Materials for Engineers and Technicians

A comprehensive yet accessible introduction to materials engineering which provides a straightforward, readable approach to the subject. The sixth edition includes a new chapter on the selection of materials, an updated discussion of new materials, and a complete glossary of key terms used in materials engineering.

This renowned text has provided many thousands of students with an easily accessible introduction to the wide-ranging subject area of materials engineering and manufacturing processes for over forty years. It avoids the excessive jargon and mathematical complexity so often found in textbooks for this subject, retaining the practical down-to-earth approach for which the book is noted. The increased emphasis on the selection of materials reflects the increased emphasis on this aspect of materials engineering now seen within current vocational and university courses.

In addition to meeting the requirements of vocational and undergraduate engineering syllabuses, this text will also provide a valuable desktop reference for professional engineers working in product design who require a quick source of information on materials and manufacturing processes.

W. Bolton was formerly Head of Research and Development and Monitoring at BTEC. He has also been a UNESCO consultant and is author of many successful engineering textbooks.

R.A. Higgins was a Senior Lecturer in Materials Science at the College of Commerce and Technology, West Bromwich, UK. He was a Chief Metallurgist at Messrs Aston Chain and Hook Ltd., Birmingham, UK, as well as an Examiner in Metallurgy for the Institution of Production Engineers, the City and Guilds of London Institute, the Union of Lancashire and Cheshire Institutes and the Union of Educational Institutes.

D0303165

Materials for Engineers and Technicians

Sixth Edition

W. Bolton

R.A. Higgins

LONDON AND NEW YORK

First published by Arnold 1972

This edition published 2015
by Routledge
2 Park Square, Milton Park, Abingdon, Oxon OX14 4RN

and by Routledge
711 Third Avenue, New York, NY 10017

Routledge is an imprint of the Taylor & Francis Group, an informa business

British Library Cataloguing-in-Publication Data
A catalogue record for this book is available from the British Library

Library of Congress Cataloging-in-Publication Data
Bolton, W. (William), 1933–
[Materials]
Materials for engineers and technicians / W. Bolton, R.A. Higgins. — 6th edition.
pages cm
Original edition published under the title: Materials.
Includes bibliographical references and index.
1. Materials. I. Higgins, Raymond A. (Raymond Aurelius), 1916– II. Title.
TA403.B64 2014
620.11—dc23
2014011242

ISBN: 9781138778757 (pbk)
ISBN: 9781315771687 (ebk)

Typeset in Times New Roman
by Florence Production Ltd, Stoodleigh, Devon

Printed and bound by CPI Group (UK) Ltd, Croydon, CR0 4YY

Contents

Preface

In revising this well-loved and established text for its sixth edition it seemed most appropriate to use the words of the author Raymond Higgins to indicate the background to his writing of the text and merely to add a footnote at the end of this preface to indicate the changes I have made for this sixth edition.

From the preface to the first edition

My grandfather used to take trout from the River Tame, which runs alongside our new college here in West Bromwich. During my boyhood, fish no longer lived there, though occasionally an adventurous water-vole would make an exploratory dive. Today the Tame is a filthy open sewer, which will support neither animal nor vegetable life, and even the lapwings have forsaken its surrounding water-meadows. The clear-water tributaries of the Tame, where as a boy I gathered watercress for my grandmother – and tarried to fish for sticklebacks – are now submerged by acres of concrete which constitute the link between the M5 and M6 motorways. On its way to the sea, the Tame spews industrial poison of the Black Country into the unsuspecting Trent, doing little to maintain the ecological stability of that river.

We Midlanders are not alone in achieving this kind of environmental despoliation. It seems that the Americans have succeeded in poisoning considerable areas of their Great Lakes. At this rate of 'progress', we must seriously consider the ultimate pollution of the sea, which, far from being 'cruel', provides Man with a great deal of his food. More important still, much of his supply of oxygen is generated by vast forests of marine kelp which is generally referred to, somewhat unkindly, as 'seaweed'.

We would do well to consider the extent to which a more effective use of materials might alleviate the pollution of our environment. Suppose, for example, a motor car were constructed of materials such that it would last much longer. Presumably, since fewer motor cars would then be produced, the total amount of environmental pollution associated with their production would also be reduced. Further, since their scrap value would be higher, less of them would be left around to disfigure – or pollute – our countryside. Obviously, if such a course were followed, the whole philosophy of consumer production would need to be rethought, since a reduced total production would mean underemployment per worker (judged by present standards, rather than those which will obtain during the twenty-first century). However, these are problems to be solved by the planner and the politician, rather than by the mechanical engineer.

Pollution of the environment is, however, only one factor in the use of materials that the engineer must consider. The supplies of basic raw materials are by no means inexhaustible, and, by the end of the present century, reserves of metals like copper and tungsten will have run out, if we continue to gobble them up at present rates. In the more distant future it may well become a punishable offence to allow iron in any of its forms to rust away and become so dispersed amongst its surroundings that it cannot conveniently be reclaimed. The motor car of the future will need to be made so that it does not corrode. Such an automobile, made to last say 40 years, would inevitably cost a lot of money, but, when worn out, would be returned to the factory for 'dismemberment', so that almost 100% of its basic materials would be reclaimed. Only in this way will looting and pollution of the environment be reduced to tolerable proportions, particularly if the world population continues to increase even at only a fraction of its present rate.

It follows that a greater appreciation of the properties of all available materials is necessary to the mechanical engineer, and, although this book has been written with the mechanical engineering technician specifically in mind, it is hoped that it will provide a useful introduction to materials science for others engaged in engineering production. The book has been so written that only a most elementary knowledge of general science is necessary in order that it can be read with advantage. As a point of interest, the original idea of the work was conceived when the author was blizzard-bound in his caravan one Easter mountaineering holiday in Skye. It was finished in high summer in the Haute Savoie, when plagues of vicious insects in the alpine meadows made it equally unpleasant to venture forth.

Preface to the sixth edition

For the sixth edition the opportunity has been taken to include a discussion of electrical conductivity in Chapter 1, expand the discussion in Chapter 27 of the new materials that are now being increasingly used, add a Chapter 28 on the selection of materials, the materials life cycle and the problems of recycling, and also add an Appendix giving a glossary of key terms.

W. Bolton

1 Engineering materials

1.1 Introduction

Figure 1.1 *An old nail.*

I found this old nail (Figure 1.1) – some 135 mm long – whilst poking about among the debris from restoration work in the medieval hill-top village of San Gimignano, Tuscany. In the Middle Ages, such nails were made by hand, a blacksmith using a hammer and an anvil, in this case containing a suitable slot so that he could form the nail head. His only additional equipment would be a charcoal fire assisted by goat-skin bellows to enable him to heat the piece of iron from which he forged and pointed the nail. Because of the way in which they were made, medieval nails were roughly square in cross-section and differed little from those with which Christ was crucified more than a thousand years earlier.

Such products were of course very labour intensive and, even allowing for the relatively low wages and long hours worked by craftsmen in those days, nails would have been quite costly items. Producing the necessary lumps of iron from the original ore was also an expensive process, adding further to the cost of the nail.

Following the Industrial Revolution, methods used to manufacture nails became increasingly mechanised. In modern processes, steel rod – or wire – is fed into a machine where a die automatically forges the head whilst almost simultaneously cutting through the stock rod at suitable angles to form the point of the resulting nail. The process is completed in a fraction of a second and the rod, or wire, immediately travels forward to produce the next nail. Since a single operator may tend several machines, the immediate labour cost of producing a nail is very low compared with medieval times. However, the costing is not quite as simple as that – things rarely are! A modern nail-making machine is a complex and fairly expensive piece of equipment which must be serviced frequently if it is to show a suitable profit before it finally becomes 'clapped out' (how often has one purchased nails whose points proved to be blunt due to lack of maintenance of the shearing blades?). There are of course many other overheads as well as running costs which must be taken into account in assessing the cost of producing a modern nail. One thing is certain: technological advances during and since the Industrial Revolution have reduced the unit cost reckoned in man-hours in producing a nail.

The above illustrates how a material, in this case iron, has been made use of for the production of a product for which there is a demand (how would structures such as wooden houses and ships have been made in ancient times without nails?). In this chapter, we take a look at the relationship between materials, their properties and the uses of the products made from them.

1.2 The requirements

The selection of a material from which a product can be manufactured is determined by:

- The *conditions under which the product is to be used,* i.e. the service requirements. These dictate the properties required of a material. For example, if a product is to be subject to forces then it might need strength and toughness and/or if subject to a corrosive environment it might require corrosive resistance.
- The *methods proposed for the manufacture of the product.* For example, if a material has to be bent as part of its processing then it must be ductile enough to be bent without breaking. A brittle material could not be used.
- The *price of the material and its availability and the cost of making the product.*

1.2.1 Properties of materials

Materials selection for a product is based upon a consideration of the properties required. These include:

- *Mechanical properties:* these include such properties as density, and the properties displayed when a force is applied to a material, e.g. yield strength, strength, stiffness, hardness, toughness, fatigue strength (how many times can it be flexed back-and-forth before it breaks?), creep strength (how will it change in length with time when subject to a constant force?).
- *Electrical properties:* these are the properties displayed when the material is used in electrical circuits or electrical components and include resistivity, conductivity and resistance to electrical breakdown.
- *Magnetic properties:* these are relevant when the material is used as, for example, a magnet or part of an electrical component such as an inductor which relies on such properties.
- *Thermal properties:* these are the properties displayed when there is a heat input to a material and include expansivity and heat capacity.
- *Optical properties:* these include transparency.
- *Surface properties:* these are, for example, relevant in considerations of abrasion and wear, corrosion and solvent resistance.
- *Aesthetic properties:* these include appearance, texture and the feel of a material.

The properties of materials are often markedly changed by the treatments they undergo. Thus, the properties of a material might be changed as a result of working, e.g. if carbon steel is permanently deformed then it will have different mechanical properties to those existing before that deformation, or heat-treatment, e.g. steels can have their properties changed by a heat-treatment such as annealing, in which the steel is heated to some temperature and then slowly cooled, or quenching, which involves heating to some temperature and then immersing the material in cold water.

1.3 The materials

The history of the human race can be divided into periods according to the materials that were predominantly used (though modern times have a multiplicity of materials in widespread use, in olden times there was no such great multiplicity):

- The *Stone Age* (about 10 000 BC–3000 BC). People could only use the materials they found around them such as stone, wood, clay, animal hides, bone, etc. The products they made were limited to what they could fashion out of these materials, thus they had tools made from stone, flint, bone and horn, with weapons – always at the forefront of technology at any time – of wood and flint.

- The *Bronze Age* (3000 BC–1000 BC). By about 3000 BC, people were able to extract *copper* from its ore. Copper is a ductile material which can be hammered into shapes, thus enabling a greater variety of items to be fashioned than was possible with stone. The copper ores contained impurities that were not completely removed by the smelting and so copper alloys were produced. It was found that when tin was added to copper, an alloy, *bronze,* was produced that had an attractive colour, was easy to form and harder than copper alone.

- The *Iron Age* (1000 BC–1620 AD). About 1000 BC the extraction of *iron* from its ores signalled another major development. Iron in its pure form was, however, inferior to bronze but by heating items fashioned from iron in charcoal and hammering them, a tougher material, called *steel,* was produced. Plunging the hot metal into cold water, i.e. quenching, was found to improve the hardness. Then reheating and cooling the metal slowly produced a less hard but tougher and less brittle material, this process now being termed *tempering*. Thus *heat-treatment processes* were developed.

- The *Cast Iron Age* (1620 AD–1850 AD). Large-scale iron production with the first coke-fuelled blast-furnace started in 1709. The use of cast iron for structures and machine parts grew rapidly after 1750, including its use for casting cannon. In 1777, the first cast iron bridge was built over the River Severn near Coalbrookdale. Cast iron established the dominance of metals in engineering. The term *Industrial Revolution* is used for the period that followed as the pace of developments of materials and machines increased rapidly and resulted in major changes in the industrial environment and the products generally available. During this period, England led the world in the production of iron.

- The *Steel Age* (1860 AD onwards). Steel was a special-purpose material during the first half of the nineteenth century. However, the year 1860 saw the development of the Bessemer and open hearth processes for the production of steel, and this date may be considered to mark the general use of steel as a constructional material. This development reinforced the dominance of metals in engineering.

- The *Light Alloys Age* (such alloys only widely used from 1940 onwards). Although aluminium was first produced, in minute quantities, by H. C. Oersted in 1825, it was not until 1886 that it was produced commercially. The high strength aluminium alloy duralumin was developed in 1909, high strength nickel–chromium alloys for high temperature use in 1931. Titanium was first produced commercially in 1948.

- The *Plastics Age* (1930 onwards). The first manufactured plastic, celluloid, was developed in 1862; in 1906, Bakelite was developed. The period after about 1930 saw a major development of plastics and their use in a wide range of products. In 1933, a Dutch scientist, A. Michels, was carrying out research into the effects of high pressure on chemical reactions when he obtained a surprise result – the chemical reaction between ethylene and benzaldehyde was being studied at a pressure of 2000 times the atmospheric pressure and a temperature of 170°C when a waxy solid was found to form. This is the material we call *polyethylene*. The commercial production of polyethylene started in England in 1941. The development of *polyvinyl chloride* was, unlike the accidental discovery of polythene, an investigation where a new material was sought. In 1936, there was no readily available material that could replace natural rubber and since, in the event of a war Britain's natural rubber supply from the Far East would be at risk, a substitute was required. In July 1940, a small amount of PVC was produced with commercial production of PVC starting in 1945.

- The *Composites Age* (from about 1950s onwards). Though composites are not new, bricks and concrete being very old examples, it is only in the second half of the twentieth century that synthetic composites became widely used. Reinforced plastics are now very widely used and carbon-fibre reinforced composites are, from their initial development in the 1960s, now becoming widely enough used to become known to the general public.

When tools and weapons were limited to those that could be fashioned out of stone, there were severe limitations on what could be achieved with them. The development of metals enabled finer products to be fashioned, e.g. bronze swords which were far superior weapons to stone weapons. The development of cast iron can be considered one of the significant developments which ushered in the Industrial Age. The development of plastics enabled a great range of products to be produced cheaply and in large numbers – what would the world be like today if plastics had not been developed? As a consequence of the evolution of materials over the years, our lifestyles have changed.

1.3.1 Materials classification

Materials can be classified into four main groups, these being determined by their internal structure and consequential properties.

1   *Metals:* these are based on metallic chemical elements. Engineering metals are generally alloys, these being formed by mixing two or more elements. For example, mild steel is an alloy of iron and carbon, stainless steel an alloy of iron, chromium, carbon, manganese and possibly other elements. The reason for such additions is to improve the properties.

The carbon improves the strength of the iron, the chromium in the stainless steel improves the corrosion resistance. In general, metals have high electrical conductivities, high thermal conductivities, have a relatively high stiffness and strength, can be ductile and thus permit products to be made by being bent into shape. For example, the body-work of a car is generally made of metal which is ductile enough to be formed into the required shape and has reasonable stiffness and strength so that it retains that shape. Metals are also classified as follows:

(a) *Ferrous alloys,* i.e. iron-based alloys, e.g. steels and cast irons.

(b) *Non-ferrous alloys,* i.e. not iron-based alloys, e.g. aluminium- and copper-based alloys.

2 *Polymers (plastics):* these are based on long chain molecules, generally with carbon backbones. In general, polymers have low electrical conductivity and low thermal conductivity (hence their use for electrical and thermal insulation), and when compared with metals have lower densities, expand more when there is a change in temperature, generally more corrosion-resistant, a lower stiffness, stretch more and are not as hard. When loaded they tend to creep, i.e. the extension gradually changes with time. Their properties depend very much on the temperature so that a polymer which may be tough and flexible at room temperature may be brittle at 0°C and show considerable creep at 100°C. They can be classified as:

(a) *Thermoplastics:* these soften when heated and become hard again when the heat is removed. Thus, they can be heated and bent to form required shapes, thermosets cannot. Thermoplastic materials are generally flexible and relatively soft. Polythene is an example of a thermoplastic, being widely used for such items as bags, 'squeeze' bottles, and wire and cable insulation.

(b) *Thermosets:* these do not soften when heated, but char and decompose. They are rigid and hard. Phenol formaldehyde, known as *Bakelite,* is a thermoset and was widely used for electrical plug casings, door knobs and handles.

(c) *Elastomers:* these are polymers which by their structure allow considerable extensions that are reversible, e.g. rubber bands.

3 *Ceramics:* these are inorganic materials, originally just clay-based materials. They tend to be brittle, relatively stiff, stronger in compression than tension, hard, chemically inert and bad conductors of electricity and heat. The non-glasses tend to have good heat and wear resistance and high-temperature strength. Ceramics include:

(a) *Glasses:* soda lime glasses, borosilicate glasses, pyroceramics.

(b) *Domestic ceramics:* porcelain, vitreous china, earthenware, stoneware, cement. Examples of domestic ceramics and glasses abound in the home in the form of cups, plates and glasses.

(c) *Engineering ceramics:* alumina, carbides, nitrides. Because of their hardness and abrasion resistance, such ceramics are widely used as the cutting edges of tools.

(d) *Natural ceramics*: rocks.

4   *Composites:* these are materials composed of two different materials bonded together, e.g. glass fibres or particles in plastics, ceramic particles in metals (cermets) and steel rods in concrete (reinforced concrete). Wood is a natural composite consisting of tubes of cellulose in a polymer called *lignin*.

The term *smart materials* is used for those materials which can sense and respond to the environment around them, e.g. the photochromic materials used in reactive spectacle lenses which become darker in response to increased light levels.

The term *nanomaterial* is used for metals, polymers, ceramics or composites that involve nanometre-sized elements (1 nm = 10^{-9} m).

1.4 Structure of materials

Materials are made up of atoms. These are very tiny particles – a pin head contains about 350 000 000 000 000 000 of them, give or take a few billions. All truly solid materials consist of atoms held together by forces of attraction which have their origin in electrical charges within each atom. A chemical *element* contains atoms all of one type. There are 92 different 'natural' chemical elements, of which over 70 are metals. Some of these metallic elements are extremely rare, whilst others are useless to the engineer, either by virtue of poor mechanical properties or because they are chemically very reactive. Consequently, less than 20 of them are in common use in engineering alloys. Of the non-metallic elements, carbon is the one which forms the basis of many engineering materials, since it constitutes the 'backbones' of all plastics. Chemically similar to carbon, the element silicon has become famous in the form of the 'silicon chip', but along with oxygen (as silicon dioxide or *silica*) it is the basis of many refractory building materials. Oxygen and silicon are by far the most common elements in the Earth's crust and account for some 75% of it in the form of clays, sands, stones and rocks like granite.

1.4.1 The chemical bonding of atoms

The main 'building blocks' of all atoms are the *electron*, the *proton* and the *neutron*. The proton and neutron are roughly equal in mass whilst the electron is only about a two-thousandth of the mass of the other two, being 9.11×10^{-31} kg. The electron carries a unit negative charge of 1.6×10^{-19} coulombs whilst the proton carries an equal but opposite charge of positive electricity. The neutron, as its name suggests, carries no electrical charge. The number of protons and electrons in any stable atom are equal, since a stable atom carries no resultant charge.

All of the relatively massive protons and neutrons are concentrated in the *nucleus* of the atom, whilst the electrons can be considered to be arranged in a series of 'orbits' or 'shells' around the nucleus. Each electron shell can only contain a specific number of electrons.

* The first shell, nearest to the nucleus, can only contain two electrons.
* The second shell can only contain eight.
* The third shell can only contain eighteen and so on.

Since the number of protons in the nucleus governs the total number of electrons in all the shells around it, it follows that there will often be insufficient electrons to complete the final outer shell. Figure 1.2 shows the electron shell structures of the first 10 elements.

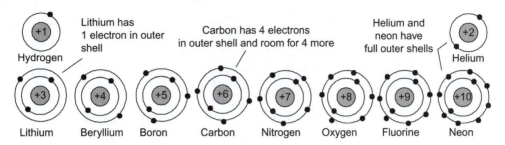

Figure 1.2 *The electron structure of the first ten elements in order of 'atomic number', i.e. the number of protons in the nucleus. Although these diagrams show electrons as being single discrete particles in fixed orbits, this interpretation should not be taken too literally. It is better to visualise an electron as a sort of mist of electricity surrounding the nucleus!*

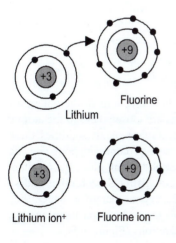

Figure 1.3 *Lithium and fluorine ions.*

1.4.2 The electrovalent (or ionic) bond

This bond between atoms occurs between metals and non-metals. Thus, the extremely reactive metal sodium combines with the non-metallic gas chlorine to form crystals of sodium chloride (common table salt). Since the atomic structures involved are simpler, we will instead consider the similar combination which occurs between the metal lithium (now an essential element in the batteries used in electronic watches, automatic cameras and the like) and the non-metallic gas fluorine (the same group of elements as chlorine).

An atom of lithium contains three electrons in orbit around its nucleus, two of these completing the first shell and leaving a lone electron in the second shell. The atom of fluorine contains nine electrons in orbit, two of these filling the first shell with seven in the second shell. The force of attraction between the positive nucleus of lithium and its outer lone electron is comparatively weak and it is easily snatched away so that it joins the outer electron shell of the fluorine atom. Because the lithium atom has lost an electron, it now has a resultant positive charge, whilst the fluorine atom has gained an electron and so has a resultant negative charge. Charged atoms of this type are called *ions* (Figure 1.3). Metals always form positively charged ions, because they are always able to easily lose electrons, whilst non-metals form negatively charged ions. As these lithium and fluorine ions carry opposite charges, they will attract each other. In a solid, the lithium and fluorine ions arrange themselves in a geometrical pattern in which each fluorine ion is surrounded by six lithium ions as its nearest neighbours, whilst each lithium ion is surrounded by six fluorine ions (Figure 1.4). The compound, lithium fluoride, forms a relatively simple cubic type of *crystal*

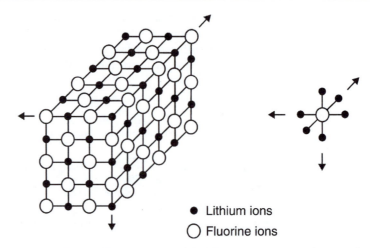

● Lithium ions
○ Fluorine ions

Figure 1.4 *The crystal structure of lithium fluoride.*

structure. Other such *salts* may form more complex crystal patterns, depending upon the relative sizes of the ions involved and the electrical charges carried by each type of ion. Here we have dealt with atoms which lose, or gain, only one electron; those which lose, or gain, two electrons will produce ions carrying twice the electrical charge.

The bonds between atoms in ceramics are frequently ionic, e.g. aluminium oxide (alumina), Al_2O_3, and these strong bonds account for the high melting points and high strengths, but low ductilities, of ceramics.

1.4.3 The metallic bond

Most metals have one, two or at the most three electrons in the outermost shell of the atom. These outer-shell electrons are loosely held to the atomic nucleus and as a metallic vapour condenses and subsequently solidifies, these outer-shell electrons are surrendered to a sort of common pool and are virtually shared between all the atoms in the solid metal. Since the resultant metallic ions are all positively charged, they will repel each other and so arrange themselves in some form of regular, crystal pattern in which they are firmly held in position by the attractive forces between them and the permeating 'cloud' of negatively charged electrons (Figure 1.5).

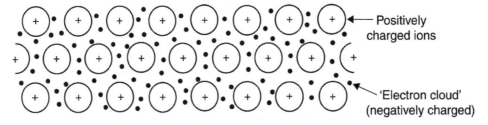

Positively charged ions

'Electron cloud' (negatively charged)

Figure 1.5 *A simple interpretation of the metallic bond concept. The positively charged ions repel each other but are held in place by the attractive force supplied by the negatively charged 'electron cloud'.*

The metallic bond explains many of the main characteristics of metallic elements:

- All metals are good conductors of electricity since they have electrons, from the 'electron cloud', which are free to move within the body of the metal when a potential difference is applied.
- Metals are good conductors of heat. The application of heat to a piece of metal causes electrons to vibrate more actively and these vibrations can be passed on quickly from one electron to another within the electron cloud which is continuous within the body of the metal.
- Most metals are ductile because layers of ions can be made to slide over each other by the application of a shearing force. At the same time, metals are strong because the attractive force provided by the electron cloud opposes the movement apart from layers of ions.
- Metals are lustrous in appearance since the free, vibrating surface electrons fling back light that falls on the surface of a metal.

1.4.4 The covalent bond

The covalent bond is formed between atoms of those non-metallic elements in which, for various reasons, there is a strong attractive force between the nucleus and the outer-shell electrons. Hence, instead of a transfer of electrons from one atom to another to give an ionic bond, there is a *sharing* of electrons between two atoms to bind them together.

An atom of carbon has four electrons in its outer shell (see Figure 1.2). This is the second electron shell and so there are four spaces for electrons in that shell and it can combine with four atoms of hydrogen. The carbon atom thus has shares of four hydrogen electrons to give a total of eight and complete the shell. Each hydrogen atom has a share in one electron of the carbon atom to complete its outer shell with two electrons (Figure 1.6). The methane molecule is thus a stable unit, being held together by, what are termed, covalent bonds.

The element carbon has the ability to form long chain-like molecules in which carbon atoms are bonded covalently to each other and to hydrogen atoms (Figure 1.7). The results are the materials we know as *plastics*. An important feature of these materials is that, unlike metals in which the outer-shell electrons can travel freely within the electron cloud, so making metals good conductors of electricity, the outer-shell electrons in these covalently

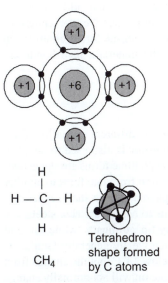

Figure 1.6 *The gas methane with covalent bonds, i.e. electron sharing between the hydrogen atoms and the carbon atom. The shape of the methane molecule is determined by the hydrogen nuclei, which being positively charged repel each other and so arrange themselves as far from each other as possible, i.e. in a tetrahedral pattern around the carbon atom to which each is bound.*

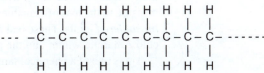

Approximately 1200 carbon atoms in the chain

Figure 1.7 *The long-chain like molecule of the plastics material polyethylene.*

bonded substances are securely held to the atoms to which they belong and so are not free to move away. Thus, the materials are excellent electric insulators. The initial use of polythene was as an electrical insulator in electronics equipment used in radar during the Second World War.

1.4.5 Intermolecular forces

However large the molecules in covalent compounds, each will contain equal numbers of protons and electrons and so will be electrically neutral and carry no resultant electric charge. So how do these molecules stick together to form coherent solids?

The short answer is that electrons and protons are not necessarily equally distributed over the molecule, often due to interaction with electrons and protons of neighbouring molecules. Consequently, due to this irregular distribution of electrons and protons, the molecule acquires a 'negative end' and a 'positive end' (Figure 1.8), rather like the north and south poles of a magnet. Since unlike charges will be attracted to each other, then the molecules will be attracted to each other. The greater the unevenness of distribution of charges within molecules, the greater the forces of attraction between them and therefore the 'stronger' the material. Such bonds are termed *van der Waals forces*.

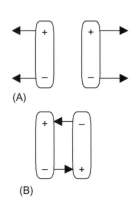

(A)

(B)

Figure 1.8 *Like charges repel (A) tending to cause realignment of molecules, so that unlike charges then attract (B) producing bonds (van der Waals forces).*

1.4.6 Polymorphism

Many solid elements can exist in more than one different crystalline form and are said to be *polymorphic* (the term *allotropy* is also used to describe this phenomenon). Generally, these different crystalline forms are stable over different temperature ranges so that the transition from one form to another takes place as the transition temperature is passed. As we shall see later, it is the polymorphism of iron which enables us to harden suitable steels.

Tin is also polymorphic, existing as 'grey tin', ordinary 'white tin' and as 'brittle tin'. White tin, the form with which we are generally familiar is stable above 13°C, whilst grey tin is stable below 13°C, but the change from one form to the other is very sluggish and white tin will not normally change to the powdery grey form unless the temperature falls well below 13°C. At one time, organ pipes were manufactured from tin and the story is told that during one particularly cold winter in St. Petersburg, the organ pipes collapsed into a pile of grey dust as the organist played his first chord. This particular affliction to which tin is prone was known as *tin pest* but was in fact a manifestation of a polymorphic change. On a more sombre note, the final tragedy which befell Scott's Polar party in 1912 was due in no small measure to the loss of valuable fuel which had been stored in containers with *tin-soldered* joints. These had failed – due to 'tin pest' – during the intensely cold Antarctic winter and the precious fuel seeped away.

Tin is one of a chemical family of elements which include carbon, silicon and germanium, all of these are polymorphic elements. As we have seen in Figure 1.6, an atom of carbon can combine with four atoms of hydrogen to form a molecule of methane. That is, an atom of carbon is capable of forming four separate bonds (*valences* as chemists call them) with other atoms.

Under conditions of extremely high temperature and pressure, carbon atoms will link up with each other to form structures in which the carbon atoms form a rigid crystalline structure, showing the same type of tetrahedral pattern that is exhibited by the distribution of hydrogen atoms in a methane molecule. This tetrahedral crystalline structure containing only carbon atoms is the substance *diamond*. Figure 1.9A indicates the arrangement of the centres of carbon atoms in the basic unit of the diamond crystal. There is an atom at the geometrical centre of the tetrahedron and one at each of the four points of the tetrahedron. The structure is, of course, continuous (Figure 1.9B) such that each atom of carbon in the structure is surrounded by four other atoms covalently bonded to it and spaced equidistant from it. Since all carbon atoms in diamond are 'joined' by strong covalent bonds to four other carbon atoms it follows that diamond is a very strong, hard material. It is in fact the hardest substance known.

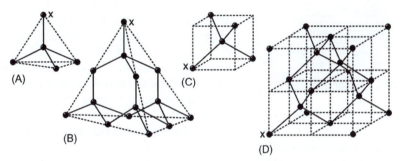

Figure 1.9 *The crystal structure of diamond.*

Under a different set of conditions of temperature and pressure, carbon atoms will combine to form layer-like molecules (Figure 1.10), a polymorph called *graphite*. The layers are held together by relatively weak van der Waals

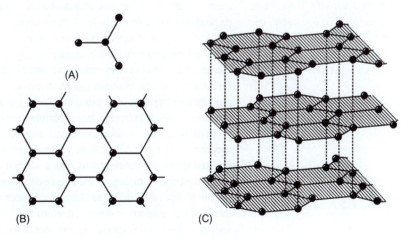

Figure 1.10 *The crystal structure of graphite.*

forces generated by the spare electrons not used by the primary bonding system. Consequently, these layers slide over each other quite easily so that graphite can be used as a lubricant. Coke, charcoal and soot, all familiar forms of carbon, contain tiny micro-crystals of graphite.

Since all of the carbon atoms in diamond are covalently bonded in such a manner that all outer-shell electrons are involved in such bonds, there are no free electrons present and so diamond does not conduct electricity. In graphite, however, spare electrons are present since not all outer-shell electrons are used up in covalent bonds. Hence, graphite is a relatively good conductor of electricity.

For centuries it was believed that there were only two allotropes of carbon, diamond and graphite, but in 1985 a third form was discovered. This form was composed of hexagons and pentagons of carbon joined together to form a completely spherical shape. Imagine a single graphite 'sheet' in which selected 'hexagons' have been replaced by 'pentagons'. These pentagons will only 'fit' if the flat sheet is allowed to form a curved surface to give a molecule which looks like a closed cage with a large empty space in the middle. The way the carbon atoms are arranged is similar to the design of geodesic domes, such as that used to house 'Science World' in Vancouver, by the architect Richard Buckminster Fuller and so this new form of carbon allotrope was named *fullerene* in his honour. There is a series of these fullerene molecules containing between 20 and 600 carbon atoms, possibly the most common containing 60 carbon atoms, i.e. C_{60}. This has a geometry similar to that of a modern soccer ball (Figure 1.11), for which reason it has become known affectionately as the *Buckyball*.

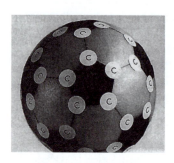

Figure 1.11 *A model of the fullerene molecule C_{60} which I constructed using a child's soccer ball, a felt pen and some small labels. Each of the 12 pentagonal panels (dark) is surrounded by 5 hexagonal panels (light). The white labels – each inscribed with a C – indicate the positions of the nuclei of carbon atoms.*

1.4.7 Electrical conductivity and materials

In terms of their electrical conductivity, materials can be grouped into three categories: *conductors*, *semiconductors* and *insulators*. Metals are conductors and the reason for this is that their atomic structure consists of atoms all of which have only one, two or three electrons in their outer shells. These outer electrons are only loosely attached to the atoms and so when the atoms are packed close together in a solid, they come under the influence of other atoms and can be pulled out of orbit and drift off, moving freely between atoms. Thus when a potential difference is applied across a metal, these 'free' electrons are able to move and give rise to a current. An increase in temperature results in a decrease in the conductivity. This is because the temperature rise does not release any more electrons but results in increased vibration of the atoms so scattering the free electrons more and hindering their movement.

Insulators, e.g. ceramics, have, however, structures in which all their electrons are tightly bound to atoms. Hence when a potential difference is applied there are no electrons free and able to give rise to a current.

Semiconductors, e.g. silicon and germanium, can be regarded as insulators at low temperatures but as they have outer electrons which do not require too much energy to remove them from their atoms, at room temperature there will be some 'free' electrons. An electron becoming free means leaving a 'hole' in the outer shell of an atom. Thus when a potential difference is applied, we can consider an electron hopping into a hole in a neighbouring atom, leaving a hole behind into which another electron can jump. The electrical conductivity of a semiconductor can be markedly changed by the presence of foreign atoms. Deliberately introducing such foreign atoms is termed doping. Some foreign atoms will make more electrons available for conduction and are thus termed donors. Such doped semiconductors are termed n-type. Other foreign atoms will supply holes and are thus termed acceptors. Such doped semiconductors are termed p-type.

1.5 Processes

The processes used for producing products from materials can be grouped as:

- *Shaping:* these include casting, moulding, powder methods, sheet forming and machining.
- *Joining:* these include adhesives, welding and fasteners.
- *Surface treatments:* these include polishing, coating, plating, etching and printing.

The choice of process is determined by the material to be used, the form of the required product, e.g. shape and size, and the economics of the process. Thus, some processes have expensive set-up costs which can only be defrayed over large production runs, while others are more economic for small runs where high cycle times are not required. For example, die-casting will produce better dimensional accuracy and finish than sand-casting and has lower labour costs per item; however, it has higher initial plant and equipment set-up costs and so is more appropriate for long production runs where the initial set-up costs can be spread over a large number of items. Sand-casting has, however, a smaller set-up cost but labour costs are incurred every time a casting is made. So as the number of items made is increased, the labour costs build up and eventually die-casting proves cheaper per item than sand-casting.

1.6 The materials society

Low material cost coupled with high labour costs have created a society that discards many products after a single use, it being cheaper to get another one than spend money on labour to get more use from the product. Milk only used to be available in glass bottles, being delivered to the doorstep and then the empty bottles being collected the next day for cleaning and refilling. Each bottle, in this way, was used many times before it became too scratched or broken and discarded. Nowadays most milk seems to come in plastics containers which are just discarded after a single use.

There is, however, a growing recognition of the waste of materials occurring as a result of the discard-after-one-use policy and recycling, the feeding of waste material back into the processing cycle so that it can be reused, is becoming important. The problem that has to be faced is that non-renewable ores are processed to give materials which are then manufactured into products which are used and then at the end of their lives are discarded with only a small percentage of the discards entering a recycling loop. At each point in the sequence, energy is used and CO_2 produced with the consequential effect on planetary warming. Reduction in CO_2 emissions is necessary for the future well-being of the planet. Also, there is also the public pressure to reduce waste and the dumping of materials in waste-fill sites.

Recycling of glass, tins and plastics is now becoming quite commonplace. Typically, of the order of 30% of the glass in a new bottle comes from recycled glass; of the order of 40% of the aluminium used in Britain is recycled from aluminium drinks cans.

2 Properties of materials

2.1 Introduction

Beer can be bought from the supermarket in aluminium cans or in glass bottles. Why aluminium? Why glass? Both materials, together with the shape adopted for the container, are able to give rigid containers which do not readily deform under the handling they receive in the supermarket and getting to it. Both materials are resistant to chemical corrosion by the contents – though some people swear by glass as they feel the aluminium does give some 'taste' to the beer. Both materials are able to keep the 'fizz' in the beer and stop it going flat, i.e. the gas in the beer is not able to escape through the walls of the container. In addition, both containers can be produced cheaply and the material cost is also low. Both materials can be recycled. The main difference between the two is that the glass bottle is significantly heavier than the aluminium can.

In this chapter, we take a look at how the properties of different types of materials compare and the sources for data on material properties.

2.2 Properties

The following are a look at some of the terms used to describe properties and comparisons of those properties for a range of materials.

2.2.1 Mechanical properties

A material subject to external forces which stretch it is said to be in *tension,* when subject to forces which squeeze to be in *compression*. In discussing the application of forces to materials the concern is the force applied per unit area, this being termed the *stress*. Stress has the units of pascal (Pa), with 1 Pa being a force of 1 newton per square metre, i.e. 1 Pa = 1 N/m². When a material is subject to tensile or compressive forces, it changes in length, the term *strain,* symbol ε, is used for the change in length/original length. Since strain is a ratio of two lengths it has no units. However, strain is frequently expressed as a percentage, i.e. the percentage change in length.

The following are some of the more frequently encountered terms that are used to describe mechanical properties:

- *Strength:* this is the ability of it to resist the application of forces without breaking. The term *tensile strength* is used for the maximum value of the tensile stress that a material can withstand without breaking; the *compressive strength* is the maximum compressive stress the material can withstand without becoming crushed. The unit of strength is that of stress and so is the pascal (Pa), with 1 Pa being 1 N/m². Strengths are often millions of pascals and so MPa is often used, 1 MPa being 10^6 Pa or 1 000 000 Pa. For example, aluminium alloys at about 20°C have

Table 2.1 *Tensile strength values at about 20°C*

Strength (MPa)	Material
2–12	Woods perpendicular to the grain
2–12	Elastomers
6–100	Woods parallel to the grain
60–100	Engineering polymers
20–60	Concrete
80–300	Magnesium alloys
160–400	Zinc alloys
100–600	Aluminium alloys
80–1000	Copper alloys
250–1300	Carbon and low alloy steels
250–1500	Nickel alloys
500–1800	High alloy steels
100–1800	Engineering composites
1000 to >10 000	Engineering ceramics

Table 2.2 *Tensile modulus values at 20°C*

Tensile modulus (GPa)	Material
<0.2	Elastomers
0.2–10	Woods parallel to grain
0.2–10	Engineering polymers
2–20	Woods perpendicular to grain
20–50	Concrete
40–45	Magnesium alloys
50–80	Glasses
70–80	Aluminium alloys
43–96	Zinc alloys
110–125	Titanium alloys
100–160	Copper alloys
200–210	Steels
80–1000	Engineering ceramics

strengths in the region of 100–600 MPa, i.e. 100–600 million pascals. Table 2.1 gives typical values of tensile strengths at about 20°C.

- *Stiffness:* this is the ability of a material to resist bending. When a strip of material is bent, one surface is stretched and the opposite face is compressed, the more it bends the greater is the amount by which the stretched surface extends and the compressed surface contracts. Thus, a stiff material would be one that gave a small change in length when subject to tensile or compressive forces, i.e. a small strain when subject to tensile or compressive stress and so a large value of stress/strain. For most materials a graph of stress against strain gives initially a straight line relationship, and so a large value of stress/strain means a steep gradient of the stress–strain graph. This gradient is called the *modulus of elasticity* (or sometimes *Young's modulus*). The units of the modulus are the same as those of stress, since strain has no units. Engineering materials frequently have a modulus of the order of 1 000 000 000 Pa, i.e. 10^9 Pa. This is generally expressed as GPa, with 1 GPa = 10^9 Pa. For example, steels have tensile modulus values in the range 200–210 GPa, i.e. 200–210 × 1 000 000 000 Pa. Table 2.2 gives some typical values of tensile modulus at about 20°C.

- *Ductility:* this is the ability of a material to suffer significant deformation before it breaks. Glass is a *brittle material* and if you drop a glass it breaks; however, it is possible to stick all the pieces together again and restore the glass to its original shape. If a car is involved in a collision, the bodywork of mild steel is less likely to shatter like the glass but more likely to dent and show permanent deformation. Ductile materials permit manufacturing methods which involve bending them or squashing into the required shape. Brittle materials cannot be formed to the required

Table 2.3 *Percentage elongation values at 20°C*

Percentage elongation	Material
0	Engineering ceramics
0	Glasses
0–18	Cast irons
1–60	Nickel alloys
2–30	Low alloy steels
6–20	Magnesium alloys
6–30	Titanium alloys
10–100	Zinc alloys
18–25	Mild steel
1–55	Copper alloys
1–70	Brasses and bronzes
1–60	Nickel alloys
10–100	Polyurethane foam
500	Natural rubber

shape in this way. A measure of the ductility of a material is obtained by determining the length of a test-piece of the material, then stretching it until it breaks and then, by putting the pieces together, measuring the final length of the test-piece. A brittle material will show little change in length from that of the original test-piece, but a ductile material will indicate a significant increase in length. The *percentage elongation* of a test-piece after breaking is thus used as a measure of ductility. Table 2.3 gives some typical values. A reasonably ductile material, such as mild steel, will have a percentage elongation of about 20%, a brittle material such as a cast iron less than 1%. Thermoplastics, generally ductile, tend to have percentage elongations of the order of 50–500%, thermosets, which are brittle, of the order of 0.1–1%.

- *Malleability:* this term is used to describe the amount of plastic deformation that occurs as a result of a compressive load. A malleable material can be squeezed to the required shape by such processes as forging and rolling.

- *Toughness:* a tough material can be considered to be one that, though it may contain a crack, resists the crack growing and running through the material. Think of trying to tear a sheet of paper. If the paper has perforations, i.e. initial 'cracks', then it is much more easily torn. In the case of, say, the skin of an aircraft there are windows or their fastenings, which are equivalent to cracks, there is a need for cracks not to propagate and so a tough material is required. *Toughness* can be defined in terms of the work that has to be done to propagate a crack through a material, a tough material requiring more energy than a less tough one. One way of obtaining a measure of toughness is in terms of the ability of a material to withstand shock loads – the so-called *impact tests* such as the Charpy and Izod (see Chapter 3). In these tests, a test-piece is struck a sudden blow and the energy needed to break it is measured.

- *Hardness:* this is a measure of the resistance of a material to abrasion or indentation. A number of scales are used for hardness, depending on the method that has been used to measure it (see Chapter 3 for discussions of test methods).

2.2.2 Electrical properties

The electrical *resistivity* ρ of a material is defined by the equation:

$$\rho = \frac{RA}{L}$$

where R is the resistance of a length L of the material of cross-sectional area A. The unit of resistivity is the ohm metre. An electrical *insulator*, such as a ceramic, will have a very high resistivity, typically of the order of 10^{10} Ωm or higher. An electrical *conductor,* such as copper, will have a very low resistivity, typically of the order of 10^{-8} Ωm. The term *semiconductor* is used for materials with resistivities roughly half-way between those of conductors and insulators, i.e. about 10^2 Ωm. Conductivity is the reciprocal of resistivity. Table 2.4 shows typical values of resistivity and conductivity for insulators, semiconductors and conductors.

Table 2.4 *Typical resistivity and conductivity values at about 20°C*

Material	Resistivity (Ωm)	Conductivity (S/m)
Insulators		
Acrylic (a polymer)	$>10^{14}$	$<10^{-14}$
Polyvinyl chloride (a polymer)	10^{12}–10^{13}	10^{-13}–10^{-12}
Mica	10^{11}–10^{12}	10^{-12}–10^{-11}
Glass	10^{10}–10^{14}	10^{-14}–10^{-10}
Porcelain (a ceramic)	10^{10}–10^{12}	10^{-12}–10^{-10}
Alumina (a ceramic)	10^{9}–10^{12}	10^{-12}–10^{-9}
Semiconductors		
Silicon (pure)	2.3×10^3	4.3×10^{-4}
Germanium (pure)	0.43	2.3
Conductors		
Nichrome (alloy of nickel and chromium)	108×10^{-8}	0.9×10^6
Manganin (alloy of copper and manganese)	42×10^{-8}	2×10^6
Nickel (pure)	7×10^{-8}	14×10^6
Copper (pure)	2×10^{-8}	50×10^6

2.2.3 Thermal properties

Thermal properties that are generally of interest in the selection of materials include:

- The *linear expansivity* α or *coefficient of linear expansion* is a measure of the amount by which a length of material expands when the temperature increases. It is defined as:

$$\alpha = \frac{\text{change in length}}{\text{original length} \times \text{change in temperature}}$$

  and has the unit of K^{-1}. For example, aluminium has a linear expansivity of 24×10^{-6} K^{-1} and so a 1 m length piece will increase in length by 24×10^{-6} m for each 1°C change in temperature.

- The *heat capacity* is the amount of heat needed to raise the temperature of an object by 1 K. The *specific heat capacity c* is the amount of heat needed per kilogram of material to raise the temperature by 1 K, hence:

$$c = \frac{\text{amount of heat}}{\text{mass} \times \text{change in temperature}}$$

  It has the unit of $J\ kg^{-1}\ K^{-1}$. Weight-for-weight, metals require less heat to reach a particular temperature than plastics, e.g. copper has a specific heat capacity of about 340 $Jkg^{-1}\ K^{-1}$ while polythene is about 1800 $Jkg^{-1}\ K^{-1}$.

- The *thermal conductivity* λ of a material is a measure of the ability of a material to conduct heat and is defined in terms of the quantity of heat that will flow per second divided by the temperature gradient, i.e.:

$$\lambda = \frac{\text{quantity of heat/second}}{\text{temperature gradient}}$$

  and has the unit of $W\ m^{-1}\ K^{-1}$. A high thermal conductivity means a good conductor of heat. Metals tend to be good conductors, e.g. copper has a thermal conductivity of about 400 $W\ m^{-1}\ K^{-1}$. Materials that are bad conductors of heat have low thermal conductivities, e.g. plastics have thermal conductivities of the order 0.3 $W\ m^{-1}\ K^{-1}$ or less. Very low thermal conductivities occur with foamed plastics, i.e. those containing bubbles of air. Foamed polymer polystyrene (expanded polystyrene) has a thermal conductivity of about 0.02–0.03 $W\ m^{-1}\ K^{-1}$ and is widely used for thermal insulation.

Table 2.5 gives typical values of the linear expansivity, the specific heat capacity and the thermal conductivity for metals, polymers and ceramics.

Table 2.5 *Thermal properties*

Material	Linear expansivity (10^{-6} K^{-1})	Specific heat capacity ($J\ kg^{-1}\ K^{-1}$)	Thermal conductivity ($W\ m^{-1}\ K^{-1}$)
Metals			
Aluminium	24	920	230
Copper	18	385	380
Mild steel	11	480	54
Alloy steel	12	510	37
Grey cast iron	11	265–460	44–53
Titanium	8	530	16
Polymers			
Polyvinyl chloride	70–80	840–1200	0.1–0.2
Polyethylene	100–200	1900–2300	0.3–0.5
Epoxy cast resin	45–65	1000	0.1–0.2
Expanded polystyrene			0.02–0.03
Natural rubber	22	1900	0.18
Ceramics			
Alumina	8	750	38
Fused silica	0.5	800	2
Glass	8	800	1

Table 2.6 *Comparative corrosion resistances*

Corrosion resistance	Material
Aerated water	
High resistance	All ceramics, glasses, lead alloys, alloy steels, titanium alloys, nickel alloys, copper alloys, PTFE, polypropylene, nylon, epoxies, polystyrene, PVC
Medium resistance	Aluminium alloys, polythene, polyesters
Low resistance	Carbon steels
Salt water	
High resistance	All ceramics, glasses, lead alloys, stainless steels, titanium alloys, nickel alloys, copper alloys, PTFE, polypropylene, nylon, epoxies, polystyrene, PVC, polythene
Medium resistance	Aluminium alloys, polyesters
Low resistance	Low alloy steels, carbon steels
UV radiation	
High resistance	All ceramics, glasses, all alloys
Medium resistance	Epoxies, polyesters, polypropylene, polystyrene, HD polyethylene, polymers with UV inhibitor
Low resistance	Nylon, PVC, many elastomers

2.2.4 Durability

Corrosion of materials by the environment in which they are situated is a major problem, e.g. the rusting of iron. Table 2.6 indicates the comparative resistance to attack of materials in various environments, e.g. in aerated water, in salt water, and to ultraviolet radiation. Thus, for example, in a salt water environment carbon steels are rated at having very poor resistance to attack, aluminium alloys good resistance and stainless steels excellent resistance. A feature common to many metals is the need for a surface coating to protect them from corrosion by the atmosphere.

2.3 Costs

Costs can be considered in relation to the basic costs of the raw materials, the costs of manufacturing, and the life and maintenance costs of the finished product. Comparison of the basic costs of materials is often on the basis of the cost per unit weight or cost per unit volume. Table 2.7 shows the relative costs per kilogram and per cubic metre of some materials. However, often a more important comparison is on the basis of the *cost per unit strength* or *cost per unit stiffness* for the same volume of material. This enables the cost of say a beam to be considered in terms of what it will cost to have a beam of a certain strength or stiffness. For example, if the cost per cubic metre is £ 900 and the tensile strength is 500 MPa, then the cost per MPa of strength is 900/500 = £1.8.

Table 2.7 *Relative costs of materials in relation to that of mild steel ingot*

Material	Relative cost/kg	Relative cost/m^3
Nickel	28	32
Chromium	26	24
Tin	19	18
Brass sheet	16	17
Al–Cu alloy sheet	14	5.3
Nylon 66	12	1.8
Magnesium ingot	9.2	2.1
Acrylic	8.9	1.4
Copper tubing	8.7	10
ABS	8.3	1.1
Aluminium ingot	4.3	1.5
Polystyrene	3.6	0.50
Zinc ingot	3.6	3.3
Polyethylene (HDPE)	3.4	0.43
Polypropylene	3.2	0.34
Natural rubber	3.1	0.50
Polyethylene (LDPE)	2.3	0.29
PVC, rigid	2.3	0.43
Mild steel sheet	1.9	1.9
Mild steel ingot	1.0	1.0
Cast iron	0.8	0.79

Materials are supplied in a number of standard forms, the forms depending on the material concerned. Thus, they may be supplied in the form of sheet and plate, barstock, pipe and tube, rolled sections, extrusions, ingots, castings and mouldings, forgings and pressings, granules and powders or liquids.

The costs of manufacturing will depend on the processes used. Some processes require a large capital outlay and then can be used to produce large numbers of the product at a relatively low cost per item. Other processes may have little in the way of setting-up costs but a large cost per unit product.

The cost of maintaining a material during its life can often be a significant factor in the selection of materials. The rusting of steels is an obvious example of this and, for example, dictates the need for the continuous repainting of ironwork.

2.4 Data sources

National standards bodies such the British Standards Institution (BS) and international bodies such as the International Organisation for Standardisation (ISO) are responsible for setting the standards to be adopted: a *standard* is a technical specification drawn up with the co-operation and general approval of all interests affected by it with the aims of benefiting all concerned by the promotion of consistency in quality, rationalisation of processes and methods of operation, promoting economic production methods, providing a means of communication, protecting consumer interests, promoting safe practices, and helping confidence in manufacturers and users. In the case of materials standards, if a material is stated by its producer to be to a certain standard, tested by the methods laid down by certain standards, then a customer need not have all the details written out to know what properties to expect of the material. There is, for example, the standard for tensile testing of metals (BS EN ISO 6892-1: 2009), this lays down such things as the sizes of the test-pieces to be used (see Chapter 3 for such details). There are standards for materials such as steel plate, sheet and strip, the plastic polypropylene, etc.

Data on the properties of materials is available from a range of sources:

- Specifications issued by bodies responsible for standards, e.g. the British Standards Institution and the International Standards Organisation.
- Trade associations which supply technical details of compositions and properties of materials.
- Data supplied, in catalogues or on websites, by suppliers of materials.
- Disc and network databases which give materials and their properties with means to rapidly access particular materials or find materials with particular properties.

Coding systems are used to refer to particular metals. Such codes are a concise way of specifying a particular material without having to write out its full chemical composition or properties. Thus, for example, steels might be referred to in terms of a code specified by the British Standards Institution. The code for wrought steels is a six symbol code. The first three digits of the code represent the type of steel:

000–199   Carbon and carbon–manganese types, the number being 100 times the manganese contents.

200–240   Free-cutting steels, the second and third numbers being approximately 100 times the mean sulphur content.

250   Silicon–manganese spring steels.

300–499   Stainless and heat-resistant valve steels.

500–999   Alloy steels with different groups of numbers within this range being allocated to different alloys. Thus, 500–519 is when nickel is the main alloying element, 520–539 when chromium is.

The fourth symbol is a letter:

A   The steel is supplied to a chemical composition determined by chemical analysis.

H   The steel is supplied to a hardenability specification.

M   The steel is supplied to mechanical property specifications.

S   The steel is stainless.

The fifth and sixth digits correspond to 100 times the mean percentage carbon content of the steel.

As an illustration of the above coding system, consider a steel with a code 070M20. The first three digits are 070 and since they are between 000 and 199 the steel is a carbon or carbon–manganese type. The 070 indicates that the steel has 0.70% manganese. The fourth symbol is M and so the steel is supplied to mechanical property specifications. The fifth and sixth digits are 20 and so the steel has 0.20% carbon.

The AISI-SAE (American Iron and Steel Institute, Society of Automotive Engineers) use a four digit code, the first and second digits indicating the type of steel and the third and fourth digits to indicate 100 times the percentage of the carbon content:

10XX, 11XX, 12XX, 14XX Plain carbon steels
13XX Manganese steels
23XX, 25XX Nickel steels
31XX, 32XX, 33XX, 34XX Nickel–chromium steels
40XX, 44XX Molybdenum steels
41XX Chromium–molybdenum steels
43XX, 47XX Nickel–chromium–molybdenum steels
46XX, 48XX Nickel–molybdenum steels
50XX, 51XX Chromium steels
61XX Chromium–vanadium steels
72XX Tungsten–chromium steels
81XX, 86XX, 87XX, 88XX Nickel–chromium–molybdenum steels
93XX, 94XX, 97XX, 98XX Nickel–chromium–molybdenum steels
92XX Silicon–manganese steels

For example, 1010 is a plain carbon steel with 0.10% carbon, 5120 is a chromium steel with 0.20% carbon.

As a further illustration of coding, the British Standards Institution coding for wrought copper and copper alloys consists of two letters followed by three digits. The two letters indicate the alloy group and the three digits the alloy within that group.

C   Very high percentage of copper.
CA Copper–aluminium alloys.
CB Copper–beryllium alloys.
CN Copper–nickel alloys (cupro-nickels).
CS Copper–silicon alloys (silicon bronzes).
CZ Copper–zinc alloys (brasses).
NS Copper–zinc–nickel alloys (nickel-silvers).
PB Copper–zinc–phosphorus (phosphor bronzes).

For cast copper alloys, letters followed by a digit are used. For example, AB1, AB2 and AB3 are aluminium bronze, LB1, LB2, LB4 and LB5 are leaded bronzes, SCB1 and SCB2 are general purpose brasses.

See Chapter 17 for the coding used with aluminium alloys.

3 Mechanical testing

3.1 Introduction

When an engineer designs some structure or machine, he or she must know the effects which the application of forces may have on the material before they can use it. Various mechanical tests have been devised over the years to determine such effects. Thus, for example, in a tensile test we measure the force required to stretch a specimen of the metal until it breaks, whilst in various hardness tests we produce a small dent in the surface by means of using a compressive force and take the force used divided by the surface area of the impression produced by it as a measure of hardness.

Complete information on the tests, and the behaviour of metals and alloys when subject to those tests, has been compiled by such bodies as the British Standards Institution, who have drawn up their series of BS specifications. Many European standards have been incorporated into our system and are designated as BSEN standards. In the USA, such matters are dealt with by the American Society for Metals (ASM) and the American Society for Testing Materials (ASTM). It is on tests and values specified by such bodies that the engineer bases his designs and accepts his materials.

In the type of tests described in this chapter, the test-piece is destroyed during the testing process. Such tests are therefore known as *destructive tests,* and can only be applied to individual test-pieces. These are taken from a batch of material which it is proposed to use for some specific purpose and they are therefore assumed to be representative of the batch. Tests of a different nature and purpose are used to examine manufactured components for internal flaws and faults, e.g. X-rays used to seek internal cavities in castings. These tests are generally referred to as *non-destructive tests* (NDT), since the component, so to speak, 'lives to tell the tale'.

3.2 The tensile test

The tensile test of a material involves a test-piece of known cross-sectional area being gripped in the jaws of a testing-machine and then subject to a tensile force which is increased in increments. For each increment of force, the amount by which the length of a known 'gauge length' on the test-piece increases is measured. This process continues until the test-piece fractures.

3.2.1 Tensile-testing machines

Tensile-testing machines of more than 50 years ago were designed on the principle of a simple first-order lever in which a massive counterpoise weight was moved along a beam to provide the necessary increasing tensile force acting on the test-piece. The great disadvantage of such a system was that once a test-piece reached its *yield point* and began to stretch rapidly, it was impossible to reduce the force quickly enough to enable the complete force–extension diagram to be plotted since fracture took place so readily once the maximum force had been reached. From a practical engineering consideration, this did not matter very much since it was the maximum force the test-piece could withstand which was regarded as important. However, in modern tensile-testing machines, force is applied by hydraulic cylinders for machines of up to 1 MN in capacity, whilst for small 'bench' machines of only 20 kN capacity the tensile force is applied by a spring beam. With such systems, force relaxes automatically as soon as rapid extension of the test-piece begins. The complete force–extension diagram can then be plotted with accuracy until fracture occurs.

3.2.2 Force–extension diagrams

Examination of a typical force–extension diagram (Figure 3.1) for an annealed carbon steel shows that at first the amount of extension is very small compared with the increase in force. Such extension, as there is, is directly proportional to the force; that is, OA is a straight line. If the force is released at any point before A is reached, the test-piece will return to its original length. Thus, the extension between O and A is *elastic* and the material is said to obey *Hooke's law,* i.e. we have the force proportional to the extension:

force ∝ extension

and so:

stress ∝ strain

stress = E × strain

where E is the tensile modulus of elasticity, also known as *Young's modulus.*

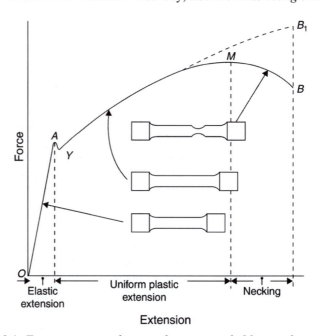

Figure 3.1 *Force–extension diagram for an annealed low-carbon steel.*

If the test-piece is stressed past point *A,* known as the *elastic limit* or the *limit of proportionality* (because the force is no longer proportional to the extension), the material suddenly 'gives', that is it suffers a sudden extension for very little increase in force. This is called the *yield point* (*Y*), and, if the force is now removed, a small permanent extension will remain in the material. The term *elastic limit* is thus used for point *A* because the material does not return to its original dimensions when stretching occurs beyond this point. Any extension which occurs past the point *A* has an element which is of a *plastic* nature. As the force is increased further, the material stretches rapidly – first uniformly along its entire length, and then locally to form a 'neck'. This 'necking' occurs just after the maximum force has been reached, at *M,* and since the cross-section decreases rapidly at the neck, the force at *B* required to break the specimen is *much less* than the maximum load at *M.*

This might be an appropriate moment to point out the difference between a force–extension diagram and a stress–strain diagram, since the terms are often loosely and imprecisely used. Figure 3.1 clearly represents a force–extension diagram, since the total force is plotted against the total extension, and, as the force decreases past the point *M,* for the reasons mentioned earlier, this decrease is indicated by the diagram. If, however, we wished to plot *stress* (force per unit area of cross-section of the specimen), we would need to measure the minimum cross-section of the specimen, as well as its length, for each increment of force. This would be particularly important for the values of force after the point *M,* since in this part of the test the cross-section is decreasing rapidly, due to the formation of the 'neck'. Thus, if the stress were calculated on this decreasing diameter, the resulting stress–strain diagram would follow a path indicated by the broken line to B_1 (Figure 3.1). In practice, however, the original cross-section of the material is used and the term *engineering stress* used for the force at any stage of the loading cycle divided by the original area of cross-section.

Thus, a *nominal* value of the *tensile strength* of a material is calculated using the maximum force (at *M*) and the original cross-sectional area of the test-piece:

$$\text{tensile strength} = \frac{\text{maximum force used}}{\text{original area of cross-section}}$$

Tensile strength is a useful guide to the mechanical properties of a material. It is primarily an aid to quality control because it is a test which can be carried out under easily standardised conditions but it is not of paramount import-ance in engineering design. After all, the engineer is not particularly interested in the material once plastic flow occurs – unless he happens to be a produc-tion engineer interested in deep-drawing, or some other forming process. In terms of structural or constructional engineering, the elastic limit *A* will be of far greater significance.

The force–extension diagram shown in Figure 3.1 is typical of a low-carbon steel in the annealed condition. Unfortunately, force–extension diagrams for heat-treated steels, and for most other alloys, do not often show

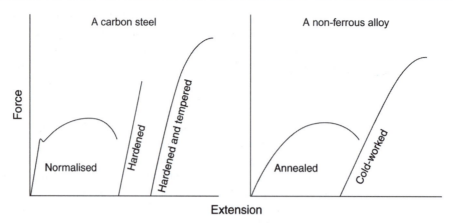

Figure 3.2 *Typical force–extension diagrams for both carbon steels and non-ferrous materials.*

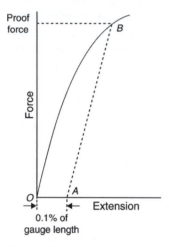

Figure 3.3 *The determination of 0.1% proof stress.*

a well-defined yield point, and the 'elastic portion' of the graph merges gradually into the 'plastic section', as shown in the examples in Figure 3.2. This makes it almost impossible to assess the yield stress of such an alloy and, in cases like this, yield stress is replaced by a value known as *proof stress*.

The 0.1% proof stress of an alloy (denoted by the symbol $R_{p0.1}$) is defined as that stress which will produce a permanent extension of 0.1% in the gauge length of the test-piece. This is very roughly equivalent to the permanent extension remaining in a normalised steel at its yield point.

The 0.1% proof stress of a material is derived as shown in Figure 3.3. The relevant part of the force–extension diagram is plotted as described earlier. A distance *OA,* equal to 0.1% of the gauge length, is marked along the horizontal axis. A line is then drawn from *A,* parallel to the straight-line portion of the force–extension diagram. The line from *A* intersects the diagram at *B* and this indicates the proof force which would produce a permanent extension of 0.1% in the gauge length of the specimen. From this value of force, the 0.1% proof stress can be calculated:

$$0.1\% \text{ proof stress} \;=\; \frac{\text{proof force}}{\text{original cross-sectional area of test-piece}}$$

It will be obvious from the above that if the yield stress (or 0.1% proof stress) is exceeded in a material, then fairly rapid plastic deformation and failure are likely to occur.

3.2.3 The percentage elongation

In addition to determining the tensile strength and the 0.1% proof stress (or alternatively, the yield stress), the percentage elongation of the test-piece at fracture is also derived. This is a measure of the ductility of the material. The two halves of the broken test-piece are fitted together (Figure 3.4) and the extended gauge length is measured, then:

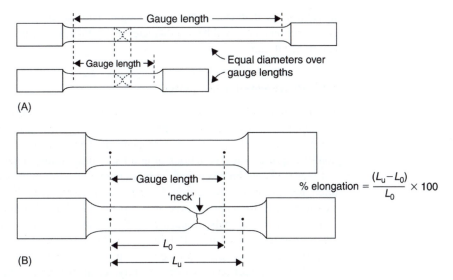

Figure 3.4 *The determination of the percentage elongation.*

$$\% \text{ elongation} = \frac{\text{increase in gauge length}}{\text{original gauge length}} \times 100$$

Mild steel is reasonably ductile and has a percentage elongation of between 18 and 25. Grey cast iron is, however, rather brittle with virtually zero percentage elongation. Table 2.3 gave some typical values of percentage elongations.

3.2.4 Proportional test-pieces

The two test-pieces in Figure 3.4A are of similar material and of equal diameter. Consequently, the dimensions and shape of both 'necked' portions will also be similar, that is the *increase* in length will be the same in each test-piece. However, since different gauge lengths have been used it follows that the percentage elongation reported on gauge length would be different for each. Therefore, in order that the values of the percentage elongation shall be comparable, it is obvious that test-pieces should be geometrically similar; that is, there must be a standard relationship or ratio between cross-sectional area and gauge length. Test-pieces which are geometrically similar and fulfil these conditions are known as *proportional test-pieces*. The British Standards Institution lays down that, for proportional test-pieces:

$$L_0 = 5.65 \sqrt{S_0}$$

where L_0 is the gauge length and S_0 the original area of cross-section. This formula has been accepted by international agreement. For test-pieces of circular cross-section, it gives a value:

$$L_0 = 5d \text{ (approximately)}$$

where d is the diameter of the gauge length. Thus, a test-piece 200 mm^2 in cross-sectional area will have a diameter of 15.96 mm (16 mm) and a gauge length of 80 mm.

3.2.5 Representative test-pieces

Test-pieces must be as representative as is possible of the material under test. This applies to test-pieces in general and not only those used in a tensile test. Many materials are far from homogeneous. Thus, because of the segregation of impurities and variations in grain size in castings, generally tensile test-pieces should be taken from more than one position in a casting. Quite often, test-pieces are made from 'runners' and 'risers' (see Section 5.3) of a casting and will generally give an adequate overall guide to quality.

In wrought materials, impurities will be more evenly distributed and the grain size more uniform but there will inevitably be a directionality of properties caused by the formation of 'fibres' of impurity in the direction of working (see Section 6.6). Consequently, tensile test-pieces should, where possible be made from material along the fibre direction (Figure 3.5A) and also at right angles to the fibre direction (Figure 3.5B). For narrow strip material this is not possible, though test-pieces of rectangular cross-section are commonly used. General methods for tensile testing of metals are covered by BS EN ISO 6892-1: 2009: *Methods of Tensile Testing Metals*, whilst the procedures for specific alloys are covered in the appropriate BS or BSEN specifications for those alloys.

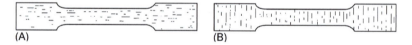

Figure 3.5 *Tensile test-pieces and fibre direction in wrought materials.*

The earlier paragraphs apply principally to the testing of metals. The tensile – and other – testing of plastics involves its own particular problems, not the least of which is the influence of temperature. The mechanical testing of plastics materials is therefore considered separately (see Section 20.3) after an introduction to the physical properties of polymers has been outlined.

3.3 Hardness tests

A true definition of surface hardness is the capacity of that surface to resist abrasion as, for example, by the cutting action of emery paper. Thus, early attempts to quantify hardness led to the adoption of Moh's scale which was used originally to assess the relative hardness of minerals. This consists of a list of materials arranged in order of hardness, with diamond the hardest of all, with a hardness index 10, at the head of the list and talc, with an index of 1, at the foot (Table 3.1). Any mineral in the list will scratch any one below it and in this way the hardness of any 'unknown' substance can be related to the scale by finding which substance on the scale will just scratch it, the substance beneath it will not, and a hardness index thus assigned to it.

Table 3.1 *Moh's scale of hardness*

Material	Hardness index	Material	Hardness index
Diamond	10	Apatite	5
Sapphire	9	Fluorspar	4
Topaz	8	Calcite	3
Quartz	7	Gypsum	2
Feldspar	6	Talc	1

Although this method of testing is useful in the classification of minerals rather than for the determination of hardness of metals, it nevertheless agrees with the definition of surface hardness as the resistance of a surface to abrasion. In the Turner sclerometer (from the Greek: skleros meaning hard), devised in Birmingham by T. T. Turner in 1886, a diamond point, attached to a lever arm, was loaded and drawn across the polished surface of the test-piece, the load being increased until a scratch was just visible. The 'hardness number' was quoted as the load in grams required to press the diamond point sufficiently to produce a 'normal' scratch. Obviously, there was considerable room for error in judging what was a 'normal' scratch. For this reason, modern methods of hardness testing really measure the material's resistance to penetration rather than to abrasion. They are therefore somewhat of a compromise on the true meaning of hardness but have the advantage of being easier to determine with accuracy.

3.3.1 The Brinell hardness test

This test, devised by a Swede, Dr. J. A. Brinell, is probably the best known of the hardness tests (BS EN 1003: *Methods for the Brinell Hardness Test*). A hardened steel ball is forced into the surface of a test-piece by means of a suitable standard load (Figure 3.6). The diameter of the impression is then measured, using some form of calibrated microscope, and the Brinell hardness number (H) is found from:

$$H = \frac{\text{load } P}{\text{area of curved surface of the impression } A}$$

If D is the diameter of the ball and d that of the impression, it can be shown that:

$$A = \frac{\pi}{2}D\left(D - \sqrt{D^2 - d^2}\right)$$

It follows that:

$$H = \frac{P}{\frac{\pi}{2}D\left(D - \sqrt{D^2 - d^2}\right)}$$

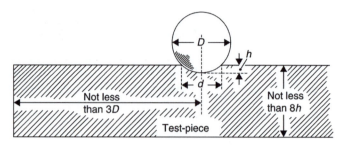

Figure 3.6 *The relationship between ball diameter D, depth of impression h and dimensions of the test-piece in the Brinell test.*

To make tedious calculations unnecessary, H is generally found by reference to tables which relate H to d – a different set of tables being used for each possible combination of P and D.

In carrying out a Brinell test, certain conditions must be fulfilled. First, the depth of impression must not be too great relative to the thickness of the test-piece, otherwise we may produce the situation shown in Figure 3.7A. Here it is the table of the machine, rather than the test-piece, which is supporting the load. Hence, it is recommended that the thickness of the test-piece shall be at least 8 times the depth of the impression. The width of the test-piece must also be adequate to support the load (Figure 3.6), otherwise the edges of the impression may collapse due to the lack of support and so give a falsely low reading.

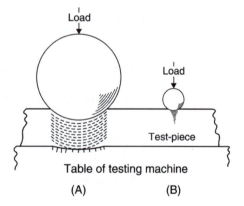

Figure 3.7 *This illustrates the necessity of using the correct ball diameter in relation to the thickness of the test-piece.*

Materials of non-uniform cross-section also present a problem. Thus, the surface skin of cold-rolled metal plate or strip will be much harder than the interior layers and the use of a small ball in the Brinell test would therefore suggest a higher hardness index than if a larger ball were used, since in the latter case the ball would be supported by the softer metal of the interior.

Similarly, case-hardened steels (see Section 14.1) consist of a very hard skin supported by softer but tougher metal beneath and the only satisfactory way to deal with such a situation is to cut a slice through the component, polish and etch it, so that the extent of the case can be seen, and then make hardness measurements at appropriate points across the section using the smallest ball or, preferably, the Vickers diamond pyramid test (see Section 3.3.2).

Balls of 10, 5 and 1 mm diameter are available; so one appropriate to the thickness of the test-piece should be chosen, bearing in mind that the larger the ball it is possible to use, the more accurate is the result likely to be. Having decided upon a suitable ball, we must now select a load which will produce an impression of reasonable proportions. If, for example, in testing a soft metal we use a load which is too great relative to the size of the ball, we shall get an impression similar to that indicated in Figure 3.8A. Here, the ball has sunk to its full diameter, and the result is meaningless. On the other hand, the impression shown in Figure 3.8B would be obtained if the load were too small relative to the ball diameter, and here the result would be uncertain. For different materials, then, the ratio P/D^2 has been standardised (Table 3.2) in order to obtain accurate and comparable results. P is still measured in 'kg force' and D in mm.

As an example, in testing a piece of steel, we can use a 10 mm ball in conjunction with a 3000 kgf load, a 5 mm ball with a 750 kgf load or a 1 mm ball with a 30 kgf load. As mentioned earlier, the choice of ball diameter D will rest with the thickness of the test-piece, whilst the load to be used with it will be determined from the appropriate P/D^2 ratio.

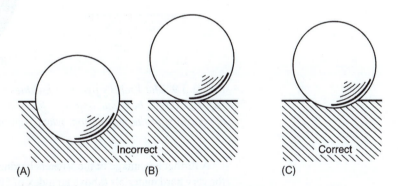

Figure 3.8 *It is essential to use the correct P/D^2 ratio for the material being tested.*

Table 3.2 *P/D^2 ratios for the Brinell test*

Material	P/D^2
Steel	30
Copper alloys	10
Aluminium alloys	5
Lead alloys and tin alloys	1

Typical Brinell hardness values are: for as-rolled carbon steel 125–295 and when annealed 110–190, for soft austenitic stainless steel (302S31) 183. White cast iron is a much harder material and has a Brinell hardness of about 500.

3.3.2 The Vickers pyramid hardness test

This test uses a square-based diamond pyramid (Figure 3.9) as the indentor. One great advantage of this is that all the impressions will be the same square shape, regardless of how big an indentation force is used. Consequently, the operator does not have to choose a P/D^2 ratio as he does in the Brinell test, though he must still observe the relationship between the depth of impression and thickness of specimen, for reasons similar to those indicated in the case of the Brinell test and illustrated in Figure 3.7. Here, the thickness needs to be at least 1.5 times the diagonal length of the indentation (BS EN ISO 6507: *Vickers Hardness Test*).

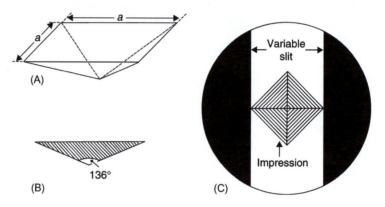

Figure 3.9 *The Vickers pyramid hardness test. (A) The diamond indentor. (B) The angle between opposite faces of the diamond is 136°. (C) The appearance of the impression, when viewed, in the microscope eyepiece.*

A further advantage of the Vickers hardness test is that the hardness values for very hard materials (above an index of 500) are likely to be more accurate than the corresponding Brinell numbers – a diamond does not deform under high pressure to the same extent as does a steel ball, and so the result will be less uncertain.

In this test, the diagonal length of the square impression is measured by means of a microscope which has a variable slit built into the eyepiece (Figure 3.9C). The width of the slit is adjusted so that its edges coincide with the corners of the impression, and the relative diagonal length of the impression is then obtained from a small instrument geared to the movement of the slit, and working on the principle of a revolution counter. The ocular reading thus obtained is converted to a Vickers pyramid hardness number (VPN) by reference to tables (in some quarters, this test is referred to as the diamond pyramid hardness (DPH) test). The size of the impression is related to hardness in the same way as is the Brinell number, i.e.

$$H = \frac{\text{load } P}{\text{surface area of indentation } A}$$

where P is in kgf. The surface area A of the indentation is $d^2/(2 \sin \theta/2)$, where θ is the angle of the diamond pyramid apex, i.e. 136°, and d is the arithmetic mean of the diagonals (in mm).

Since the impression made by the diamond is generally much smaller than that produced by the Brinell indentor, a smoother surface finish is required on the test-piece. This is produced by rubbing with fine emery paper of about '400 grit'. Surface damage is negligible, making the Vickers test more suitable for testing finished components. The specified time of contact between the indentor and the test-piece in most hardness tests is 15 seconds. In the Vickers testing machine (Figure 3.10), this period of contact is timed automatically by a piston working in an oil dashpot. Typical Vickers hardness values are: brasses in the half hard condition 80 and the full hard condition 130.

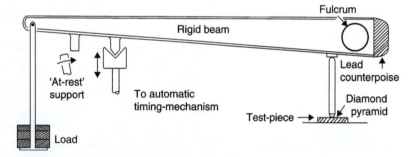

Figure 3.10 *The loading system for the Vickers pyramid hardness machine. It is essentially a second-order lever system. The 15-second period of load application is timed by an oil dashpot system.*

3.3.3 The Rockwell hardness test

The Rockwell test was devised in the USA. It is particularly useful for rapid routine testing of finished material, since the hardness number is indicated on a dial, and no subsequent measurement of the diameter is involved. The test-piece, which needs no preparation save the removal of dirt and scale from the surface, is placed on the table of the instrument and the indentor is brought into contact with the surface under 'light load'. This takes up the 'slack' in the system and the scale is then adjusted to zero. 'Full load' is then applied, and when it is subsequently released (timing being automatic), the test-piece remains under the 'light load', whilst the hardness index is read directly from the scale. Although the penetration depth h (Figure 3.11) of the impression is measured by the instrument, this is converted to hardness values which appear on the dial. The Rockwell hardness number is given by:

Rockwell hardness $= E - h$

where E is a constant determined by the form of the indentor; for a diamond cone indentor E is 100, for a steel ball 130 (BS EN 10109: *Rockwell Hardness Test;* BS 4175: *Rockwell Superficial Hardness Test*)

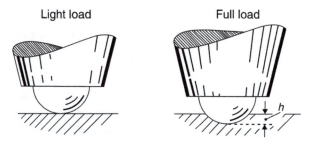

Figure 3.11 *The Rockwell test.*

There are several different scales on the dial, the scale being determined by the indentor and 'full load' used. The most important scales are:

1   Scale B, which is used in conjunction with a 1/16 inch diameter steel ball and a 100 kgf load. This is used mainly for softer metals, such as copper alloys, aluminium alloys, normalised steel and mild steel.
2   Scale C, which is used in conjunction with a diamond cone of 120° angle and a 150 kgf load. This is used mainly for hardened steels and other very hard materials such as hard cast irons. A typical value on this scale for a water-hardening tool steel at 20°C is 63.
3   Scale A, which is used in conjunction with the diamond cone and a 60 kgf load. This is used for extremely hard materials, such as tool steels.

The Rockwell machine is very rapid in action, and can be used by relatively unskilled operators. Since the size of the impression is also very small, it is particularly useful for the routine testing of stock or individual components on a production basis.

3.3.4 The Shore scleroscope

This is a small portable instrument which can be used for testing the hardness of large components such as rolls, drop-forgings, dies, castings and gears. Such components could not be placed on the table of one of the more orthodox machines mentioned earlier. The scleroscope embodies a small diamond-tipped 'tup', or hammer, of mass approximately 2.5 g, which is released so that it falls from a standard height of about 250 mm inside a graduated glass tube placed on the test surface. The height of rebound is taken as the hardness index.

Soft materials absorb more of the kinetic energy of the hammer, as they are more easily penetrated by the diamond point, and so the height of rebound is less. Conversely, a greater height of rebound is obtained from hard materials.

3.3.5 Wear resistance

It was suggested at the beginning of this section that the well-established hardness tests do not represent the classical idea of surface hardness, that is the resistance of a surface to abrasion as is recorded in Moh's test for minerals or by the long obsolete Turner sclerometer. Instead, modern 'hardness' tests measure the resistance to penetration which, however, is closely connected to resistance to abrasion. In some instances it is desirable to obtain a more accurate assessment of the *wear resistance* of a material. Then specific tests can be devised which estimate the amount of wear which occurs when two surfaces rub together under standardised conditions. Thus, a disc of the material under examination can be rotated against a standard abrasive disc for a given number of revolutions under a standard pressure. The loss in mass suffered by the test disc is a direct measure of the abrasion it has suffered.

Obviously, the coefficient of friction between the rubbing surfaces is involved and hard materials have a relatively low coefficient. When two surfaces are pressed into contact there is likely to be contact made at only a few 'high spots'. However, since metals can deform plastically under pressure, the areas of contact at the 'high spots' will increase. Interatomic bonding forces in metals are high, and so increasing the 'real' areas of contact will resist the local removal of metal by friction. Moreover, due to this deformation, *work-hardening* occurs and this further resists abrasion. Surface abrasion, in particular, is important in the design of bearing surfaces.

3.4 Impact tests

These tests are used to indicate the toughness of a material, and particularly its capacity for resisting mechanical shock. Brittleness, resulting from a variety of causes, is often *not* revealed during a tensile test. For example, nickel–chromium constructional steels suffer from a defect known as *temper brittleness* (see Section 13.2). This is caused by faulty heat-treatment, yet a tensile test-piece derived from a satisfactorily treated material and one produced from a similar material but which has been incorrectly heat-treated might both show approximately the same tensile strengths and elongations. In an impact test, however, the difference would be apparent; the unsatisfactory material would prove to be extremely brittle as compared with the correctly treated one, which would be tough.

With the Izod and Charpy impact tests, a heavy pendulum, mounted on ball-bearings, is allowed to strike a test-piece after being released from a fixed height (Figure 3.12). The tests differ in the form of test-pieces used and height from which the pendulum swings. The striking energy is partially absorbed in breaking the test-piece, and as the pendulum swings past, it carries with it a drag pointer which it leaves at its highest point of swing. This indicates the amount of mechanical energy used in fracturing the test-piece.

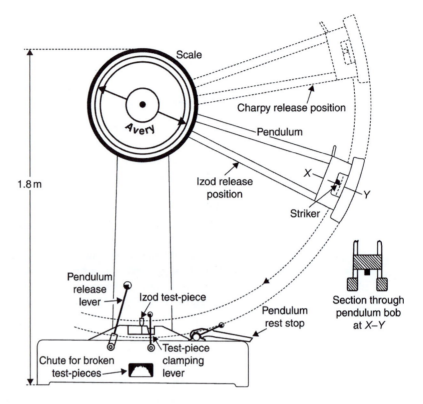

Figure 3.12 *The Avery–Denison universal impact-testing machine. This machine can be used for either Charpy or Izod impact tests. For Izod tests, the pendulum is released from the lower position, to give a striking energy of 170 J and for the Charpy test it is released from the upper position to give a striking energy of 300 J. The scale carries a set of graduations for each test. The machine can also be used for impact-tension tests.*

3.4.1 The Izod impact test

This test employs a standard notched test-piece (Figure 3.13) which is clamped firmly in a vice. The striking energy is approximately 170 J. The test-piece is notched so that there is an initial 'crack' to initiate fracture. It is essential, however, that this notch always be standard, for which reason a standard gauge is supplied to test the dimensional accuracy of the notch, both in this and other impact tests dealt with below (BS 131: *Notched-bar Tests*). Typical Izod impact values are: for hard drawn 4% phosphor bronze 62 J, 7% annealed aluminium bronze 32 J, 070M20 carbon steel quenched and tempered 40 J, 120M19 manganese steel quenched and tempered 40 J.

Examination of the fractured cross-section of the test-piece reveals further useful information. In the most ductile materials the fractured surface is likely to be of a 'fibrous' nature and will be rather dull and 'silky' in appearance (Figure 3.14A) since plastic flow of the crystalline structure has occurred. With very brittle materials, the fractured surface will be relatively bright,

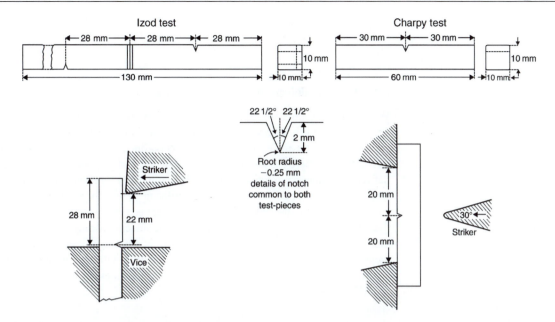

Figure 3.13 *Details of standard test-pieces used in both the Izod and Charpy tests.*

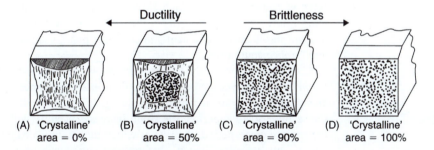

Figure 3.14 *The nature of the fractured surface in the Izod test.*

sparkling and 'crystalline' since crystals have not been plastically deformed. As fracture has followed the crystal boundaries, each small crystal reflects a bright point of light (Figure 3.14D). Many metals will fall somewhere between these extremes and the fractured surface will show a combination of ductile and 'crystalline' areas. For steels in particular, it is possible to estimate quite accurately the percentage crystalline area of the fractured surface of the test-piece and to use this as a measure of the notch-ductility.

3.4.2 The Charpy impact test

This test is of continental origin, and differs from the Izod test in that the test-piece is supported at each end (Figure 3.13), whereas the Izod test uses a test-piece held cantilever fashion. Here, the load on the pendulum can be varied so that the impact energy is either 150 J or 300 J. Typical Charpy impact test values are: for cartridge brass (70% Cu, 30% Zn) annealed 88 J and 3/4 hard 21 J, grey cast iron 3 J, carbon steel with 0.2% carbon as rolled 50 J.

3.4.3 Effect of defects on impact properties

Whilst the presence of minute cavities, slag particles, grain-boundary segregates and other flaws is likely to have a limited effect on the tensile strength of a metal and relatively none on its hardness, it will seriously impair its impact properties. Defects such as these will act as stress raisers, particularly under conditions of shock loading, by introducing points at which a sudden concentration of stress builds up. Thus, there will tend to be a much greater 'scatter' of results in the impact test on a single cast material than in the tensile test for the same material, since a casting is likely to be much less homogeneous in its structure than is a wrought material. Nevertheless, in wrought materials directionality of fibre will have a much more significant effect on impact toughness than it does on either tensile strength or ductility (see Table 6.2).

3.5 Creep

When stressed over a long period of time, some metals extend very gradually and may ultimately fail at a stress *well below* the tensile strength of the material. This phenomenon of slow but continuous extension under a steady force is known as *creep*. Such slow extension is more prevalent at high temperatures, and for this reason the effects of creep must be taken into account in the design of steam and chemical plant, gas and steam turbines and furnace equipment.

Creep occurs generally in three stages (Figure 3.15), termed *primary*, *secondary* and *tertiary* creep. During the primary creep period, the strain is changing but the rate at which it is changing with time decreases. During the secondary creep period, the strain increases steadily with time. During the tertiary creep period, the rate at which the strain is changing increases and eventually failure occurs. A family of curves can be produced which show the creep for different initial stresses and temperatures. At low stress and/or low temperature (curve I in the figure) some *primary* creep may occur but this falls to a negligible amount in the *secondary* stage when the creep curve becomes almost horizontal. With increased stress and/or temperature (curves II and III), the rate of secondary creep increases, leading to *tertiary* creep and inevitable catastrophic failure.

Creep tests are carried out on test-pieces which are similar in form to ordinary test-pieces. A test-piece is enclosed in a thermostatically controlled electric tube furnace which can be maintained accurately at a fixed temperature over the long period of time occupied by the test. The test-piece is statically stressed, and some form of sensitive extensometer is used to measure the extremely small extensions at suitable time intervals. A set of creep curves, obtained for different static forces at the same temperature, is finally produced, and from these the limiting creep stress is derived.

The stress which will produce a limiting amount of creep in a fixed time, say 2 or 3 days, at a particular temperature can be determined. Thus, the

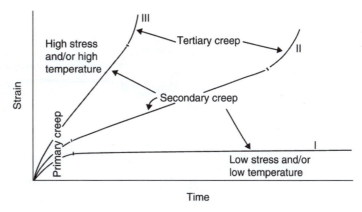

Figure 3.15 *The variation of creep rate with stress and temperature.*

stress/temperature combination which will produce a limiting value of creep strain of, say, 10^{-3} in a time of 1000 hours might be determined. The term *stress to rupture* of a material is used for the initial stress which will result in failure at a particular temperature after some period of time. For example, for a 0.2% plain carbon steel at 400°C, the stress to rupture in 1000 hours is 295 MPa; at 500°C it has dropped to 118 MPa.

3.6 Fatigue

Whenever failure of some structure or machine leads to a disaster which is worthwhile reporting, the press invariably hints darkly at 'metal fatigue'. Unfortunately, the technical knowledge of these gentlemen of the press is often suspect, as is evidenced by the oft-repeated account of the legendary cable 'through which a *current* of 132 000 V was flowing'.

Although the public was first made aware of the phenomenon of fatigue following investigations of the Comet airline disasters in 1953 and 1954, the underlying principles of metal fatigue were appreciated more than a century ago by the British engineer Sir William Fairbairn, who carried out his classical experiments with wrought iron girders. He found that a girder which would support a static load of 12 tonf for an indefinite period would nevertheless *fail* if a load of only 3 tonf were raised and lowered on it some 3 million times. Fatigue is associated with the effects which a fluctuating or an alternating force may have on a member – or, in everyday engineering terms, be subjected to the action of a 'live load'. If you take a paper clip and repeatedly flex it back and forth, it does not last very long before it breaks, even though you have not applied a large stress – you would find it very difficult to break it by direct tensile pulling!

3.6.1 *S/N* curves

Following the work of Fairbairn, the German engineer Wöhler, with native inventiveness, produced the well-known fatigue-testing machine which still bears his name. This is a device (Figure 3.16A) whereby alternations of stress can be produced in a test-piece very rapidly, and so reduce to a reasonable

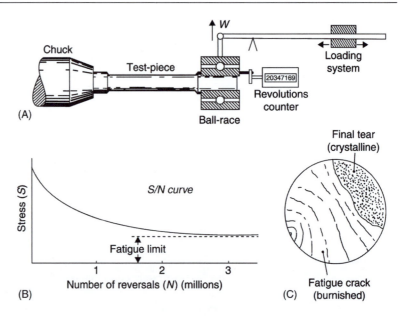

Figure 3.16 *(A) The principle of a simple fatigue-testing machine. (B) A typical S/N curve obtained from a series of tests. (C) The appearance of the fractured surface of a shaft which has failed due to fatigue.*

period the time required for a fatigue test. As the test-piece runs through 180°, the force W acting on the specimen falls to zero, and then increases to W in the opposite direction. To find the fatigue limit, a number of similar specimens of the material are tested in this way, each at different values of W, until failure occurs, or, alternatively, until about 20 million reversals have been endured (it is of course not possible to subject the test-piece to the ideal infinite number of reversals). From these results, an S/N curve is plotted; that is, stress S against the number of cycles endured (Figure 3.16B). The curve becomes horizontal at a stress which will be endured for an infinite number of reversals. This stress is the *fatigue limit* or *endurance limit*. Some non-ferrous materials do not show a well-defined fatigue limit; that is, the S/N curve slopes gradually down to the horizontal axis. With these materials it is only possible to design for a limited life, since eventually, regardless of the stress, they will fail through fatigue.

A fatigue fracture has a characteristic type of surface and consists of two parts (Figure 3.16C). One is smooth and burnished and shows ripple-like marks radiating outwards from the centre of crack formation, whilst the other is coarse and crystalline, indicating the final fracture of the remainder of the cross-sectional area which could no longer withstand the load.

Any feature which increases stress concentrations may precipitate fatigue failure. Thus, a fatigue crack may start from a keyway, a sharp fillet, a microstructural defect or even a bad tool mark on the surface of a component which has otherwise been correctly designed with regard to the fatigue limit of the material from which it was made.

3.6.2 Fatigue failure case study

Fatigue failure can be due simply to bad design and lack of understanding of fatigue, but is much more likely to be due to the presence of unforeseen high-frequency vibrations in a member which is stressed above the fatigue limit. This is possible since the fatigue limit is well below the tensile strength of all materials. Fatigue failure will ultimately occur in any member in which the operating stress fluctuates or alternates to a stress which is above its fatigue limit.

Some time ago, the author's opinion was sought on the possible cause of failure of copper tube which had held, cantilever fashion, a small pressure gauge in the manner shown in Figure 3.17. The broken tube was sent with the information that it had broken off at XX_1 under the action of no other force than the weight of the pressure gauge P, which was very small. The fracture appeared, on examination, to be of a typical fatigue nature. This tentative suggestion brought forth derisory comments from the engineer in charge, who claimed that the copper tube must be of pure quality, since failure had taken place after only a few days of service in *static* conditions. On ultimately visiting the site, the author's suspicions were soon confirmed – the pipe T was connected to a large air-compressor, the nerve-shattering vibrations from which were apparent at a distance of several hundred metres.

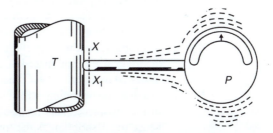

Figure 3.17 *Fatigue failure with a cantilevered pressure gauge.*

3.7 Other mechanical tests

Most of these tests are designed to evaluate some property of a material which is not revealed adequately during any of the preceding tests.

3.7.1 The Erichsen cupping test

Materials used for deep-drawing are inevitably those of very high ductility. However, a simple measurement of ductility in terms of percentage elongation (obtained during the tensile test) does not always give a complete assessment of deep-drawing properties. A test which imitates the conditions present during a deep-drawing operation is often preferable. Such a test is the Erichsen cupping test (Figure 3.18) and it is commonly used to assess the deep-drawing quality of soft brass, aluminium, copper or mild steel.

In this test, a hardened steel ball is forced into the test-piece, which is clamped between a die-face and a blank-holder. When the test-piece splits, the height of the cup which has been formed is measured, and this height

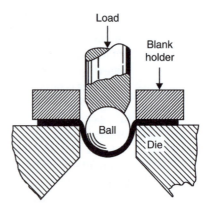

Figure 3.18 *The principle of the Erichsen cupping machine.*

(in mm) is taken as the Erichsen value. Unfortunately, the results from such a test can be variable, even with material of uniform quality. The depth of cup that can be drawn depends largely upon the pressure between the blank-holder and the die-face. Light pressure will allow metal to be drawn between the die and the holder, and a deeper cup is formed. The continentals tend to use firm pressure between die and holder, whereas in Britain we 'cheat' by using lighter pressure. Nevertheless, the British Standards Institution recognises the Erichsen test in BS EN ISO 20482; test-piece dimensions, testing apparatus, lubricants and precautions are dealt with.

Probably, the most useful aspect of the test is that it gives some idea of the grain size of the material, and hence its suitability for deep-drawing, as indicated in Figure 3.19. Coarse grain, always associated with poor ductility in a drawing operation, will show up as a rough rumpled surface in the dome of the test-piece. This resembles the 'lumpy' outside skin of an orange – hence the description as the 'orange-peel effect'.

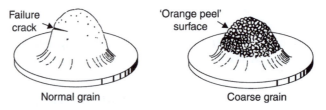

Figure 3.19 *Erichsen test-pieces.*

3.7.2 Bend tests

Bend tests are often used as a means of judging the suitability of a metal for similar treatment during a production process. For example, copper, brass or bronze strip used for the manufacture of electrical switch-gear contacts by simple bending processes may be tested by a bending operation which is somewhat more severe than that which will be experienced during production.

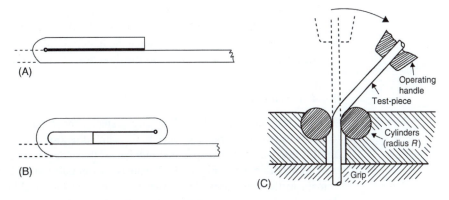

Figure 3.20 *Simple bend tests. (A) The material is bent back upon itself. (B) It is doubled over its own thickness, the second bend being the test bend. (C) A specific radius R is used.*

Simple bend tests are probably more closely related to the fundamental ideas of toughness than are the widely used impact tests. The latter are more easily measured in numerical terms and in any case deal specifically with conditions of *shock* loading. Figure 3.20A and B illustrates simple tests requiring little equipment other than a vice, whilst C represents a more widely accepted test in which the wire is bent through 90° over a cylinder of specified radius *R,* then back through 90° in the opposite direction. This is continued until the test-piece breaks, the number of bending cycles being counted. The surface affected by the bending process is also examined for cracks, and, if necessary, for coarse grain ('orange peel').

3.7.3 Compression tests

These tests are used mainly in connection with cast iron and concrete, since these are materials more likely to be used under the action of compressive forces than in tension. A cylindrical block, the length of which is twice its diameter, is used as a test-piece (Figure 3.21A). This is compressed (using a tensile-testing machine running in 'reverse') until it fails.

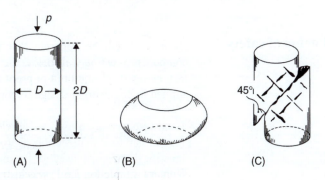

Figure 3.21 *The behaviour of brittle and ductile materials during a compression test.*

Malleable metals do not show a well-defined point of failure (Figure 3.21B), but with brittle materials (Figure 3.21C) the ultimate compressive stress can be measured accurately, since the material fails suddenly, usually by multiple shear at angles of 45° to the direction of compression.

3.7.4 Torsion tests

Torsion tests are often applied to wire – steel wire in particular. The test consists of twisting a piece of wire in the same direction round its own axis until it breaks, or until a specified number of twists has been endured. The simple machine used (Figure 3.22) consists essentially of two grips which remain in the same axis. The test-piece is held in the machine so that its longitudinal axis coincides with the axis of the grips and so that it remains straight during the test. This is achieved by applying a constant tensile force just sufficient to straighten it but not exceeding 2% of the nominal tensile strength of the specimen. One grip remains stationary, whilst the other is rotated at a constant speed which shall be sufficiently slow to prevent any rise in temperature of the test-piece, the distance between the grips being adjustable so that test-pieces of different lengths can be tested. This rotation continues until the test-piece breaks or until the specified number of turns is registered. The number of complete turns of the rotating grip is counted.

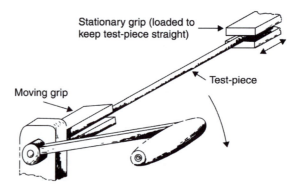

Figure 3.22 *A very simple 'twist' testing machine.*

3.8 Factor of safety

By now it will be apparent that the working stress in an engineering component must be well below the tensile strength of the material and, in fact, below its elastic limit or proof stress. Factors of safety have long been applied in engineering design to cope with this situation:

$$\text{factor of safety} \; = \; \frac{\text{tensile strength}}{\text{safe working stress}}$$

Wöhler (see Section 3.6.1) was instrumental in establishing factors of safety based on the earlier criteria. For 'dead loads' he suggested a factor of 4; for 'live loads' a factor of 6 or 8 and for alternating loads a safety factor between

12 and 18. In modern practice, a factor of safety is influenced by many features such as cost, uncertainty of working conditions and consequences of possible failure. Thus, in structural steelwork a factor of 3 is commonly used, whilst in steam boilers where creep, fluctuating stresses and the dire consequences of failure must be considered, a much higher factor of safety of 8 or 10 may be involved. Nevertheless, in aircraft construction where it is important to keep weight to a minimum, the factor of safety may be as low as 2 in some cases. In view of the catastrophic consequences of failure, this means that rigorous quality control must be applied to materials and components used as well as stress analysis to all members involved. In addition, periodic non-destructive investigation is used to detect the onset of such faults as fatigue cracks.

It was mentioned earlier that the elastic limit (or proof stress) is in reality a much more useful value than the tensile strength in engineering design since, once the elastic limit has been exceeded, the metal has already begun to fail. Thus, it has become more frequent practice to define factors of safety in terms of the elastic limit or proof stress. Obviously, it becomes necessary to specify whether the factor of safety is based on the *tensile strength* or on *proof stress*. Thus:

$$\text{proof factor of safety} = \frac{\text{proof stress}}{\text{safe working stress}}$$

Whichever factor of safety is adopted, a mixture of mathematical design criteria and *experience* of the behaviour of materials is generally used to enable a reliable estimate of the safe working stress to be made.

4 The crystal structure of metals

4.1 Introduction

All metals – and other elements, for that matter – can exist as either gases, liquids or solids. The 'state' in which a metal exists depends upon the conditions of temperature and pressure which prevail at the time. Thus, mercury will freeze to form a solid, rather like lead, if cooled to $-38.8°C$ and will boil to form a gas or vapour if heated to $357°C$ at atmospheric pressure. At the other end of the scale, tungsten melts at $3410°C$ and boils at $5930°C$.

This chapter is about the regular patterns adopted by metal atoms when metals solidify, such regular patterns being termed *crystalline* structures, and how they solidify into such patterns.

4.2 From gas to solid

In any gas, the particles (either atoms or molecules, depending upon the constitution of the gas) are in a state of constant motion and the impacts which these particles make with the walls of the containing vessel constitute the pressure exerted by the gas. As the temperature increases so their kinetic energy increases and hence the velocity of the particles increases. They thus travel faster between the container walls and hence the number of impacts per unit time with the walls increases and therefore the pressure within the vessel increases. This is the basis of the kinetic theory of gases.

In a metallic gas, the particles consist of single atoms (metallic gases are said to be *monatomic*, with most non-metallic gases such as oxygen and nitrogen consisting of molecules each of which contains two atoms) which are in a state of continuous motion. As the temperature falls, condensation occurs at the boiling-point and in the resultant liquid metal the atoms are jumbled together willy-nilly. Since they are held together only by weak forces of attraction at this stage, the liquid lacks cohesion and will flow. When the metal solidifies, the energy of each atom is reduced. This energy is given out as latent heat during the solidification process, which for a pure metal occurs at a fixed temperature (Figure 4.1). During solidification, the atoms arrange themselves according to some regular pattern, or 'lattice structure' and so each atom becomes firmly bonded to its neighbours by stronger forces of attraction; so the solid metal acquires strength.

Since the atoms are now arranged in a regular pattern, they generally occupy less space. Thus, most metals shrink during solidification, and the foundryman must allow for this, not only making the wooden pattern a little larger than the required size of the casting, but also by providing adequate runners and risers, so that molten metal can feed into the body of a casting as it solidifies and so prevent the formation of internal shrinkage cavities.

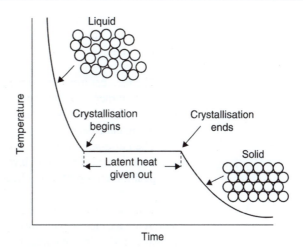

Figure 4.1 *A pure metal solidifies at a fixed single temperature, and the atoms arrange themselves in some regular pattern.*

4.3 Metal crystals

Most of the important metals crystallise into one of three different patterns as solidification takes place (Figure 4.2). The (A) diagrams represent the simplest building block from which much larger structures are built, like a

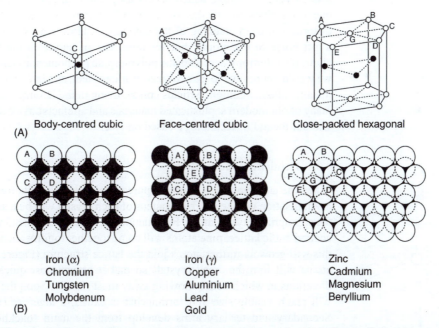

Figure 4.2 *The principal types of lattice structure occurring in metals. (A) indicates the positions of the centres of the atoms only, in the simplest unit of the structure, whilst (B) shows how these occur in a continuous crystal structure viewed in 'plan'. In each case, the letters A, B, C, etc. indicate the appropriate atoms in both diagrams. Although the atoms here are shown in black or white, to indicate in which layer they are situated, they are, of course, all of the same type. The lattice structures were derived by X-ray analysis.*

brick from which buildings are made. Only the positions occupied by the centres of the atoms are shown. Each outer 'face' of the figure is also part of the next adjacent unit, as shown in (B).

- The *face-centred cubic* arrangement can be considered to have atoms at each corner of a cube and also one in the centre of each face.
- The *body-centred cubic* has atoms at each corner of a cube and also one in the cube centre.
- The *close-packed hexagonal* structure is the arrangement which would be produced if a second layer of snooker balls were allowed to fall into position on top of a set already packed in the triangle.

Of these 'space lattices', the close-packed hexagonal arrangement represents the closest packing of atoms. The face-centred cubic arrangement is also a fairly close packing of atoms, but the body-centred cubic form is relatively 'open'; that is, the lattice contains relatively more space.

Iron is a *polymorphic element,* i.e. an element which exists in more than one crystalline form, and exists in two principal crystalline forms. The body-centred cubic form (α-iron) exists up to 910°C when it changes to the face-centred cubic form (γ-iron). On cooling again, the structure reverts to the body-centred cubic. It is this fact which enables steel to be heat-treated in its own special way, as we shall see later in Chapter 12. Unfortunately, the sudden volume change which occurs as γ-iron changes to α-iron on being quenched gives rise to the formation of internal stresses, and sometimes distortion or even cracking of the component. However, we must not look a gift horse in the mouth: but for this freak of nature in the form of the allotropic (allotropy describes the polymorphic phenomenon) change in iron, we would not be able to harden steel. Without steel, as Man's most important metallic alloy, we might perhaps still be living in the Bronze Age, for could many of our modern sophisticated materials and artefacts have been produced without the aid of technology based on steel?

4.3.1 Dendritic solidification

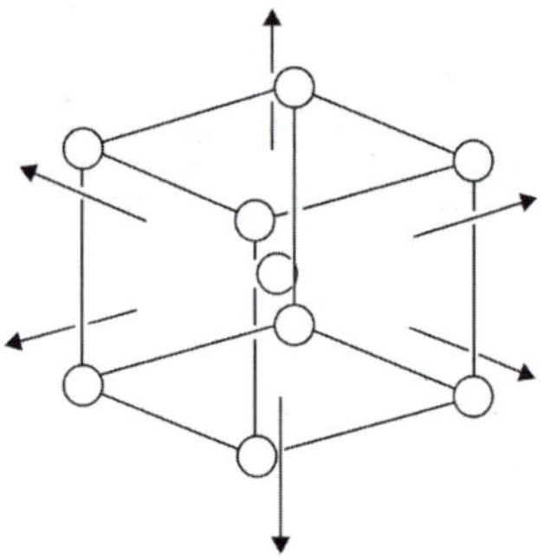

When the temperature of a molten pure metal falls below its freezing point, crystallisation will begin. The nucleus of each crystal will be a single unit of the appropriate crystal lattice. For example, in the case of a metal with a body-centred cubic lattice, nine atoms will come together to form a single unit and this will grow as further atoms join the lattice structure (Figure 4.3). These atoms will join the 'seed crystal' so that it grows most quickly in those directions in which heat is flowing away most rapidly. Soon the tiny crystal will reach visible size and form, what is called, a *dendrite* (Figure 4.4). Secondary and tertiary arms develop from the main 'backbone' of the dendrite – rather like the branches and twigs which develop from the trunk of a tree, except that the branches in a dendrite follow a regular geometrical pattern. The term *dendrite* is, in fact, derived from the Greek *dendron* – a tree.

The arms of the dendrite continue to grow until they make contact with the outer arms of other dendrites growing in a similar manner nearby. When

Figure 4.3 *The nucleus of a metallic crystal (in this case a body-centred cubic structure).*

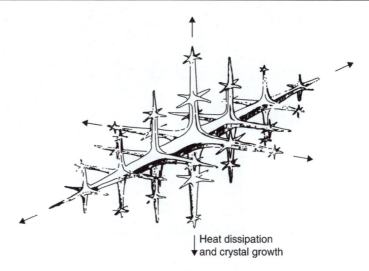

Heat dissipation
and crystal growth

Figure 4.4 *The early stages in the growth of a metallic dendrite.*

the outward growth is thus restricted, the existing arms thicken until the spaces between them are filled, or, alternatively, until all the remaining liquid is used up (Figure 4.5). As mentioned earlier in this chapter, shrinkage usually accompanies solidification, and so liquid metal will be drawn in from elsewhere to fill the space formed as a dendrite grows. If this is not possible, then small shrinkage cavities are likely to form between the dendrite arms.

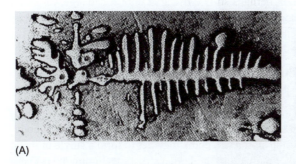

(A)

(B)

Figure 4.5 *Metallic dendrites. In (A) a molten solution of iron in copper was allowed to freeze. The iron, being of higher melting-point, crystallised first and was all used up before it could complete the formation of a crystal, so leaving a dendrite 'skeleton' of primary and secondary 'arms'. The molten copper then solidified, filling in the spaces. To be more accurate, the iron dendrite contains a small amount of dissolved copper, whilst the solid copper matrix contains a small amount of dissolved iron. In (B) a zinc dendrite was grown by electrolysis from a water solution of zinc sulphate.*
   Note that the secondary 'arms' in the iron dendrite (A) are at right angles to the 'backbone' because the crystal structure in iron is basically cubic, whilst in the zinc dendrite (B) the secondary 'arms' are at angles of 60° to the backbone, reflecting the close-packed hexagonal structure of zinc.

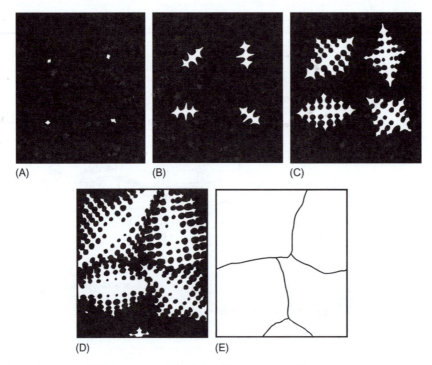

Figure 4.6 *The dendritic solidification of a metal. (A) Nuclei form. (B) Dendrites begin to grow from these nuclei. (C) Secondary arms form, and meet others growing in the opposite direction. (D) Dendrites grow until their outer arms touch. Existing arms thicken and – (E) when the metal is completely solid, only the grain boundaries are visible.*

Figure 4.7 *A section through a small aluminium ingot in the cast condition, etched in hydrofluoric acid (2%). The crystals are all of the same composition – pure aluminium. Since each crystal has developed independently, its lattice structure is 'tilted' at a random angle to the surface section, and so each crystal reflects light of a different intensity.*

Since each dendrite forms independently, it follows that outer arms of neighbouring dendrites are likely to make contact with each other at irregular angles and this leads to the irregular overall shape of crystals, as indicated in Figures 4.6 and 4.7.

4.4 Impurities in cast metals

If the metal we have been considering is pure, then there will be no hint of the dendritic process of crystallisation once solidification is complete, because all atoms in a pure metal are identical. However, if impurities were dissolved in the molten metal, these would tend to remain in solution until solidification was almost complete. They would therefore remain concentrated in that metal which solidified last, i.e. between the dendrite arms. They would thus reveal the dendritic pattern (Figure 4.8) when a suitably prepared specimen was viewed under the microscope.

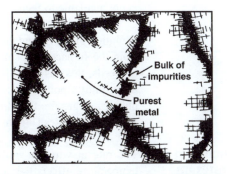

Figure 4.8 *The local segregation of impurities. The heavily shaded region near the crystal boundaries contains the bulk of the impurities.*

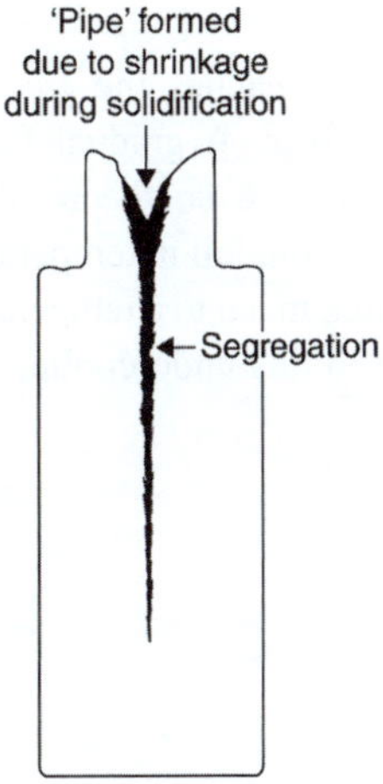

This concentration or *segregation* of impurities at crystal boundaries explains why a small amount of impurity can have such a devastating effect on mechanical properties, making the cast metal brittle and likely to fail along the crystal boundaries. In addition to this local segregation at all crystal boundaries, there is a general accumulation of impurities in the central 'pipe' of a cast ingot (Figure 4.9). This is where metal solidifies last of all, becoming most charged with impurities, relatively pure metal having crystallised during the early stages of solidification.

Some impurities are insoluble in the molten metal and these generally float to the surface to form a slag. They include manganese sulphide in steels, residual from the combined deoxidisation and desulphurisation processes (see Chapter 11), and oxide dross in brass, bronzes and aluminium alloys. Some of the material may be carried down into the casting as a result of turbulence caused by pouring the charge and become entrapped by the growing dendrite arms. Such inclusions may occur anywhere in the crystal structure. Particles of manganese sulphide in steel are dove-grey in colour.

The formation of shrinkage cavities was mentioned earlier. They will form only between dendrite arms (Figure 4.10) and along crystal boundaries during the final stages of solidification when, due to poor moulding design

'Pipe' formed due to shrinkage during solidification

←Segregation

Figure 4.9 *The segregation of impurities in the central 'pipe' of an ingot. In casting steel ingots, the central pipe is confined by using a fire clay collar on top of the mould (see Figure 5.1). The mould is tapered upwards, so that it can be lifted off the solid ingot.*

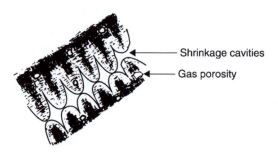

Figure 4.10 *The distribution of shrinkage cavities and gas porosity.*

or adverse casting conditions, pockets of molten metal remain isolated. As these solidify and shrink, small cavities remain into which molten metal is not available to feed.

Other cavities may be found in cast metals and are generally due to the evolution of gas during the solidification process. This gas may be dissolved from the furnace atmosphere during the melting process. As solidification proceeds, this gas tends to remain dissolved in the still molten part of the metal until a point is reached where the solubility limit is exceeded. Tiny bubbles of gas are then liberated and become trapped between the rapidly growing dendrites (Figure 4.10).

4.5 The influence of cooling rates on crystal size

The rate at which a molten metal is cooling as it reaches the freezing temperature affects the size of the crystals which form. A gradual fall in temperature results in the formation of few nuclei and so the resultant crystals grow unimpeded to a large size. However, a rapid fall in temperature will lead to some degree of *undercooling* of the molten metal to a temperature *below* its actual freezing-point (Figure 4.11). Due to this undercooling and

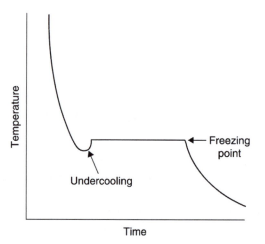

Figure 4.11 *Undercooling of a molten metal due to a steep temperature gradient.*

the instability associated with it, a sudden shower of nuclei is produced and, because there are many nuclei, the ultimate crystals will be tiny. As the foundryman says, 'Chilling causes *fine-grain* castings' (in metallurgy, the term *grain* is often used to mean 'crystal' and does not imply any directionality of structure as does the term grain in timber technology).

Because of the difference in the rates of cooling, the resultant grain size of a die-casting is small as compared with that of a sand-casting. This is an advantage, since fine-grained castings are generally tougher and stronger than those with a coarse grain size.

When a large ingot solidifies, the rate of cooling varies from the outer skin to the core during the crystallisation process. At the onset of crystallisation, the cold ingot-mould chills the molten metal adjacent to it, so a layer of small crystals is formed. Due to the heat flow outwards, the mould warms up and so chilling becomes less severe. This favours the growth inwards of elongated or columnar crystals. These grow inwards more quickly than fresh nuclei can form and this results in their elongated shape. The residue of molten metal, at the centre, cools so slowly that very few nuclei form and so the crystals in that region are relatively large. They are termed *equi-axed crystals* – literally 'of equal axes' (Figure 4.12).

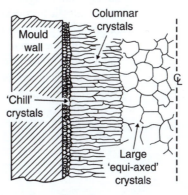

Figure 4.12 *Zones of different crystal forms in an ingot.*

A cast material tends to be rather brittle, because of both its coarse grain and the segregation of impurities mentioned earlier. When cast ingots are rolled or forged to some other shape, the mechanical properties of the material are vastly improved during the process, for reasons discussed later (see Section 6.6). Frequently, the engineer uses sand- or die-castings as integral parts of a machine. Often no other method is available for producing an intricate shape, though in many cases a sand-casting will be used because it involves the cheapest method of producing a given shape. The choice rests with the production engineer: he must argue the mechanical and physical properties required in a component with the Spectre of Production Costs – who is forever looking over his shoulder.

4.5.1 Rapid solidification processing (RSP)

Rapid solidification processing has been widely investigated with the aim of producing even smaller crystals by increasing the rate of cooling as the freezing range of the metal or alloy is reached. In fact, metallurgists have devised rapid solidification processes involving cooling rates in the region of 10^{6}°C/second. Now cooling a molten metal so that its temperature falls at a rate equal to a rate of a million degrees Celsius per second taxes the ingenuity of the R and D staff (see Section 22.7) but, under such conditions, extremely tiny *micro-crystals* are produced and in some cases *amorphous,* i.e. non-crystalline, solid metals are obtained. Thus, in extreme cases, the non-crystalline liquid structure is retained since the metal is given no opportunity to crystallise.

A most important feature of this development is that an amorphous solid metal contains no heavily segregated areas (Section 4.4) since any impurity would be dispersed evenly throughout the liquid metal and will remain so in the amorphous solid. Segregation only occurs during crystallisation. In RSP, the crystals are tiny as there is a much increased area of grain-boundary per unit volume of metal over which impurities are spread. This thinner spreading of impurities means that the metal will be more uniform in composition and materials which contain no regions of brittle segregate are stronger and tougher as well as being malleable and ductile.

Further, it becomes possible to make alloys from metals which do not normally mix (dissolve in each other) as liquids, by quenching a mixture of them from the *vapour* state. RSP powders can also be produced in which the alloy is in a micro-crystalline or amorphous state. The powder is then consolidated under pressure and subsequently shaped by extrusion (see Section 7.2).

5 Casting process

5.1 Introduction

Casting is the process in which metal is heated to make it molten and then poured into a mould. Most metallic materials pass through a molten state at some stage during the shaping process. Generally, molten metal is cast into ingot form prior to being shaped by some mechanical working process. However, some metals and alloys – and a number of non-metallic products such as concrete – can be shaped only by casting since they lack the necessary properties of either malleability or ductility which make them amenable to working processes. This chapter is about the casting processes used with metals.

A few metallic substances are produced in powder form. The powder is then compressed and sintered to provide the required shape. This branch of technology is termed *powder-metallurgy* (see Chapter 7).

5.2 Ingot-casting

Many alloys, both ferrous and non-ferrous, are cast in the form of *ingots* which are then rolled, forged or extruded into strip, sheet, rod, tube or other sections. When produced as single ingots, steel is generally cast into large iron moulds holding several tonnes of metal. These moulds generally stand on a flat metal base and are tapered upwards very slightly so that the mould can be lifted clear of the solid ingot. A 'hot top' (Figure 5.1) is often used in order that a reservoir of molten metal shall be preserved until solidification of the body of the ingot is complete. The reservoir feeds metal into the 'pipe' which is formed as the main body of the ingot solidifies and contracts. Many non-ferrous metals are cast as slabs in cast-iron 'book-form' moulds, whilst some are cast as cylindrical ingots for subsequent extrusion.

5.2.1 Continuous casting

This process is used to produce ingots of both ferrous and non-ferrous alloys (Figure 5.2). Here, the molten metal is cast into a short water-cooled mould A which has a retractable base B. As solidification begins, the base is withdrawn downwards at a rate which will keep pace with that of pouring.

With ordinary ingots, a portion of the top must always be cropped off and rejected, since it contains the pipe, and is a region rich in impurities. In continuous casting, however, there is little process scrap, since very long ingots are produced and consequently there is proportionally less rejected pipe. In fact, pipe is virtually eliminated in modern steel-making since the continuous-cast ingot is conveyed by a system of guide rolls direct to the rolling mill from which the emerging product (plate, sheet, rod, etc.) is cut to the required lengths by a flying saw. The bulk of steel made in the developed world is produced by 'continuous' plant.

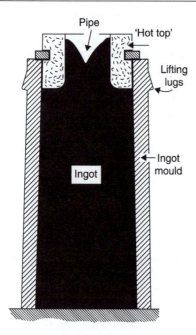

Figure 5.1 *A typical mould for producing the steel ingots. The 'hot top' restricts the formation of the 'pipe' which is caused by contraction of the metal as it solidifies.*

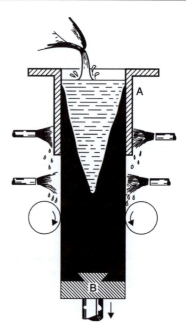

Figure 5.2 *A method for continuous casting. The molten metal is cast into mould A which has a retractable base B.*

5.3 Sand-casting

The production of a desired shape by a *sand-casting process* first involves moulding foundry sand around a suitable pattern in such a way that the pattern can be withdrawn to leave a cavity of the correct shape in the sand. To facilitate this, the sand pattern is split into two or more parts which can be separated so that the wooden pattern can be removed.

The production of a very simple sand mould is shown in Figure 5.3.

(A) The pattern of the simple gear blank is first laid on a moulding-board, along with the 'drag' half of a moulding-box.
(B) Moulding sand is now riddled over the pattern and rammed sufficiently for its particles to adhere to each other. When the drag has been filled, the sand is 'cut' level with the edge of the box.
(C) The assembly is turned over.
(D) A layer of parting sand (dry, clay-free material) is now sifted onto the sand surface, so that the upper half of the mould will not adhere to it when this is subsequently made. The 'cope' half of the moulding-box is now placed in position, along with the 'runner' and 'riser' pins which are held steady by means of a small amount of moulding sand pressed around them. The purpose of the runner in the finished mould is to admit the molten metal, whilst the riser provides a reservoir from which molten metal can feed back onto the casting as it solidifies and shrinks.

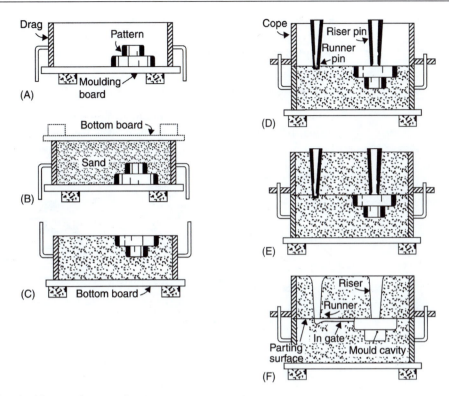

Figure 5.3 *Moulding with a simple pattern.*

(E) Moulding sand is now riddled into the cope and rammed around the pattern, runner and riser.

(F) The cope is then gently lifted off, the pattern is removed and the cope is replaced in position so that the finished mould is ready to receive its charge of molten metal.

The foregoing description deals with the manufacture of a sand mould of the simplest type. In practice, a pattern may be of such a complex shape that the mould must be split into several sections and consequently a multipart box is used. Cores may be required to form holes or cavities in the casting.

5.3.1 Uses of sand-casting

Sand-casting is a very useful process since very intricate shapes can be produced in a large range of metals and alloys. Moreover, relatively small numbers of castings can be made economically since the necessary cost of the simple equipment required is low. Wooden patterns are cheap to produce, as compared with the metal die which is necessary in die-casting processes. However, the process is labour intensive and so labour costs can be high. The process can be economical for low production runs or where one-offs are required. Typical uses of sand-casting are to produce engine blocks, machine tool bases and pump housings.

5.4 Die-casting

In *die-casting,* a permanent metal mould is used and the charge of molten metal is either allowed to run in under the action of gravity (gravity die-casting) or is forced in under pressure (pressure die-casting).

5.4.1 Pressure die-casting

A number of different types of machines are employed in pressure die-casting, but possibly the most widely used is the *cold-chamber machine* (Figure 5.4). Here, a charge of molten metal is forced into the die by means

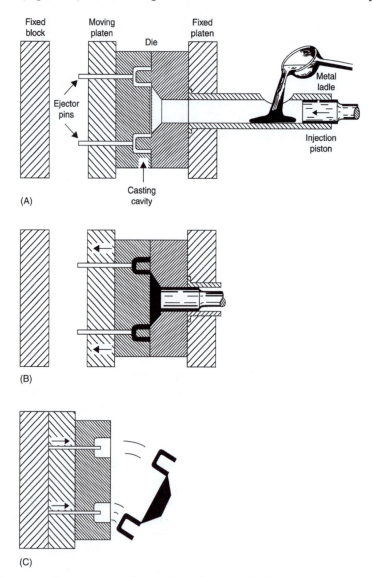

Figure 5.4 *A 'cold-chamber' pressure die-casting machine. The molten metal is forced into the die by means of the piston. When the casting is solid, the die is opened, and the casting moves away with the moving platen, from which it is ejected by 'pins' passing through the platen.*

of a plunger. As soon as the casting is solid, the moving pattern is retracted and, as it comes against a fixed block, ejector pins are activated so that the casting is pushed out of the mould. Cycling time is rapid.

5.4.2 Gravity die-casting

In gravity die-casting, or 'permanent-mould casting' as it is now generally called, the die is of metal and may be a multi-part design if the complexity of shape of the casting demands it. Metal cores of complex shapes must be split in order to allow their removal from the finished castings; otherwise sand cores may have to be used. The die cavity is filled under gravity, and the charge may be poured by hand or it may be fed in automatically in modern high-speed plant.

5.4.3 Uses of die-casting

The product of die-casting is metallurgically superior to that of sand-casting in that the internal structure is more uniform and the grains much finer, because of the rapid cooling rates which prevail. Moreover, output rates are much higher when using a permanent metal mould than when using sand moulds. Greater dimensional accuracy and a better surface finish are also obtained by die-casting. However, some alloys which can be sand-cast cannot be die-cast, because of their high shrinkage coefficients. Such alloys would inevitably crack due to their contraction during solidification within the rigid, non-yielding metal mould. Die-casting is confined mainly to zinc- or aluminium-base alloys.

Typical applications of gravity die-casting are for the production of cylinder heads, pistons, gear and die blanks and pressure die-casting to engine and pump parts, domestic appliances and toy parts.

5.5 Centrifugal-casting

This process has a permanent cylindrical mould, without any central core, which is spun at high speed and molten metal then poured into it (Figure 5.5). Centrifugal force flings the metal to the surface of the mould, thus producing a hollow cylinder of uniform wall thickness. The product has a

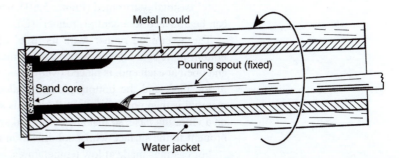

Figure 5.5 *The principle of the centrifugal casting of pipes. The mould rotates and also moves to the left, so that molten metal is distributed along the mould surface.*

uniformly fine-grained outer surface and is considered superior to a similar shape which has been sand-cast.

5.5.1 Uses of centrifugal-casting

Steel, iron, aluminium, copper and nickel alloys can be cast in this process. This process was widely used in the manufacture of cast-iron pipes for water, sewage and gas mains, though for such purposes cement- or plastic-based materials are now more commonly used. It is now used for such applications as pipes, gun-barrels and engine-cylinder liners.

5.6 Investment-casting

Investment-casting is in fact the most ancient casting process in use. It is thought that prehistoric man had learned to fashion an image from beeswax, and then knead a clay mould around it. The mould was then hardened by firing, a process which also melted out the wax pattern, leaving a mould cavity without cores or parting lines.

The process was rediscovered in the sixteenth century by Benvenuto Cellini, who used it to produce many works of art in gold and silver. Cellini kept the process secret, so it was lost again until, during the latter part of the nineteenth century, it was rediscovered for a second time and became known as the *cire perdue,* or *lost wax,* process. Since an expendable wax pattern is required for the production of each mould, it follows that a permanent mould must be first produced to manufacture the wax pattern – assuming of course that a large number of similar components is required. This *master mould* could be machined in steel or produced by casting a low melting-point alloy around a master *pattern.*

5.6.1 Production of castings

Figure 5.6 shows the sequence of operations involved in producing castings by this process.

To produce wax patterns, the two halves of the mould are clamped together and the molten wax is injected at a pressure of about 3.5 MPa. When the wax pattern has solidified, it is removed from the mould and the wax 'gate' is suitably trimmed (Figure 5.6B), using a heated hand-tool, so that it can be attached to a central 'runner' (C).

The assembled runner, with its 'tree' of patterns is then fixed to a flat bottom-plate by a blob of molten wax. A metal flask, lined with waxed paper and open at each end, is placed over the assembly. The gap between the end of the flask and the bottom plate is sealed with wax and the investment material is then poured into the flask. This stage of the process is conducted on a vibrating base, so that any entrapped air bubbles or excess moisture are brought to the surface whilst solidification is taking place (D).

For castings made at low temperatures, an investment mixture composed of very fine silica sand and plaster of Paris is still sometimes used. A more refractory investment material consists of a mixture of fine sillimanite sand and ethyl silicate. During moulding and subsequent firing of the mould, ethyl

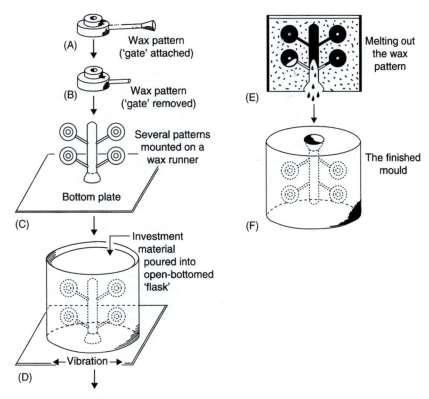

Figure 5.6 *The investment-casting (or 'lost-wax') process.*

silicate decomposes to form silica which knits the existing sand particles together to give a strong, rigid mould.

The investment is allowed to dry in air for some 8 hours. The baseplate is then detached and the inverted flask is passed through an oven at about 150°C, so that the wax melts and runs out, leaving a mould cavity in the investment material (E).

When most of the wax has been removed, the mould is preheated prior to receiving its charge of molten metal. The preheating temperature varies with the metal being cast, but is usually between 700°C and 1000°C. The object of the preheating is to remove the last traces of wax by volatilisation, to complete the decomposition of the ethyl silicate bond to silica, and also to ensure that the cast metal will not be chilled, but will flow into every detail of the mould cavity.

Molten metal may be cast into the mould under gravity, but if thin sections are to be formed then it will be necessary to inject the molten metal under pressure.

5.6.2 Applications of investment-casting

Tolerances in the region of 3–5 mm/m are obtained industrially by the investment-casting process, but it can also be used to produce complicated shapes which would be very difficult to obtain by other casting processes. A further advantage is the absence of any disfiguring parting line, which always appears on a casting made by any process involving the use of a two-part mould.

The main disadvantages of investment-casting are its high cost and the fact that the size of components is normally limited to 2 kg or so.

The process is particularly useful in the manufacture of small components from metals and alloys which cannot be shaped by forging and machining operations. Amongst the best-known examples of this type of application are the blades of gas-turbine and jet engines.

In addition to jet-engine blades, other typical investment-cast products include special alloy parts used in chemical engineering, valves and fittings for oil-refining plant, machine parts used in the production of modern textiles, tool and die applications such as milling-cutters, precision gauges and forming and swaging dies and parts for various industrial and domestic equipment, such as cams, levers, spray nozzles, food-processing plant, parts for sewing machines and washing machines. The materials from which many of these components are made are extremely hard and strong over a wide range of temperatures, making it impossible or, at best, expensive to shape them by orthodox mechanical methods. Investment-casting provides a means of producing the required shapes whilst retaining the required dimensional accuracy.

5.7 Full-mould process

This somewhat novel process resembles investment-casting in that a single-part mould is used, so that no parting lines, and hence no fins, appear on the finished casting. However, it is essentially a 'one-off' process, since the consumable pattern is made from expanded polystyrene. This is a polymer derived from benzene and ethylene, and in its 'expanded' form it contains only 2% actual solid polystyrene. Readers will be familiar with the substance which is widely used as a packaging material for fragile equipment. An expendable pattern, complete with runners and risers, is cut from expanded polystyrene (Figure 5.7A) and is then completely surrounded with silica sand containing a small proportion of thermosetting resin (B). The molten metal is then poured *onto the pattern,* which melts and burns very quickly, leaving a cavity which is immediately occupied by molten metal. No solid residue is formed and the carbon dioxide and water vapour evolved in the combustion of the polystyrene do not dissolve in the molten metal, but escape through the permeable mould.

The small amount of resin added to the sand is sufficient to form a bond between the sand grains, preventing premature collapse of the mould. In some cases, the resin is omitted and as the expanded polystyrene burns it produces a tacky bond between the sand grains for long enough for a skin of metal to form.

5.7.1 Use of full-mould casting

A well-known example of full-mould casting is Geoffrey Clarke's cross which originally adorned the summit of the post-war Coventry Cathedral.

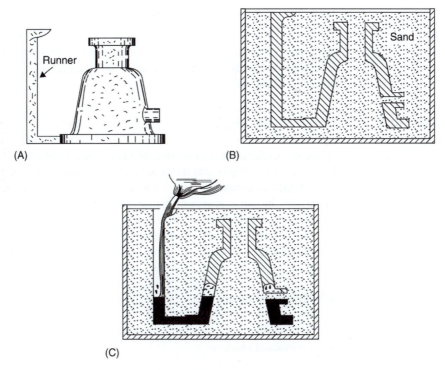

(A)

(B)

(C)

Figure 5.7 *The 'full-mould' process, which uses an expendable polystyrene pattern. (A) Polystyrene pattern. (B) Pattern moulded in single-part flask. (C) Casting in progress.*

In the engineering industries, the process is used in the manufacture of large press-tool die-holders, and similar components in the 'one-off' category.

5.7.2 The lost-foam process

This is a later modification of the full-mould process to adapt it to mass production. The expanded polystyrene patterns are made by injecting polystyrene beads into a heated slit aluminium die. Steam is then passed through the mould to fuse and expand the polystyrene to form the expanded solid pattern. The pattern can then be used in the same way as in full-mould casting involving a polystyrene pattern carved from solid expanded polystyrene.

5.8 Semi-solid metal processing

This is a process whereby suitable alloys can be shaped to something near final form from a mushy, part solid–part liquid state (see Section 9.2). The semi-molten alloy (containing about 50% solid) is stirred vigorously so that the dendrites tend to break up as they form and the viscosity of the slurry is greatly reduced. On standing, the slurry will begin to stiffen again, becoming more viscous, but the viscosity can be reduced again by stirring. This reversible behaviour is similar to that of thixotropic (non-drip) paint. Stirred slurries of this type can be die-cast in a normal cold-chamber machine (Figure 5.4) provided relatively simple shapes are attempted.

5.8.1 Uses of semi-solid metal processing

This type of process is used principally for manufacturing automobile components in aluminium alloys of suitable wide freezing range. It is an environment-friendly process since up to 35% fuel saving is achieved over normal die-casting processes.

5.9 The choice of casting process

Compared with other metal forming processes, casting is likely to be the optimum method if a part has a large or complex internal cavity. In such a situation, machining might be impossible or would certainly give rise to a lot of waste. Casting is also likely to be the optimum method if the material used is difficult to machine, e.g. white cast iron which is very hard. Casting might also be more economical than other methods for complex shaped objects.

If dimensional accuracy is the main criterion in choosing a casting method, then those processes described earlier could be considered in the following order:

1   Investment-casting
2   Die-casting
3   Sand-casting
4   Full-mould process

Relative costs per casting too will be roughly in that order.

Also to be taken into account would be the cost of any subsequent machine (metal cutting) operations. With very hard alloys, machining is either very expensive or impossible. In such cases, investment-casting comes into its own. A further advantage of the process is that there is no intrusive parting line. In two-part moulds or dies, pressure applied during the casting process may cause some separation of the mould parts and give rise to dimensional inaccuracy in the resulting casting. Cored holes which must be an accurate distance apart should never be separated by a parting line.

At the other end of the scale, the full-mould process is less suitable for producing castings of high dimensional accuracy but is particularly suitable where a one-off feature is to be cast, e.g. a statue or other architectural object in aluminium. Simple one-off machine parts for subsequent machining can also be made economically by this process. Thus, investment-casting may be chosen where a high dimensional accuracy is required in the resulting casting, whilst the full-mould process is a useful method where single castings of no great accuracy are required.

5.9.1 'Long run' production

'Long run' production of castings usually follows either the well established sand-casting or die-casting route. In general, die-casting will provide a better surface finish and greater dimensional accuracy, as well as superior mechanical properties (due to a smaller crystal size) than are available with sand-castings and one or more of these features may determine the choice

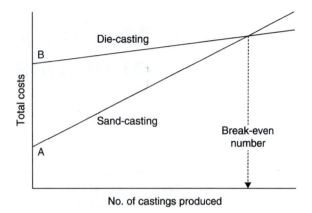

Figure 5.8 *The initial costs of plant and equipment are higher in the case of die-casting (B) than for sand-casting (A). Labour costs, however, are continuous during operation of the process and are higher for sand-casting. The break-even value, i.e. the number of castings which must be made before the use of die-casting is justified, may be of the order of several thousands. Once the break-even value has been reached, the cost per die-casting becomes less than the cost per sand-casting.*

of die-casting. However, where none of these features are important then the manufacturer will obviously choose the cheaper process. This in turn will depend upon a forecast of the number of castings required. Generally speaking, due to the high cost of producing metal dies, it is not economical to use die-casting unless a large number of castings are required. On the other hand, the cost of making a wooden pattern (for sand-casting) is relatively low but the labour costs of an experienced sand moulder are higher than those of a die-casting machine operator. Consequently, all other things being equal, the number of castings required will govern the choice of process (Figure 5.8).

6 Mechanical deformation of metals

6.1 Introduction

When the shape of a metal is changed by the application of forces, deformation takes place in two stages as is shown during a tensile test. At first, crystals within the metal are distorted in an elastic manner and this distortion increases proportionally with the increase in stress. If the stress is removed during this stage, the metal returns to its original shape, illustrating the *elastic* nature of the deformation. On the other hand, if the stress is increased further, a point is reached (the yield point) where forces which bind together the atoms in the lattice structure are overcome to the extent that layers or planes of atoms begin to slide over each other. This process of '*slip*', as metallurgists call it, is not reversible; so, if the stress is now removed, permanent deformation remains in the metal (Figure 6.1). This type of deformation is termed *plastic*.

This chapter is a look at how metals behave when mechanically deformed. Chapter 7 then follows on from this and looks at how mechanical deformation methods can be used to shape metals and obtain products.

6.2 Slip

Visual evidence of the nature of slip is fairly easily obtained. If a piece of pure copper or aluminium is polished and etched, and then squeezed in a vice with the polished face uppermost, so that it is not directly damaged, examination under the microscope shows a large number of parallel, hair-like lines on the polished and etched surface. These lines are in fact shadows cast by minute ridges formed when blocks or layers of atoms within each crystal slide over each other (Figure 6.2). These hair-like lines are known as *slip bands*.

Slip of this type can occur along a suitable plane, until it is prevented by some fault or obstacle within the crystal. A further increase in the stress will

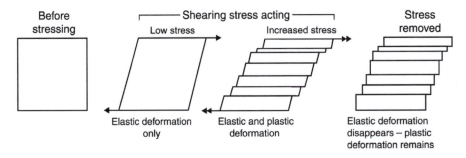

Figure 6.1 *Deformation of a metal by both elastic and plastic means.*

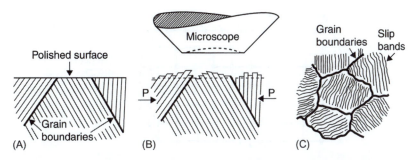

Figure 6.2 *The formation of slip bands: (A) indicates the surface of the specimen before straining and (B) the surface after straining. The relative slipping along the crystal planes produces ridges which cast shadows on the surface of the specimen when this is viewed through the microscope (C).*

then produce slip on another plane or planes, and this process goes on until all the available slip planes in the piece of metal are used up. The metal is then said to be *work-hardened,* and any further increase in stress will lead to fracture. Microscopic examination will show that individual crystals have become elongated and distorted in the direction in which the metal was deformed. In this condition, the metal is hard and strong; but it has lost its ductility, and, if further shaping is required, must be softened by annealing.

6.2.1 Dislocations

We know the distance apart of the atoms (or more properly ions) within a metallic crystal, as it can be measured with great accuracy using X-ray methods. We also know the magnitude of the forces of attraction between these positively charged ions and the negatively charged electrons within the metallic bond and how it varies as the ions are pulled further apart. It is therefore possible to calculate the *theoretical strength* of a metal, i.e. the force needed to overcome the sum total of all the net forces of attraction of the metallic bond across a given plane in a metal. Under controlled laboratory conditions, a single crystal of a metal can be grown and if this is subject to a tensile test its *true strength* is found to be only about one-thousandth of the theoretical when the latter is calculated as outlined earlier.

It is now realised that instead of slip taking place simultaneously by one block of atoms sliding wholesale over another block across a complete crystal plane, it occurs *step-by-step* by the movement of faults of discontinuities within the crystallographic planes. These faults, where in effect half-planes of atoms are missing within a crystal (Figure 6.3), are known as *dislocations*.

When the crystal is stressed to its yield point, these dislocations will move step-by-step along the crystallographic plane until they are halted or 'pegged' by some obstruction such as the atom of some alloy metal which is larger (or smaller) than those of the parent metal. Alternatively, the dislocation will move along until it is stopped by a crystal boundary of another dislocation moving across its path.

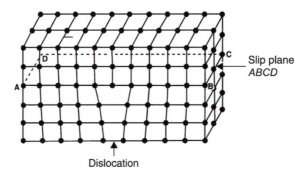

Figure 6.3 *An 'edge' dislocation. This is the most simple form of dislocation and can actually be seen in some materials when they are examined under very high magnifications in electron microscopy.*

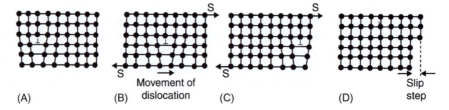

Figure 6.4 *Movement of an edge dislocation under the action of stress S. Dislocations are denoted by ⊥.*

If the surface of the metal cuts through the slip plane a minute 'step' will be formed there (Figure 6.4). If several hundreds of these dislocations follow each other along the same slip plane then the step at the surface may be large enough to be visible as a 'slip band' (Figure 6.2C) using an ordinary optical microscope.

The force necessary to move dislocations *step-by-step* through a metallic crystal is much less than the force which would be necessary to overcome the total forces of attraction between ions and electrons along the complete slip plane if this slip occurred *simultaneously in a single jerk*. Thus, the true strength of a metal is only a small fraction of that which was calculated on the assumption that slip took place by a single sudden movement.

6.2.2 The carpet analogy

Professor N. Mott's well-known analogy with the means available for smoothing the wrinkles from a carpet helps to explain this process of slip in metallic crystals. Imagine a heavy wall-to-wall carpet which, when laid, proves to be wrinkled near one wall (Figure 6.5A). Attempts to remove the wrinkle by pulling on the opposite edge (Figure 6.5B) will be fruitless and lead only to broken finger nails and utterance of unseemly expletives, since an attempt is being made to overcome the *sum* of the forces of friction between the *whole* of the carpet and the floor and so to produce the slip over the whole surface of the carpet at the same instant. Instead, it requires very little force to ease the wrinkle along with one's toe (Figure 6.5C) until it has

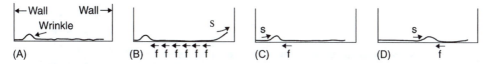

Figure 6.5 *Professor Mott's 'carpet analogy' in which a wrinkle in the carpet is likened to a 'dislocation' in a metallic crystal.*

been moved progressively to the opposite edge of the carpet (Figure 6.5D). In this way, only the small force of friction between the carpet and the floor *adjacent to the wrinkle* has to be overcome during any instant when that small part of the carpet is moving.

Similarly, the step-by-step movement of a dislocation through the crystal structure of a metal requires a much smaller force than would the movement of a large block of ions by simultaneous slip. The real strength of a metal is therefore much less than that calculated from consideration of a perfect crystal.

The movement of dislocations through a metal is hindered by the grain boundaries. So the more grain boundaries there are in a metal the more difficult it is to produce yielding. Thus, a treatment which reduces grain size makes a material stronger, whilst increasing the grain size makes it weaker.

6.2.3 Work-hardening

A metal becomes *work-hardened* when all the dislocations present have become jammed against each other or against various other obstacles within the crystal or indeed against the crystal boundaries themselves. Many dislocations will move out to the surface of the metal and be lost as slip bands.

6.2.4 Deformation by twinning

Whilst most metals deform plastically by just the process of slip described earlier, a few metals can also deform by what is known as *twinning*. Whereas during slip all ions in a block have moved the same distance when slip is complete, in deformation by twinning (Figure 6.6) ions within each successive plane in a block will have moved different distances. When twinning is complete, the lattice direction will have altered so that one half of the twinned crystal is a mirror image of the other half, the twinning plane corresponding to the position of the mirror. Twinning may be regarded as a special case of slip movement.

The stress required to cause deformation by twinning is generally higher than that needed to produce deformation by slip. Twinning is more commonly found in close-packed hexagonal (CPH) metals such as zinc and tin. When a bar of tin is bent quickly, the sound of the layers of ions moving relative to each other is emitted as an audible 'squeak'. This was known by the Ancients as 'The cry of tin'.

These twins formed during the mechanical deformation are not to be confused with 'annealing twins' occurring in copper, brasses, bronzes and

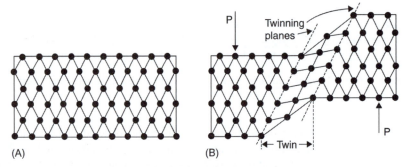

Figure 6.6 *Deformation by twinning: (A) before stressing and (B) after stressing. In the deformed crystal one half is a mirror image of the other half about the twin plane.*

austenitic stainless steels. In these instances, the twin crystals grow from a plane during recrystallisation of a cold-worked metal instead of a point source. Annealing twins are thus not produced as a result of deformation but form as part of grain growth.

6.3 Annealing

A cold-worked metal, i.e. one which has been deformed without the application of heat, is in a state of considerable internal stress due to the presence of elastic strains internally balanced within the distorted crystal structure. During annealing these stresses are removed as the thermal activation supplied allows ions to migrate into positions approaching nearer to their equilibrium positions and, as a consequence, the original ductility of the metal returns. The changes which accompany an annealing process occur in three stages:

1   The relief of stress
2   Recrystallisation
3   Grain growth

6.3.1 The relief of stress

As the temperature of a cold-worked metal is gradually raised, some of the internal stresses disappear as atoms and probably whole dislocations move through small distances into positions nearer their equilibrium positions. At this stage, there is no alteration in the generally distorted appearance of the structure and, indeed, the strength and hardness produced by cold-working remain high. Nevertheless, some hard-drawn materials, such as 70–30 brass, are given a low-temperature anneal in order to relieve internal stresses, as this reduces the tendency towards 'season cracking' during service, i.e. the opening up of cracks along grain boundaries due to the combined effects of internal stresses and surface corrosion.

6.3.2 Recrystallisation

With further increase in temperature, a point is reached where new crystals begin to grow from nuclei which form within the structure of the existing distorted crystals. These nuclei are generally produced where internal stress is greatest, i.e. where dislocations are jammed together at grain boundaries and on slip planes. As the new crystals grow they take up atoms from the old distorted crystals which they gradually replace. Unlike the old crystals, which had become elongated in one direction by the cold-working process, these new crystals are small and equi-axed.

This phenomenon, known as *recrystallisation* (Figure 6.7), is used to obtain a tough fine-grained structure in most non-ferrous metals. The minimum temperature at which it will occur is called the *recrystallisation temperature* for that metal (Table 6.1), though it is not possible to determine

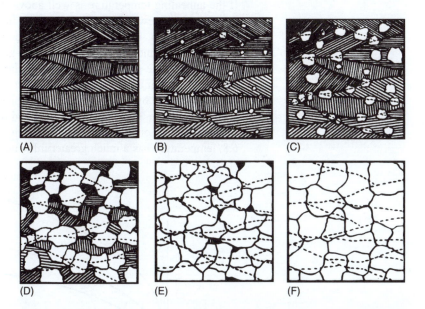

(A)   (B)   (C)

(D)   (E)   (F)

Figure 6.7 *Stages in the recrystallisation of a metal during an annealing process. (A) represents the metal in the cold-rolled state. At (B) recrystallisation has just begun, with the formation of new crystal nuclei. These grow by absorbing the old crystals, until at (F) recrystallisation is complete.*

Table 6.1 *Recrystallisation temperatures of some metals*

Metal	Recrystallisation temperature (°C)
Tungsten	1200
Nickel	600
Pure iron	450
Copper	190
Aluminium	150
Zinc	20
Lead, Tin	Below room temperature; hence, they cannot be 'cold-worked'

this temperature on the Kelvin scale precisely because it varies with the amount of cold-work to which the metal has been subjected before the annealing process. The more heavily the metal is cold-worked, the greater the internal stress, and the lower the temperature at which recrystallisation will begin. Alloying, or the presence of impurities, raises the recrystallisation temperature of a metal. Typically, the recrystallisation temperature is between 0.3 and 0.5 times the melting temperature.

6.3.3 Grain growth

If the annealing temperature is well above that for the recrystallisation of the metal, the new crystals will increase in size by absorbing each other cannibal fashion, until the resultant structure becomes relatively coarse-grained. This is undesirable, since a coarse-grained material is generally less ductile than a fine-grained material of similar composition. Moreover, if the material is destined for deep-drawing, coarse-grain tends to disfigure a stretched surface by giving it a rough, rumpled appearance known as *orange peel* (see Section 3.7.1). Both the time and temperature of annealing must be controlled, in order to limit grain growth; though, as indicated by Figure 6.8, temperature has a much greater influence than time.

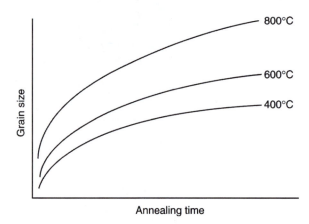

Figure 6.8 *The effects of time and temperature during annealing on the grain size of a previously cold-worked metal.*

The amount of cold-work the material receives prior to being annealed also affects the grain size produced. In heavily cold-worked metal, the amount of locked-up stress is great and so, when thermal activation is supplied during annealing, a large number of new crystal nuclei will form instantaneously as the recrystallisation temperature is reached and so the resultant crystals will be small since there will be less space in which individual crystals can grow.

On the other hand, very little cold-work (the 'critical' amount) gives rise to few nuclei, because the metal is not highly stressed. Consequently, the grain size will be large, and the ductility poor.

Alloy additions are often made to limit the grain growth of metals and alloys during the heat-treatment process. Thus, up to 5% nickel is added to case-hardening steels in order to reduce grain growth during the carburising process. When we say that nickel toughens a steel, we must remember that it does so largely by limiting its grain growth during heat-treatment.

6.4 Cold-working process

Most metals and alloys are produced in wrought form by hot-working processes, because they are generally softer and more malleable when hot, and consequently require much less energy to shape them. Hot-working processes make use of *compressive* forces to shape the work-piece. The reason for this is that metals become weak in tension at high temperatures so their ductility decreases. Any attempt to pull a metal through a die at high temperature would fail, because the metal would be so weak as to tear. Consequently, those shaping processes in which *tension* is employed are cold-working processes. Since most metals work-harden quickly during cold-working operations, frequent inter-stage annealing is necessary. This increases the cost of the process and so, as far as possible, operations involving the use of hot-working by compression are used, rather than cold-working processes which make frequent annealing stages necessary. Wire used to be made by drawing down a previously rolled bar. This involved many drawing operations, interspersed with annealing stages. Now, a cast billet is hot-extruded as thick wire in a single operation; this thick wire is then drawn down in a minimum of cold-working stages, through dies.

Reasons for using cold-working processes are:

- To obtain the necessary combination of strength, hardness and toughness for service. Mild steel and most non-ferrous materials can be hardened only by cold-work.
- To produce a smooth, clean surface finish in the final operation. Hot-working generally leaves an oxidised or scaly surface, which necessitates 'pickling' the product in an acid solution.
- To attain greater dimensional accuracy than is possible in hot-working processes.
- To improve the machinability of the material by making the surface harder and more brittle (see Section 7.5).

6.4.1 Uses of cold-working

The main disadvantage of cold-working is that the material quickly work-hardens, making frequent inter-stage annealing processes necessary. A typical sequence of operations for the manufacture of a cartridge-case by deep-drawing 70–30 brass sheet is shown in Figure 6.9. Here, the interstage annealing processes will be carried out in an oxygen-free atmosphere (bright annealing), in order to make acid-pickling unnecessary.

Other cold-working processes include:

- The drawing of wire and tubes through dies.
- The cold-rolling of metal plate, sheet and strip.

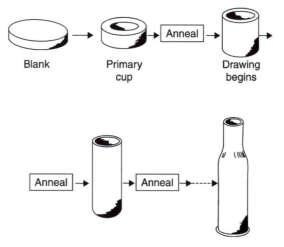

Figure 6.9 *Typical stages in the deep-drawing of a cartridge-case. In practice, many more intermediate operations may be necessary, only the principles being indicated here.*

- Spinning and flow-turning, as in the manufacture of aluminium kitchenware.
- Stretch forming, particularly in the aircraft industry.
- Cold-heading, as in the production of nails and bolts.
- Coining and embossing.

6.5 Hot-working processes

A hot-working process is one which is carried out at a temperature well above the recrystallisation temperature of the metal or alloy. At such a temperature, recrystallisation will take place simultaneously with deformation and so keep pace with the actual working process (Figure 6.10). For this reason, the metal will not work-harden and can be quickly and continuously reduced to its required shape, with the minimum of expended energy. Not only is the metal naturally more malleable at a high temperature, but it remains soft, because it is recrystallising continuously during the working process.

Thus, hot-working leads to a big saving both in energy used and in production time. It also results in the formation of a uniformly fine grain in the recrystallised material, replacing the original coarse grain cast structure. For this reason, the product is stronger, tougher and more ductile than was the original cast material.

The main disadvantage of hot-working is that the surface condition is generally poor, due to oxidation and scaling at the high working temperature. Moreover, accuracy of dimensions is generally more difficult to attain because the form tools need to be of simpler design for working at high temperatures. Consequently, hot-working processes are usually followed by some surface-cleaning process, such as water quenching (to detach scale), shot-blasting and/or acid-pickling prior to at least one cold-working operation which will improve the surface quality and accuracy of dimensions.

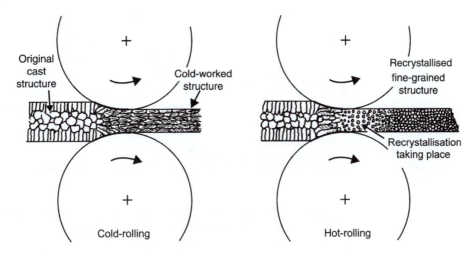

Figure 6.10 *During cold-rolling a metal becomes work-hardened, but during a hot-rolling process recrystallisation can take place.*

6.5.1 Uses of hot-working

The principal hot-working processes are:

- Hot-rolling, for the manufacture of plate, sheet, strip and shaped sections such as rolled-steel joists.
- Forging and drop-forging, for the production of relatively simple shapes, but with mechanical properties superior to those of castings.
- Extrusion, for the production of many solid and hollow sections (tubes) in both ferrous and non-ferrous materials.

Many metal-working processes involve preliminary hot-working, followed by cold-forming or finishing. Mild-steel sheet destined for the production of a motor-car body is first hot-rolled down from the ingot to quite thin sheet. This is cooled, and then pickled in acid, after which it is lightly cold-rolled to give a dense smooth surface. Finally, it is cold-formed to shape by means of presswork. This final shaping also hardens and strengthens the material, making it sufficiently rigid for service. Similarly, a drop-forged connecting rod will be finally 'sized' by 'tapping' it in a cold die, in order to improve both dimensions and finish.

6.6 Grain flow and fibre

During any hot-working process, the metal is moulded in a plastic manner with a result similar to that produced by a baker kneading his dough (though a machine now does this for him). Segregated impurities in the original casting are mixed in more uniformly, so that brittle films no longer coincide with grain boundaries. Obviously, these impurities do not disappear completely, but form 'flow lines' or 'fibres' in the direction in which the

Table 6.2 *The effects of hot-working on mechanical properties*

Condition	Tensile strength (MPa)	Percentage elongation	Izod impact (J)
As cast	540	16	3
Hot-rolled (in the direction of rolling)	765	24	100
Hot-rolled (at right angles to the direction of rolling)	750	15	27

material has been deformed. Since new crystals grow independently of these fibres, the latter weakens the structure to a much smaller extent than did the original intercrystalline films. Consequently, the material becomes stronger and tougher, particularly along the direction of the fibres. At right angles to the fibres, the material is weaker (Table 6.2), since it tends to pull apart at the interface between the metal and each fibre.

Stages in the manufacture of a simple bolt are shown in Figure 6.11. The bolt has been produced from steel rod in which a fibrous structure (A) is the result of the original hot-rolling process. The rod will be either hot- or cold-headed, and this operation will give rise to plastic flow in the metal, as indicated by the altered direction of the fibre in the head (B). The thread will then be rolled in, producing a further alternation in the direction of the fibre in the region (C). The mechanical properties of a bolt manufactured in this way will be superior to those of one machined from a solid bar (D). The latter would be much weaker, and it is highly likely that, in tension, the head may shear off, due to weakness along the flow lines – even supposing the thread had not already stripped for the same reason. This bolt (D) would be extremely anisotropic (this term means a substance having different physical properties in different directions) in its mechanical properties – in this instance strong parallel to the fibre direction but weak at right angles to the

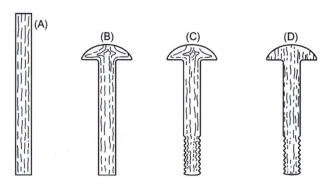

Figure 6.11 *The direction of 'fibre' in a cold-headed or forged bolt as compared with that of a bolt machined from a solid bar (D).*

fibre direction. For the mass production of bolts, the forging/thread-rolling method would be less costly in any case; so the choice of process here is very simple.

It is thus bad engineering practice to use any shaping process which exposes a cut fibrous structure such that shearing forces can act *along* exposed fibres and so cause failure. Similarly, in a drop-forging, it is essential that fibres are formed in those directions in which they will give rise to maximum strength and toughness.

6.6.1 The macro-examination of fibre direction

It is often necessary for the drop-forger to examine a specimen forging to ensure that grain flow is taking place in the desired direction. This will be indicated by fibre direction. Similarly, the engineer may wish to find out whether or not a component has been manufactured by forging – assuming that any such evidence as 'flash lines' and the like have been obliterated by subsequent light machining. Whereas *micro-examination* involves the use of a microscope to view the prepared surface (see Chapter 10), *macro-examination* implies that no such equipment is necessary, other than, possibly, a hand magnifier. Hence, the macro-examination of flow lines or fibre is a fairly simple matter.

A section is first cut so that it reveals fibre on a suitable face – in the case of the bolt mentioned earlier, a section cut symmetrically along its axis is used. The section is then filed perfectly flat, or, better still, ground flat with a linishing machine. It is then rubbed on successively finer grades of emery paper, the 'grits' most generally used being 120, 180, 220 and 280. Finer grades than these are not necessary, though the method of grinding requires a little care to ensure that deep scratches are eliminated from the surface. This is achieved by grinding the surface such that, on passing from the linisher to the coarsest paper, the specimen is turned through 90°. The new set of grinding marks will then be at right angles to the previous ones, and in this way it is easy to see when the old grinding marks have been removed. The same procedure is adopted in passing from one paper to the next. Small specimens are ground by rubbing the specimen on paper supported on a piece of plate glass. Large specimens are more difficult to manipulate, and it is often more convenient to grind such specimens by rubbing a small area at a time, using a wooden block faced with emery paper.

When a reasonably smooth surface with no deep scratches has been produced, the specimen is wiped free of swarf and dust. It is then immersed in a 50% solution of hydrochloric acid, and gently heated. Considerable effervescence takes place as hydrogen gas is liberated and the surface layers of steel are dissolved. Those areas containing the most impurity tend to dissolve more quickly and this helps to reveal the fibrous structure. The specimen is examined from time to time, being lifted from the solution with laboratory tongs. This deep-etching process may take up to 20 minutes or so, depending upon the quality of the steel.

When the flow lines are suitably revealed (Figure 6.12), the specimen is washed in running water for a few seconds, and then dried as quickly as

Figure 6.12 *The fibrous structure of a hot-forged component. The 'flowlines' indicate the direction in which the metal moved during the forging operation. Etched in boiling 50% hydrochloric acid for 15 minutes.*

possible, by immersion in 'white' methylated spirit, followed by holding in a current of warm air from a hairdryer.

If a section has been really deeply etched, it is possible to take an ink print of the surface. A blob of printer's ink is rolled on a flat glass plate, so that a thin but even film is formed on the surface of a rubber roller or squeegee. The roller is then carefully passed over the etched surface, after which a piece of paper is pressed on to the inked surface. Provided the paper does not slip in the process, a tolerable print of the flow lines will be revealed.

6.6.2 Mechanical deformation or casting

Compared with casting, products produced by mechanical deformation methods tend to have a greater degree of uniformity and reliability of mechanical properties. However, they do tend to have a directionality of properties which is not the case with casting. For this reason, casting may sometimes be preferred.

6.7 Metallurgical furnaces

These vary in size from small bench units used for the heat-treatment of dies and tools to the iron blast-furnaces some 60 m high (Section 11.1). Designs also differ widely, depending upon factors like fuel economy, operational temperature and chemical purity required in the charge. Either gas, oil, coke or electricity may be used as fuel.

Metallurgical furnaces can be conveniently classified into three groups according to the degree of contact between the fuel and the charge:

1   Furnaces in which the charge is in direct contact with the burning fuel and its products of combustion. Such furnaces include the blast-furnace and the foundry cupola. Due to intimate contact between the charge and the burning fuel and the fact that operation is continuous, the fuel efficiency is very high but impurities in the fuel may contaminate the charge.

2   Furnaces in which the charge is out of contact with the fuel but is in contact with the products of combustion. This group includes a large variety of gas- and oil-fired installations used in the melting, smelting and heat-treatment of metals. Heat efficiency is still reasonably high but the charge may still be contaminated by gaseous impurities originating in the burning fuel.

3   Furnaces in which the charge is totally isolated from the fuel and from its combustion products. Here, we include the crucible furnaces used for melting metals, the molten surface of which is often protected by a layer of molten flux; and the gas- or oil-fired muffle furnaces where the solid charge is heated in a totally enclosed chamber. Fuel efficiency is obviously lower than in the previous groups but the purity of the charge can be maintained and there is complete control over the furnace atmosphere so that bright annealing is employed.

Electricity has the obvious advantage as a metallurgical fuel in that it can be used to generate heat either by electrical resistance methods, by high- or low-frequency alternating current, or by direct-arc methods and at the same time produce no chemical contamination of the charge or the natural environment. Unfortunately, electricity is still rather more expensive than most other fuels.

6.7.1 Furnace atmospheres

Most common engineering metals become badly scaled during heat-treatment unless preventative precautions are taken. Scaling, i.e. oxidation, occurs when a metal is heated to a temperature which allows it to react with any oxygen in the surrounding atmosphere. In many cases, this can be prevented by burning the fuel gas with a restricted air supply so that *no free oxygen* remains in the gas which circulates within the furnace chamber (assuming that a fairly simple furnace is being used in which the burning fuel is in direct contact with the charge).

The principal fuel gases are based on hydrocarbons, i.e. methane (natural gas) and propane. When these burn in a mixture with air, the result is a mixture of nitrogen (N_2) (from the air used for combustion), carbon dioxide (CO_2), carbon monoxide (CO) and water vapour (H_2O). Depending on the air/gas ratio, the atmospheres can be:

•   *Exothermic atmospheres* are produced by combustion without the addition of external heat. Hydrocarbon gases are incompletely burned, the air supply being restricted so that no free oxygen remains in the atmosphere produced, e.g. air/methane ratio of 6/1 to 10/1 or propane 14/1 to 24/1. Since the product contains large amounts of water vapour and carbon dioxide, these are removed to give an atmosphere containing mainly carbon monoxide, hydrogen and nitrogen.

- *Endothermic atmospheres* are provided by combustion when external heat must be supplied. The initial gas, usually propane, is mixed with a controlled amount of air (air/propane ratio of 7.1/1) and passed through a heated gas generator. The so-called *carrier gas* (containing mainly carbon monoxide, hydrogen and nitrogen) produced is useful as an atmosphere in which many steels can be heat-treated without damage or change in carbon content of the surface. Moreover, if extra propane is added to this carrier gas an atmosphere is produced which will *carburise* the steel (see Chapter 14).

Possibly, the ideal furnace atmosphere for the prevention of surface deterioration would be the gas argon (argon is used to surround the tungsten filament of an electric lamp – such a heated filament would oxidise or burn out very quickly if surrounded by air). Unfortunately, argon is too expensive to be used in the large volumes required, so nitrogen is the most widely used, reasonably inert gas. It can be produced by the decomposition of liquid ammonia. From the resultant nitrogen/hydrogen mixture, hydrogen is burned to form water vapour which is removed by condensation, leaving almost pure nitrogen but allowing about 1% hydrogen to remain, this ensuring that no excess oxygen has been admitted to the resulting mixture. This gas is still comparatively expensive to produce and is used mainly where *bright annealing* is required. Even nitrogen is reactive to some metals at high temperatures, notably titanium which must be heat-treated in a complete vacuum, thus increasing the production costs even further.

Whatever atmosphere is used to protect components from oxidation, it is obvious that they must not be withdrawn from this atmosphere whilst still hot and can react with the oxygen in the air. Consequently, furnaces must be either of the batch type (in which the charge cools down whilst still surrounded by a protective atmosphere), or the continuous type (where the charge is carried on a moving hearth through a cooling tunnel – both of which contain the protective atmosphere – before being discharged, cold, from the exit). Steel components to be hardened are tipped from the end of the moving hearth into a quench bath sited within the protective atmosphere.

7 The mechanical shaping of metals

7.1 Introduction

It has been suggested that the Romans conquered their known world with the aid of the bronze sword. Be that as it may, it is certain that before that time Man had found that hacking his enemy apart with sword and spear was a more efficient way of killing him than by the earlier method of clubbing him to death. The smith who shaped the sword and spear – and later the ploughshare (early ones were wood) – was a very important member of society. Using his skin bellows to raise the temperature of his hearth, he was able to form metals to a variety of shapes. Almost as soon as he discovered bronze, he found that a tolerable cutting edge could be developed by cold-forging to harden the alloy. Shaping processes for metals have thus been an integral element in the technological development of man.

Modern shaping processes for metals can be divided into hot- and cold-working operations. The former tend to be used wherever possible since less power is required and working can be carried out more rapidly. Cold-working processes, on the other hand, are used in the final stages of shaping some materials, so that a high-quality surface finish can be obtained or suitable strength and hardness developed in the material.

7.2 Hot-working processes

A hot-working process is one which is carried out at a temperature above that of recrystallisation for the material. Consequently, deformation and recrystallisation take place at the same time, so the material remains malleable during the working process. Intermediate annealing processes are therefore not required, so working takes place very rapidly.

7.2.1 Forging

The simplest and most ancient metal-working process is that of hand-forging, mentioned earlier. With the aid of simple tools called *swages*, the smith can produce relatively complex shapes using either a hand- or power-assisted hammer.

If large numbers of identically shaped components are required, then it is convenient to mass-produce them by *drop-forging*. A shaped two-part die is used, one half being attached to the hammer, whilst the other half is carried by a massive anvil. For complicated shapes, a series of dies may be required. The hammer, working between two vertical guides, is lifted either mechanically or by steam pressure and is then allowed to fall, or is driven down (Figure 7.1), onto the metal to be forged. This consists of a hot bar of

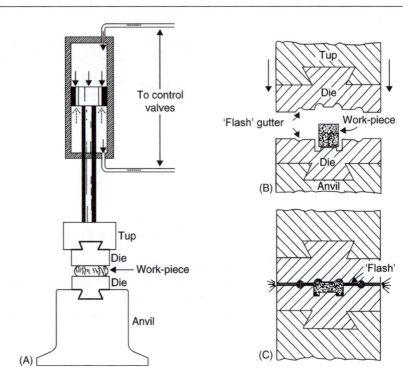

Figure 7.1 *(A) Drop-forging using a double-acting steam hammer. (B) A typical die arrangement. (C) To ensure that the die is completely filled, the work-piece must contain excess metal. This excess is forced into the 'flash gutter' and is trimmed off.*

metal, held on the anvil by means of tongs. As the hammer comes into contact with the metal, it forges it between the two halves of the die.

Hot-pressing is a development of drop-forging which is generally used in the manufacture of simpler shapes. The drop hammer is replaced by a hydraulic driven ram, so that, instead of receiving a rapid succession of blows, the metal is gradually squeezed by the static pressure of the ram. The downwards thrust is sometimes as great as 500 MN.

The main advantage of hot-pressing over drop-forging is that the mechanical deformation takes place more uniformly throughout the work-piece and is not confined to the surface layers, as it is in drop-forging. This is important when forging large components like marine propeller shafts, which may otherwise suffer from having a non-uniform internal structure.

Hot forging is typically used for the production of connecting rods, crankshafts, axle shafts and tool bodies. It is mainly used with carbon alloy and stainless steels and alloys of aluminium, copper and magnesium. The material must be ductile at the forging temperature.

7.2.2 Hot-rolling

Until the Renaissance in Europe, hand-forging was virtually the only method available for shaping metals. Rolling seems to have originated in France in about 1550, whilst in 1680 a sheet mill was in use in Staffordshire. In 1783, Henry Cavendish adapted the rolling mill for the production of wrought-iron bar. The introduction of mass-produced steel by Bessemer, in 1856, led to the development of bigger and faster rolling mills. The first reversing mill was developed at Crewe, in 1866, and this became standard equipment for the initial stages in 'breaking-down' large ingots to strip, sheet, rod and sections.

Traditionally, a steel-rolling shop consists of a powerful 'two-high' reversing mill (Figure 7.2) to reduce the section of the incoming white-hot ingots, followed by trains of rolls which are either plain or grooved according to the product being manufactured. Now that the bulk of steel produced in the Developed World originates from continuous-cast ingots (see Section 5.2), the two-high reversing mill is no longer suitable. Instead, the continuous ingot leaves the casting unit via a series of guide rolls which convey it to a train of reduction rolls followed by a set of finishing rolls. The finished strip passes into a coiling machine and is cut by a flying saw as required.

Hot rolling is also applied to most non-ferrous alloys in the initial shaping stages but the finishing is more likely to be a cold-working operation.

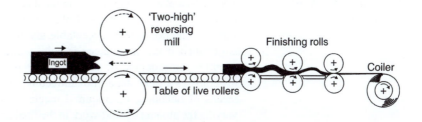

Figure 7.2 *The rolling of steel strip. The ingot is first 'broken down' by the two-high reversing mill (the piped top is usually cropped after several passes through the mill). The work-piece is then delivered to the train of rolls which roll it down to strip.*

7.2.3 Extrusion

The extrusion process is used for shaping a variety of both ferrous and non-ferrous alloys. The most important feature of the process is that, in a single operation from a cast billet, quite complex sections of reasonably accurate dimensions can be obtained.

The billet is heated to the required temperature (350–500°C for aluminium alloys, 700–800°C for brasses and 1100–1250°C for steels) and placed in the container of the extrusion press (Figure 7.3). The ram is then driven forward hydraulically, with sufficient force to extrude the metal through a hard alloy-steel die. The solid metal section issues from the die in a manner similar to the flow of toothpaste from its tube.

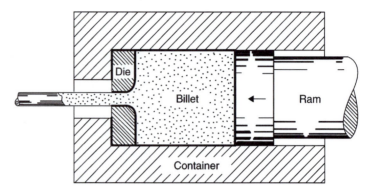

Figure 7.3 *The extrusion process.*

A wide variety of sections can be extruded, including round rod, hexagonal brass rod (for parting off as nuts), brass curtain rail, small diameter rod (for drawing down to wire), stress-bearing sections in aluminium alloys (mainly for aircraft construction), and tubes in carbon and stainless steels, as well as in aluminium and copper alloys. Here, some sort of mandrel (see Section 7.3.2) is used to support the bore of the tube.

7.3 Cold-working processes

The surface of a hot-worked component tends to be scaled, or at least heavily oxidised; so it needs to be sand-blasted or 'pickled' in an acid solution if its surface condition is to be acceptable. Even so, a much better surface quality is obtained if the work-piece is cold-worked *after* being pickled. Consequently, some degree of cold-work is applied to most wrought metallic materials as a final stage in manufacture. However, cold-working is also a means of obtaining the required mechanical properties in a material. By varying the amount of cold-work in the final operation, the degree of hardness and strength can be adjusted. Moreover, some operations can be carried out only on cold metals and alloys. Those processes which involve *drawing* – or pulling – the metal must generally be carried out on cold material, since ductility is usually less at high temperatures. This is because tensile strength is reduced, so the material tears apart very easily when heated. So, whilst malleability is increased by a rise in temperature, ductility is generally reduced. Hence, there are more cold-working operations than there are hot-working processes, because of the large number of different final shapes which are produced in metallic materials of varying ductility. Finally, with cold-working alloys much greater accuracy of dimensions is obtained in the finished material.

7.3.1 Cold-rolling

Cold-rolling is used during the finishing stages in the production of both strip and section, and also in the manufacture of foils. The types of mill used in the manufacture of the latter are shown in Figure 7.4. To roll very thin materials, small-diameter rolls are necessary and, if the material is of great

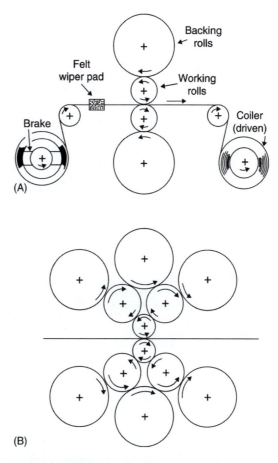

Figure 7.4 *Roll arrangement in mills used for the production of thin foil. (A) A four-high mill. (B) A 'cluster' mill.*

width, this means that the working rolls must be supported by backing rolls, otherwise they will bend to such an extent that reduction in thickness of very thin material becomes impossible. For rolling thicker material, ordinary two-high mills are generally used. The production of mirror-finished metal foil necessitates the use of rolls with a highly polished surface; only by working in perfectly clean surroundings with highly polished rolls can really high grade foil be obtained.

7.3.2 Drawing

Drawing is exclusively a cold-working process, because it relies on high ductility of the material being drawn. Rod, wire and hollow sections (tubes) are produced by drawing them through dies. In the manufacture of wire (Figure 7.5), the material is pulled through the die by winding it onto a rotating drum or 'block'.

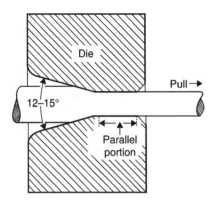

Figure 7.5 *A wire-drawing die.*

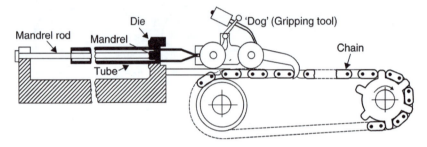

Figure 7.6 *A draw-bench for the drawing of tubes. Rod could also be drawn at such a bench, the mandrel then being omitted.*

In the production of tube, the bore is maintained by the use of a mandrel. Rods and tubes are drawn at a long draw-bench, on which a power-driven 'dog' pulls the material through the die (Figure 7.6). In each case, the material is lubricated with oil or soap before it enters the die aperture.

Drawing dies are made from high-carbon steel, tungsten–molybdenum steels and tungsten carbide, and those used for the production of very fine-gauge copper wire from diamond.

7.3.3 Cold-pressing and deep-drawing

These processes are so closely allied to each other that it is often difficult to define each separately; however, a process is generally termed *deep-drawing* if some thinning of the walls of the component occurs under the application of tensile forces. Thus, the operations range from making a pressing in a single-stage process to cupping followed by a number of drawing stages. In each case, the components are produced from sheet stock, and range from pressed mild-steel motor-car bodies to deep-drawn brass cartridge-cases (see Figure 6.9), cupro-nickel bullet-envelopes and aluminium milk-churns.

Only very ductile materials are suitable for deep-drawing. The best known of these are 70–30 brass, pure copper, cupro-nickel, pure aluminium and

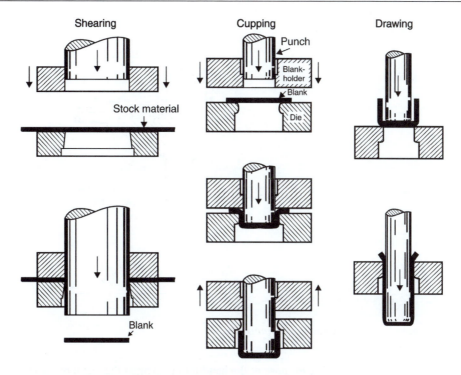

Figure 7.7 *Stages in the deep-drawing of a component.*

some of its alloys, and some of the high-nickel alloys. Mild steel of excellent deep-drawing quality is produced in the 'oxygen process' (see Section 11.2) and is generally used in a large number of motor-car parts as well as domestic equipment. It has in fact replaced much of the very expensive materials such as 70–30 brass for deep-drawn articles.

Typical stages in a pressing and deep-drawing process are shown in Figure 7.7. Wall-thinning may or may not take place in such a process. If wall-thinning is necessary, then a material of high ductility must be used. Although Figure 7.7 shows the process of shearing and cupping being carried out on different machines, usually a combination tool is used so that both processes take place on one machine.

Simple cold-pressing is widely used and alloys which are not sufficiently ductile to allow deep-drawing are generally suitable for shaping by single presswork. Much of the bodywork of a motor car is produced in this way from mild steel.

7.3.4 Spinning

This is one of the oldest methods of shaping sheet metal, and is a relatively simple process, in which a circular blank of metal is attached to the spinning chuck of a lathe. As the blank rotates, it is forced into shape by means of hand-operated tools of blunt steel or hardwood, supported against a fulcrum pin (Figure 7.8).

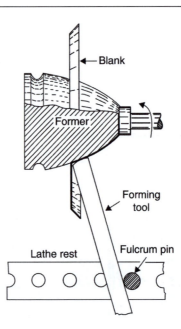

Figure 7.8 *Spinning.*

The purpose of the hand-tool is to press the metal blank into contact with a former of the desired shape. The former, which is also fixed to the rotating chuck, corresponds to the internal shape of the finished component, and may be made from a hardwood such as maple, or, in some cases, from metal. Formers may be solid; if the component is of re-entrant shape (as in Figure 7.8), then the former must be segmented, to enable it to be withdrawn from the finished product. Adequate lubrication is necessary during spinning. For small-scale work, beeswax or tallow are often used, whilst for larger work, soap is the usual choice.

Large reflectors, components used in chemical plant, stainless-steel dairy-utensils, bells, light shades and aluminium teapots, ornaments in copper and brass and other domestic hollow-ware are frequently made by spinning.

A mechanised process somewhat similar to spinning, but known as *flow-turning,* is used for the manufacture of such articles as stainless-steel or aluminium milk-churns. In this process, thick-gauge material is made to flow plastically, by pressure-rolling it in the same direction as the roller is travelling, so that a component is produced in which the wall thickness is much less than that of the original blank (Figure 7.9). Aluminium cooking utensils in which the base is required to be thicker than the side walls are made in this way.

7.3.5 Stretch-forming

Stretch-forming was introduced in the aircraft industry just before the Second World War, and soon became important in the production of metal-skinned

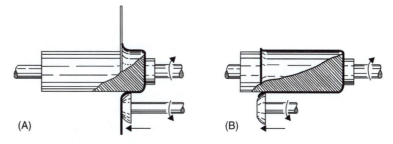

Figure 7.9 *The principles of flow-turning.*

aircraft. The process is also used in the coach-building trade, where it is one of the principal methods of forming sheet metal and sections.

In any forming process, permanent deformation can only be achieved in the work-piece if it is *stressed beyond the elastic limit*. In stretch-forming, this is accomplished by applying a tensile load to the work-piece such that the elastic limit is exceeded, and plastic deformation takes place. The operation is carried out over a form-tool or stretch-block, so that the component assumes the required shape.

In the 'rising-table' machine (Figure 7.10), the work-piece is gripped between jaws, and the stretch-block is mounted on a rising table which is actuated by a hydraulic ram. Stretching forces of up to 4 MN are obtained with this type of machine. Long components, for example aircraft fuselage panels up to 7 m long and 2 m wide, are stretch-formed in a similar machine in which the stretch-block remains stationary and the jaws move tangentially to the ends of the stretch-block.

Stretch-blocks are generally of wood or compressed resin-bonded plywoods, other materials, such as cast synthetic resins, zinc-base alloys or reinforced concrete, are also used. Lubrication of the stretch-block is, of course, necessary.

Although the process is applied mainly to the heat-treatable light alloys, stainless steel and titanium are also stretch-formed on a commercial scale.

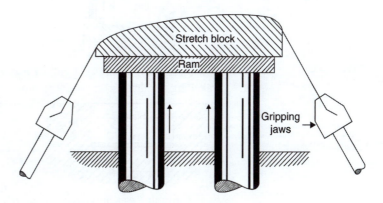

Figure 7.10 *The principles of stretch-forming.*

7.3.6 Coining and embossing

Coining is a cold-forging process in which the deformation takes place entirely by compression. It is confined mainly to the manufacture of coins, medals, keys and small metal plaques. Frequently, pressures in excess of 1500 MPa are necessary to produce sharp impressions, and this limits the size of work which is possible.

The coining operation is carried out in a closed die (Figure 7.11). Since no provision is made for the extrusion of excess metal, the size of the blanks must be accurately controlled to prevent possible damage to the dies, due to the development of excessive pressures.

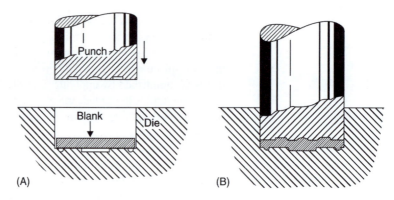

Figure 7.11 *Coining.*

Embossing differs from coining in that virtually no change in thickness of the work-piece takes place during pressing. Consequently, the force necessary to emboss metal is much less than in coining, since little, if any, lateral flow occurs. The material used for embossing is generally thinner than that used for coining, and the process is effected by using male and female dies (Figure 7.12). Typical embossed products include badges and military buttons.

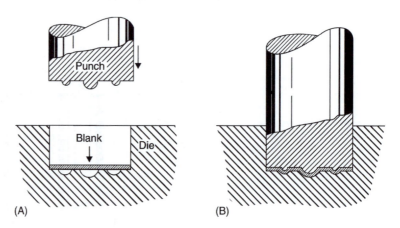

Figure 7.12 *Embossing.*

7.3.7 Impact-extrusion

Extrusion as a hot-working process was described earlier in this chapter (Section 7.2.3). A number of cold-working processes also fall under the general description of extrusion. Probably, the best known of these is the method by which disposable collapsible tubes are manufactured. Such tubes were produced in lead, for containing artists' colours, as long ago as 1841. A few years later, similar tubes were produced in tin, but it was not until 1920 that the impact-extrusion of aluminium was established on a commercial scale.

Heavily built mechanical presses are used in the impact-extrusion of these collapsible tubes. The principles of die and punch arrangement are illustrated in Figure 7.13. A small unheated blank of metal is fed into the die cavity. As the ram descends, it drives the punch very rapidly into the die cavity, where it transmits a very high pressure to the metal, which then immediately fills the cavity. Since there is no other method of exit, the metal is forced upwards through the gap between punch and die, so that it travels along the surface of the punch, forming a thin-shaped shell. The threaded nozzle of the collapsible tube may be formed during the impacting operation, but it is more usual to thread the nozzle in a separate process.

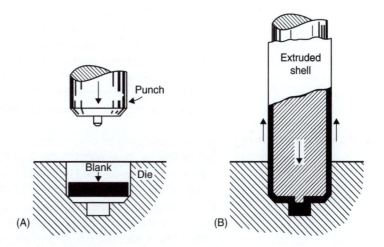

Figure 7.13 *The impact-extrusion of a disposable tube.*

The impact-extrusion of tin and lead is carried out on cold metal, but aluminium blanks may be heated to 250°C for forming. Zinc, alloyed with 0.6% cadmium and used for the extrusion of dry-battery shells, is first heated to 160°C, so that the alloy becomes malleable.

Disposable tubes in lead, tin and aluminium are used as containers for a wide range of substances, e.g. shaving cream, toothpaste, medicines and adhesives, though in recent years thermoplastic materials have replaced aluminium in many such containers. Impact-extrusion is also used for the production of many other articles, principally in aluminium, e.g. canisters and capsules for food, medical products and photographic film and shielding cans for radio components.

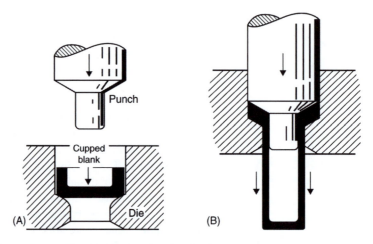

Figure 7.14 *Impact-extrusion by the Hooker process.*

The Hooker process (Figure 7.14) closely resembles hot-extrusion in so far as the flow of metal in relation to the die and punch is concerned; it is, however, a cold-working operation of the impact type. Its products include small brass cartridge-cases, copper tubes for radiators and heat exchangers and other short tubular products. Fat 'slugs' are sometimes used, but cupped blanks are usually considered to be more satisfactory. The blank is placed in the die and, as the punch descends, metal is forced down between the body of the punch and the die, producing a tubular extrusion.

7.4 Powder metallurgy

Powder-metallurgy processes were originally developed to replace melting and casting for those metals – the so-called *refractory* metals – which have very high melting-points. For example, tungsten melts at 3410°C, and this is beyond the softening temperatures of all ordinary furnace lining materials. Hence, tungsten is produced from its ore as a fine powder. This powder is then 'compacted' in a die of suitable shape at a pressure of approximately 1500 MPa. Under such high pressure, the particles of tungsten become joined together by 'cold-welding' at the points of contact between particles.

The compacts are then heated to a temperature above that of recrystallisation – about 1600°C in the case of tungsten. This treatment causes recrystallisation to occur, particularly in the highly deformed regions where cold-welding has taken place, and in this way the crystal structure becomes regular and continuous as grain-growth takes place across the original boundaries where cold welds have formed between particles. The heating process is known as *sintering* (Figure 7.15). At the end of this process, the slab of tungsten has a strong, continuous structure, though it will contain a large number of cavities. It might then be rolled, and drawn down to wire, which is used in electric lamp filaments. Most of the cavities are welded up by the working process.

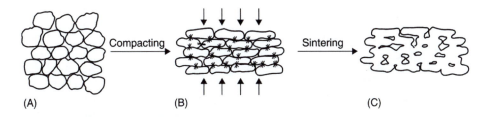

Figure 7.15 *Stages in a powder-metallurgy process. (A) Particles of tungsten powder. (B) Cold-welding between particles. (C) Grain-growth across particle boundaries.*

7.4.1 Uses of powder-metallurgy

Although powder-metallurgy was originally used to deal with metals of very high melting-point, its use has been extended for other reasons, such as:

- To produce metals and alloys of *controlled porosity,* e.g. stainless-steel filters for corrosive liquids and oil-less bronze bearings.
- To produce 'alloys' of metals which do not mix in the molten state, e.g. copper and iron for use as a cheap bearing material.
- For the production of small components, such as the G-frame of a micrometer screw gauge where the negligible amount of process scrap makes the method competitive.

7.4.2 Cemented carbides

One well-known use of powder-metallurgy is in the manufacture of cemented carbides for use as tool materials. Here, tungsten powder is heated with carbon powder at about 1500°C to form tungsten carbide. This is ground in a ball mill, to produce particles of a very small size (about 20 μm) and the resultant tungsten carbide powder is mixed with cobalt powder, so that the particles of tungsten carbide become coated with powdered cobalt. The mixture is then compacted in hardened steel dies at pressure of about 300 MPa, to cause cold-welding between the particles of cobalt. The compacts are then sintered at about 1500°C, to cause recrystallisation and grain growth in the cobalt, so that the result is a hard, continuous structure consisting of particles of very hard tungsten carbide in a matrix of hard, tough cobalt.

7.4.3 Sintered-bronze bearings

The principles involved in the manufacture of sintered-bronze bearings are slightly different. Here, the main reason for using a powder-metallurgy process is to obtain a bearing with a controlled amount of porosity, so that it can be made to absorb lubricating oil. Powders of copper and tin are mixed in the correct proportions (about 90% copper, 10% tin), and are then compacted in a die of suitable shape (Figure 7.16). The compacts are then sintered at 800°C for a few minutes. This, of course, is *above* the

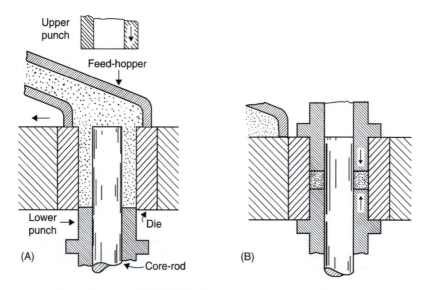

Figure 7.16 *The compacting process used in the manufacture of an 'oil-less' bronze bearing-bush.*

melting-point of tin, so the process differs from the true powder-metallurgy processes in which no fusion occurs. As the tin melts, it percolates between the copper particles and alloys with them to produce a continuous mass. The compact, however, still retains a large amount of its initial porosity and, when it is quenched into lubricating oil, oil is drawn into the pores of the bearing as it cools down. Sometimes the bearings are then placed under vacuum, whilst still in the oil bath. This causes any remaining air to be drawn out of the bearing, being replaced by oil when the pressure returns to that of the atmosphere.

The final structure resembles that of a metallic sponge which, when saturated with lubricating oil, produces a self-oiling bearing. In some cases, the amount of oil in the bearing lasts for the lifetime of the machine. Such bearings are not only used in the motor-car industry, but also particularly in many domestic machines such as vacuum cleaners, refrigerators, electric clocks and washing machines, in all of which long service with a minimum of attention is desirable.

7.5 Machining metals

Metal-cutting processes are expensive, so it is usual to shape a component to as near its final form as possible by casting, forging or powder-metallurgy processes. The need for accuracy of dimensions will nevertheless often demand some amount of machining. The ease with which a metal can be cut depends not only upon the design of tools, method of lubrication and so on, but also upon the microstructure of the work-piece itself and it is this aspect which concerns us briefly here. After all, complete books are available dealing with the various techniques of metal-cutting processes!

Machining is a cold-working process in which the cutting edge of a tool forms shavings or chips of the material being cut. During cold-working processes, heat is invariably generated as mechanical energy is being absorbed. In a machining process, heat arises from this source as well as from energy dissipated by ordinary friction between tool and work-piece. Not only is energy wasted in this way, but the heat generated tends to soften the cutting tool so that its 'edge' is lost. This involves expensive tool maintenance or the use of expensive high-speed steels (see Section 13.3) or ceramic tool materials (see Section 23.2).

As Figure 7.17A indicates, ductile materials will generally machine badly. Not only does the material tend to tear under the pressure of the tool, giving a poor surface finish, but, since the displaced metal 'flows' around the cutting edge, heat generated by friction between the two will be greater and the tool life short. Brittle materials, on the other hand, generally machine well, producing a fine powdery swarf (Figure 7.17B), in contrast to the curly slivers from a ductile metal. Unfortunately, metals which are brittle overall are unsuitable for most engineering purposes, but it is possible to improve machinability by introducing a degree of *local* brittleness in the material without impairing unduly its overall toughness. This local brittleness then causes minute chip cracks to form just in advance of the edge of the cutting tool.

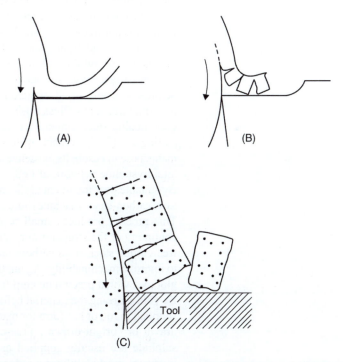

Figure 7.17 *The influence of microstructure upon machinability. In (A) a soft ductile material flows around the cutting tool and heat builds up at the cutting edge due to friction between the metal and the tool. (B) indicates a fragmented swarf from a brittle material with consequent lower friction losses. (C) A typical free-cutting alloy containing isolated particles which initiate chip cracks in advance of the cutting edge.*

Such local brittleness can be imparted in a number of ways:

- By the presence of a separate constituent in the microstructure. Small isolated particles in the structure behave as tiny holes and have the effect of setting up local stress concentrations just ahead of the advancing cutting edge. These stress raisers cause minute fractures to travel between the particles and the tool edge, thus reducing friction between tool and work-piece (Figure 7.17C). As well as a reduction in tool wear, there will be a reduction in power consumption. Moreover, since a fine powdery swarf is generated, there will be a smaller remelting loss.
- Some alloys contain microconstituents which put them in the 'free-cutting' category. The graphite flakes in grey cast iron (Section 15.5) and the particles of hard compound in tin bronzes (Section 16.6) are examples. Generally, however, small alloy additions are made to improve machinability in this way. Thus up to 2.0% lead, which is insoluble in both liquid and solid copper alloys, is added to brasses and bronzes (Sections 16.5 and 16.6) to improve machinability. Being insoluble, this lead is present as tiny particles in the microstructure. 'Ledloy' steels contain no more than 0.2% lead, yet this improves machinability by up to 25% whilst causing very little loss in mechanical properties due to the presence of the tiny lead globules.
- In the cheaper free-cutting steels, the presence of a high sulphur content is utilised by adding an excess of manganese so that isolated globules of manganese sulphide, MnS, are formed in preference to the brittle intercrystalline network of ferrous sulphide, FeS, which would otherwise be present. These small isolated globules of MnS, scattered throughout the structure, act as stress raisers during a cutting process. So, whilst a good-quality steel contains a maximum of only 0.06% sulphur, a free-cutting steel may contain up to 0.3% sulphur along with up to 1.5% manganese to ensure its existence as MnS globules rather than as a brittle intercrystalline network of FeS.
- By suitable heat-treatment of the material prior to the machining process. Low carbon-steels machine more easily when normalised (Section 11.5). This process produces small patches of pearlite which break up the continuity of the structure in much the same way as MnS globules in a free-cutting steel. High-carbon steels can be spheroidised (Section 11.5) to improve machinability. Again, this has the effect of providing isolated globules which precipitate chip formation.
- By cold-working the material before the machining operation. Materials supplied in rod or bar form for machining operations are generally cold-drawn or 'bright-drawn'. Free-cutting steels containing manganese sulphide are usually supplied as 'bright-drawn bar'. This treatment improves machinability by reducing the ductility of the surface layers and introducing local brittleness there.

7.5.1 Machinability

The term *machinability* is used to describe the ease of machining. Material with good machinability will produce small chips and need low cutting force

Table 7.1 *Machinability index values for steels*

Steel	Machinability index
Plain carbon steel 070M20, 0.20% carbon	100
Plain carbon steel 080M40, 0.40% carbon	70
Plain carbon steel 070M55, 0.55% carbon	50
Alloy steel 120M36, 1.2% manganese steel	65
Alloy steel 530M40, chromium steel	40
Alloy steel 708M40, chromium–molybdenum steel	40
Free-cutting steel 210M15	200
Free-cutting steel 214M15	140

and energy expenditure. It will be capable of being machined quickly and give good tool life. A relative measure of machinability is given by the *machinability index*. It is only a rough guide to machinability, but the higher the value the better the machinability. With British Standards, for steels the plain carbon steel 070M20 is given a value of 100% and others compared with it. Table 7.1 shows some typical values for steels.

8 Alloys

8.1 Introduction

I still have my grandmother's copper kettle. It was a wedding present from her young brother in 1877. At that time, copper was used for the manufacture of kettles, pots and pans because it was ductile and reasonably corrosion-resistant. Now aluminium, unknown commercially when my grandmother was married, has replaced copper for the manufacture of such kitchen-ware. In the meantime, the production of increasingly high-purity dead-mild steel – which is as near as we get to pure iron commercially – has confirmed its use in the manufacture of bodywork of motor cars, domestic refrigerators, washing machines and a multitude of other items of everyday equipment.

In all these products, strength is developed in otherwise soft metals by cold-work. When greater strength or hardness is required, it can only be achieved by alloying, i.e. by adding an element or elements which, by modifying the crystal structure in some way, will oppose the process of slip when stress is applied.

An alloy is a mixture of two or more metals, made with the object of improving the properties of one of these metals, or, in some cases, producing new properties not possessed by either of the metals in the pure state. For example, pure copper has a very low electrical resistance and is therefore used as a conductor of electricity; but, with 40% nickel, an alloy 'constantan' with a relatively high electrical resistance is produced. Again, pure iron is a ductile though rather weak material; yet, the addition of less than 0.5% carbon will result in the exceedingly strong alloy we call steel. In this chapter, we shall examine the internal structures of different types of alloy and show to what extent the structures of these alloys influence their mechanical properties.

8.1.1 Solutions

Oil and water do not 'mix'. If we shake a bottle containing a quantity of each liquid, droplets of one may form in the other but they soon separate into two distinct layers – oil floating on top of water. Alcohol and water, however, will 'mix' – or 'dissolve' – in each other completely in all propor-tions. Beer is a solution of about 5% alcohol in water, while whisky is a solution of some 40% alcohol in water. In each case, various flavourings are also present but the result is a single homogeneous *solution* in the glass.

Molten metals behave in much the same way. Thus, molten lead and molten zinc are, like oil and water, *insoluble* in each other and would form two separate layers in a crucible, molten zinc floating on top of the more dense molten lead. Clearly such metals, as they do not dissolve in each other as liquids, are unlikely to form a useful alloy since in such an alloy the atoms of each metal must mingle intimately together. This can only occur when they form a homogeneous liquid solution. However, it is to the manner in which this liquid solution subsequently solidifies that we must turn our attention in this chapter.

8.2 Eutectics

Sometimes on solidification, the two metals cease to remain dissolved in one another, but separate instead, each to form its own individual crystals. In a similar way, salt will dissolve in water. When the solution reaches its freezing-point, separate crystals of pure ice and pure salt are formed. Another fact we notice is that the freezing-temperature of the salt solution is much lower than that of pure water (which is why salt is scattered on roads after a night of frost). This phenomenon is known as *depression of freezing-point,* and it is observed in the case of many metallic alloys. Thus, the addition of increasing amounts of the metal cadmium to the metal bismuth will cause its freezing-point to be depressed proportionally; whilst, conversely, the addition of increasing amounts of bismuth to cadmium will have a similar effect on its melting-point as shown in Figure 8.1.

It will be noticed that the two lines in Figure 8.1 meet at the point *E*, corresponding to an alloy containing 60% bismuth and 40% cadmium. This alloy melts and freezes at a temperature of 140°C, and is the lowest melting-point alloy which can be made by mixing the metals cadmium and bismuth. The point *E* is called the *eutectic point* (pronounced 'yoe-tek-tick', this word is derived from the Greek *eutektos,* meaning 'capable of being melted easily'), and the composition 60% bismuth with 40% cadmium is the *eutectic*

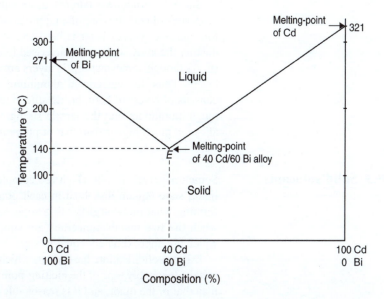

Figure 8.1 *The freezing-points (melting-points) of both bismuth and cadmium are depressed by adding each to the other. A minimum freezing-point – or 'eutectic point' – is produced.*

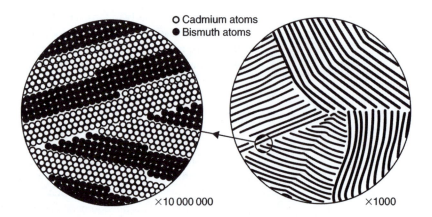

Figure 8.2 *If we had a microscope which gave a magnification of about 10 million times, the arrangement of atoms of cadmium and bismuth would look something like that in the left-hand part of the diagram, except that the bands in the eutectic would each be many thousand atoms in width.*

mixture. If a molten alloy of this composition is allowed to cool, it will remain completely liquid until the temperature falls to 140°C, when it will solidify by forming alternating thin layers of pure cadmium and pure bismuth (Figure 8.2) until solidification is complete. The metallic layers in this eutectic structure are extremely thin, and a microscope with a magnification of at least ×100 is generally necessary to be able to see the structure.

Since the structure is laminated, something like plywood, the mechanical properties of eutectics are often quite good. For example, when the material forming one type of layer is hard and strong, whilst the other is soft and ductile, the alloy will be characterised by strength and toughness, since the strong though somewhat brittle layers are cushioned between soft but tough layers. Thus, the eutectic in aluminium–silicon alloys (see Section 17.6) consists of layers of hard, brittle silicon sandwiched between layers of soft, tough aluminium and the tensile strength of these cast aluminium–silicon alloys is much higher than that of pure aluminium.

8.3 Solid solutions

Sometimes, two metals which are completely soluble in each other in the liquid state remain dissolved in each other during and after solidification, forming what metallurgists call a *solid solution*. This is generally the case when the two metals concerned are similar in properties and have atoms which are approximately equal in size.

During solidification, the crystals which form are built from atoms of both metals. Inevitably, one of the melting-points will have a melting-point higher than that of the other, and it is reasonable to expect that this metal will tend to solidify at a faster rate than the one of lower melting-point. Consequently, the core of a resultant dendrite contains rather more of the metal with the higher melting-point; whilst the outer fringes of the crystal will contain

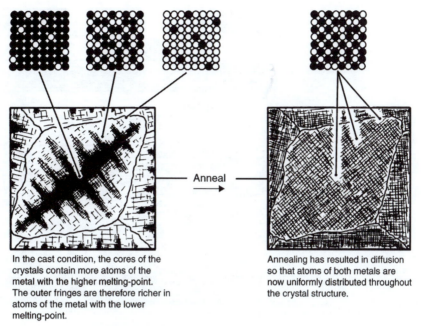

In the cast condition, the cores of the crystals contain more atoms of the metal with the higher melting-point. The outer fringes are therefore richer in atoms of the metal with the lower melting-point.

Annealing has resulted in diffusion so that atoms of both metals are now uniformly distributed throughout the crystal structure.

● Atoms of the metal with the higher melting-point
○ Atoms of the metal with the lower melting-point

Figure 8.3 *The variations in composition in a cored solid solution. The coring can be dispersed by annealing.*

correspondingly more of the metal with the lower melting-point (Figure 8.3). This effect, known as *coring,* is prevalent in all solid solutions in the cast condition.

8.3.1 Brick analogy

Many students appear to find the basic idea of solid solution difficult to understand; so an analogy, in which bricks replace atoms as the building units, will be drawn here.

Suppose that, in building a high wall, a team of bricklayers is given a mixture of red and blue bricks with which to work. Further, suppose that, as building proceeds, each successive load of bricks which arrives at the site contains a slightly higher proportion of red bricks than the previous load. We will also assume that the 'brickies', being paid on a piece-work basis, lay whatever brick comes to hand first. Thus, there will be no pattern in the laying of individual bricks; although, as the wall rises, there will be more red bricks and less blue ones in successive courses. By standing close to the wall, one would observe a small section, possibly like that shown in Figure 8.4A, and this would give no indication of the overall distribution of red and blue bricks in the wall. On standing further away from the wall, the lack of any pattern in the laying of individual bricks would still be obvious, though the general relationship between the numbers of red and blue bricks at the

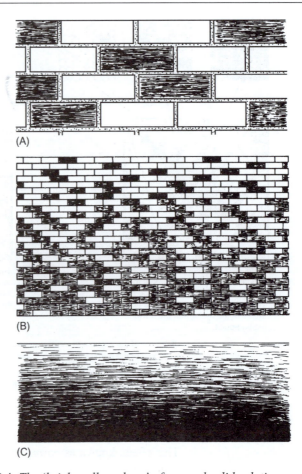

Figure 8.4 *The 'brick-wall analogy' of a cored solid solution.*

top and bottom of the wall would become apparent (Figure 8.4B). Provided that the red and blue bricks were of roughly equal size and strength, the wall would be perfectly sound, though it might look a little odd. Let us assume that we now view the wall from a distance of about half a kilometre. Individual bricks will no longer be visible, though a gradual change in colour from blue at the bottom, through various stages of purple, to red at the top will indicate the relative numbers of each type of brick at various levels in the wall (Figure 8.4C).

In many metallic solid solutions, the distribution of the two different types of atom follows a pattern similar to that of the bricks in the unlikely wall just described, i.e. whilst the atoms in general conform to some overall geometrical arrangement – as do the bricks in the wall – there is usually no rule governing the order or frequency in which atoms of each type will be present within the overall pattern. The brick-wall analogy also illustrates the folly of viewing a microstructure at a high magnification, without first examining it with a low-power lens in the microscope. When using a high magnification, one will have only a restricted field of view of a very small

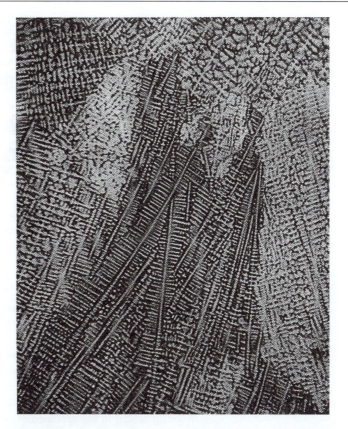

Figure 8.5 *The dendritic structure of a cast alloy. This is a photomicrograph of cast 70–30 brass at a magnification of ×39. The dendrites would not be visible were it not for the coring of the solid solution.*

part of the structure, so that no overall pattern is apparent; whereas, the use of a low-power lens may reveal the complete dendritic structure, and show beyond doubt the nature of the material (Figure 8.5).

8.3.2 Substitutional solid solution

In the earlier discussion, it was assumed that the type of solid solution formed was one in which the atoms of two metals concerned were of roughly equal size. This type of structure (Figure 8.6A) is termed a *substitutional solid solution,* since atoms of one metal have, so to speak, been substituted for atoms of the other.

Many pairs of metals, including copper/nickel, silver/gold, chromium/iron and many others, form solid solutions of this type in all proportions. A still greater number of pairs of metals will dissolve in each other in this way but in limited proportions. Notable examples include copper/tin, copper/zinc, copper/aluminium, aluminium/magnesium and a host of other useful alloys.

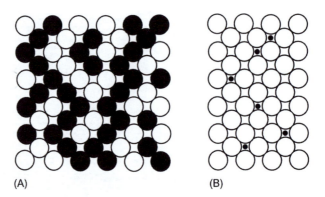

(A)   (B)

Figure 8.6 *The two main types of solid solution: (A) a substitutional solid solution, (B) an interstitial solid solution.*

8.3.3 Interstitial solid solution

There is, however, another type of solid solution, which is formed when the atoms of one element are so much smaller than those of the other that they are able to fit into the *interstices* (or spaces) between the larger atoms. Accordingly, this is known as an *interstitial solid solution* (Figure 8.6B). Carbon dissolves in face-centred cubic iron in this way, since the relatively small carbon atoms fit into the spaces between the much larger iron atoms, this explains why a piece of *solid* steel can be carburised (see Section 14.2) by being heated in the presence of carbon at a temperature high enough to make the steel face-centred cubic in structure. The carbon atoms 'infiltrate', so to speak, through the face-centred cubic ranks of iron atoms.

8.3.4 Diffusion

When a solid solution of either type is heated to a sufficiently high temperature, the atoms become so thermally activated that a process of *diffusion* takes place and atoms exchange positions. As a result, coring gradually disappears and the structure becomes more uniform in composition throughout. This is achieved by atoms moving from one part of the crystal to another. It is easy to see that this can happen in an interstitial solution where tiny solute (a solution consists of a *solute* dissolved in *solvent*) atoms can move through gaps between the larger solvent atoms, but in a substitutional solid solution such movement would be impossible if crystal structures were as perfect as that shown in Figure 8.6A. In the real world, however, crystal structures are never perfect and gaps exist where atoms or groups of atoms are missing. Examination of a piece of cast metal under the microscope will reveal many tiny cavities, even in high-quality material. Each of these cavities represents some thousands of absent atoms, so it is reasonable to assume that points where a single atom is missing are numerous indeed. Such *vacant sites,* as they are called, will allow the movement of individual atoms through the crystal structure of a substitutional solid solution.

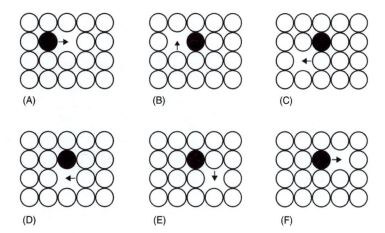

Figure 8.7 *This is the way in which metallurgists believe diffusion takes place in a solid solution. A series of five 'moves' is necessary in order that the 'black' atom (and an accompanying 'vacant site') can go forward by one space.*

It is assumed that the process of diffusion takes place by a series of successive 'moves', as suggested in Figure 8.7 for a substitutional solid solution. It is likely that a solute atom alongside a vacant site will first move; in this way, stress in the structure will be kept to a minimum. By following the series of five 'moves' shown in Figure 8.7, the solute atom (black) advances to the right by one lattice space. The series of moves will, of course, be repeated every time the solute atom and its accompanying vacant site move forward by one lattice space.

As, even with a substitutional solid solution, the atoms are not the same size, the driving force behind this process of diffusion is the lattice strains (and the stresses they produce) caused by irregularities in the lattice due to the presence of atoms of different sizes. During diffusion, atoms move into positions of lower strain and consequently lower stress.

8.3.5 Solid solutions and strength

Solid solutions are the most important of metallic alloy structures since they provide the best combination of strength, ductility and toughness. The increase in strength is a result of distortions arising in the crystal structure when atoms of different sizes accommodate themselves into the same lattice pattern (Figure 8.8). Slip then becomes more difficult as dislocations must follow a less direct path and so a greater force must be used to produce it – which is another way of saying that the yield stress has increased. Since slip is nevertheless still ultimately able to take place, much of the ductility of the original pure metal is retained. Most of our useful metallic alloys are basically solid solution in structure.

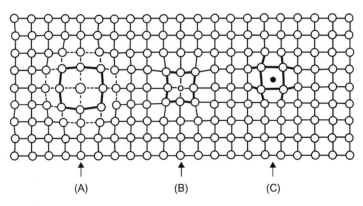

Figure 8.8 *Crystal lattice distortions caused by the presence of solute atoms: (A) a large substitutional atom, (B) a small substitutional atom, (C) an interstitial atom. In each case, the distortion produced will oppose the passage of a dislocation through the system.*

8.4 Intermetallic compounds

When heated, many metals combine with oxygen to form compounds we call oxides, whilst some metals are attacked by sulphur gases in furnace atmospheres to form sulphides. This is a general pattern in which metals as positive ions combine with non-metals as negative ions to form ionic compounds which usually bear no physical resemblance to the elements from which they are formed. Thus, the very reactive sodium (which is silvery-white in appearance and gives positive ions) combines with the very reactive gas chlorine (which is greenish in colour and forms negative ions) to form sodium chloride, i.e. table-salt. Metallic oxides, sulphides and chlorides are examples of ionic compounds formed by the attraction between positive and negative ions. However, sometimes two metals when melted together will combine to form a chemical compound called an *intermetallic compound*. This often happens when one of the two metals has strongly positive ions and the other weakly positive ions.

When a solid solution is formed, it usually bears at least some resemblance to the parent metals, as far as colour and other physical properties are concerned. For example, the colour of a low-tin bronze (see Section 16.6) is a blend of the colours of its parent metals, copper and tin, as one might expect; but if the amount of tin is increased, so that the alloy contains 66% copper and 34% tin, a hard and extremely brittle substance is produced which bears no resemblance whatever to either copper or tin. What is more, this intermetallic compound – for such it is – is of a pale blue colour. This is due mainly to the fact that an intermediate compound generally crystallises in a different pattern to that of either of the parent metals. This particular intermetallic compound has the chemical formula $Cu_{31}Sn_8$. Since an intermetallic compound is always of fixed composition, in common with all chemical compounds, there is never any coring in crystals of such a substance in the cast state.

Because of the excessive brittleness of most of these intermetallic compounds, they are used to only a limited extent as constituents of engineering alloys, and then only in the form of small, isolated particles in the microstructure. Since many intermetallic compounds are very hard, they also have very low coefficients of friction. Consequently, a typical use of these compounds is as a constituent in bearing metals (see Section 18.6). If present in an alloy in large amounts, an intermetallic compound will often form brittle intercrystalline networks. The strength and toughness of such an alloy would be negligible.

8.5 Summary: alloys

In a study of metallurgy, the term *phase* refers to any chemically stable, single homogeneous constituent in an alloy. Thus, in a solid alloy, a phase may be a solid solution, an intermetallic compound or, of course, a pure metal. A homogeneous liquid solution from which an alloy is solidifying also constitutes a phase. Any of the solid phases form the basic units of which metallic alloys are composed. It may be helpful therefore to summarise their properties:

- *Solid solutions* are formed when one metal is very similar to another, both physically and chemically, and is able to replace it, atom for atom, in a crystal structure or if the atoms of the second element are very small and able to fit into the spaces between the larger atoms of the other metal. Solid solutions are stronger than pure metals, because the presence of atoms of the second metal causes some distortion of the crystal structure, thus making slip more difficult. At the same time, solid solutions retain much of the toughness and ductility of the original pure metal.
- *Intermetallic compounds* are formed by chemical combination, and the resultant substance generally bears little resemblance to its parent metals. Most intermetallic compounds are hard and brittle, and of limited use in engineering alloys.
- *Eutectics* are formed when two metals, soluble in each other in the liquid state, become insoluble in each other in the solid state. Then, alternate layers or bands of each metal form, until the alloy is completely solid. This occurs at a fixed temperature, which is below the melting-point of either of the two pure metals. When two metals are only partially soluble in each other in the solid state, a eutectic may form consisting of alternate layers of two solid solutions. In some cases, a eutectic may consist of alternate layers of a solid solution and an intermetallic compound.

Engineering alloys often contain more than two elements and many contain six or even more, each contributing its own special effect. This structure of such an alloy is often much more complex. For example, ternary or quaternary eutectics may be formed in which layers of three or four different solid phases are present. In general, however, it is more likely that a uniform solid solution will be present, one metal acting as the solvent of all the other additions. Thus, the stainless steel 347S17 (Table 13.9) is composed almost entirely of a solid solution in which iron has dissolved 18% chromium, 10% nickel, 1% niobium and 0.8% manganese – a residual 0.04% carbon existing as a few scattered undissolved carbide particles.

9   Equilibrium diagrams

9.1 Introduction

Amongst some engineering students, the prospect of having to study the topic of equilibrium diagrams seems to be received with some dismay. Nevertheless, the topic need cause the reader no undue alarm, since for most purposes we can regard the equilibrium diagram as being no more than a graphical method of illustrating the relationship between the composition, temperature and structure, or state, of any alloy in a series.

Much useful information can be obtained from these diagrams, if a simple understanding of their meaning has been acquired. For example, an elementary knowledge of the appropriate equilibrium diagram enables us to decide upon a suitable heat-treatment process to produce the required properties in a carbon-steel. Similarly, a glance at the equilibrium diagram of a non-ferrous alloy system will often give us a pretty good indication of the structure – and hence the mechanical properties – a particular composition is likely to have. In attempting to assess the properties of an unfamiliar alloy, the modern metallurgist invariably begins by consulting the thermal equilibrium diagram for the series. There is no reason why the engineering technician should not be in a position to do precisely the same.

9.2 Obtaining equilibrium diagrams

How are equilibrium diagrams obtained? Purely by a great deal of laborious routine laboratory work accompanied by some experience of the behaviour of metals in forming alloys, but it is only possible to predict the general 'shape' of an equilibrium diagram with any certainty in the case of a limited number of alloy systems. Altogether there are some 70 different metallic elements and if these are taken in *pairs* to form *binary alloys,* quite a large number of binary alloy systems is involved. Of course, it would be extremely difficult to make alloys from some pairs of metals, e.g. high melting-point tungsten with very reactive caesium. Nevertheless, a high proportion of the metallic elements have been successfully alloyed with each other and with some of the non-metallic elements like carbon, silicon and boron.

9.2.1 Lead–tin alloys

From Roman times until the middle of the twentieth century (omitting that period of almost a thousand years between the fall of Rome and the Renaissance) much domestic and industrial pipework carrying water was of lead but, when its potentially toxic nature was realised, lead was replaced by copper and PVC. Nevertheless, he who installs or repairs our pipework is still known as a *plumber;* a plumber was originally one who worked in lead – the word being derived from the Latin word *plumbum* for lead. In the days of lead piping, the plumber joined lengths of pipe, or repaired fractures

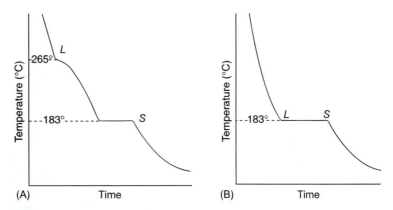

Figure 9.1 *Cooling curves for plumber's solder and for tinman's solder.*

in them, using 'plumber's solder'. This contains roughly two parts by weight
of lead to one of tin. On cooling, it begins to solidify at about 265°C, but it
is not completely solid until its temperature has fallen to 183°C (Figure 9.1A).
It thus passes through a pasty, part solid–part liquid range of some 80°C,
enabling the plumber to 'wipe' a joint using, traditionally, a moleskin 'glove'
whilst the solder is in this 'mushy' state between 265°C and 183°C.

In the case of tinman's solder, used to join pieces of suitable metal, rather
different properties are required. The solder must of course 'wet' (alloy with)
the surfaces being joined; but it will be an advantage if its melting-point is
low and, more important still, if it freezes quickly over a small range of
temperatures, so that the joint is less likely to be broken by rough handling
in its mushy state. A solder with these properties contains 62% tin and 38%
lead. It freezes, as does a pure metal, at a single temperature – 183°C in this
case (Figure 9.1B). Since the cost of tin is more than 10 times that of lead,
tinman's solder often contains less than the ideal 62%. It will then freeze
over a range of temperatures which will vary with its composition. Thus,
'coarse' tinman's solder contains 50% tin and 50% lead. It begins to solidify
at 220°C and is completely solid at 183°C.

The freezing-range of any tin–lead solder can be determined by melting
a small amount of it in a clay crucible and then taking temperature readings
of the cooling alloy every 15 seconds (Figure 9.2). A thermocouple is
probably the best temperature-measuring instrument to use for this, though
a 360°C thermometer will suffice, provided it is protected by a fireclay sheath.
Failure to use the latter will probably lead to the fracture of the thermometer
as the solder freezes onto it, contracting in the process.

A temperature/time cooling curve can now be plotted in order to determine
accurately the temperature at which freezing of the alloy begins (L) and
finishes (S) (Figure 9.1). This procedure is repeated for a number of tin–lead
alloys of different compositions. Representative values of L and S for some
tin–lead alloys are shown in Table 9.1.

The information obtained from Table 9.1 can now be plotted on a simple
diagram, as shown in Figure 9.3, in order to relate freezing-range to
composition of alloy. The line LEL_1 (called the *liquidus*) joins all points (L)

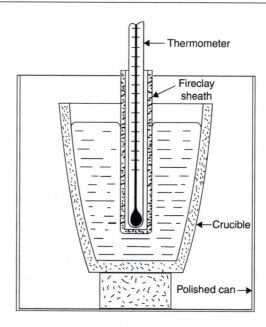

Figure 9.2 *Simple apparatus for determining the freezing-range of a low-temperature alloy. The polished can prevents the alloy from cooling too quickly and also protects the crucible from draughts.*

Table 9.1 *Freezing ranges for tin–lead alloys*

Composition		Temperature at which solidification begins (L) (°C)	Temperature at which solidification ends (S) (°C)
Lead (%)	Tin (%)		
67	33	265	183
50	50	220	183
38	62	183	183
20	80	200	183

at which solidification of the various alloys begins; whilst the line SES_1 (called the *solidus*) joins all points (*S*) at which solidification of the alloys has finished. What we have plotted is only *part* of the lead–tin thermal equilibrium diagram. The complete diagram contains other lines – or 'phase boundaries', as they are called. To determine these lines, other, more complex methods have to be used; but we are not concerned here with advanced metallurgical practice.

Even this restricted portion of the lead–tin equilibrium diagram provides us with some useful information. We can, for example, use it to determine the freezing-range of any alloy containing between 67% lead/33% tin and 20% lead/80% tin. Thus, reading from the diagram (Figure 9.3), a solder containing 60% lead/40% tin would solidify between 245°C and 183°C.

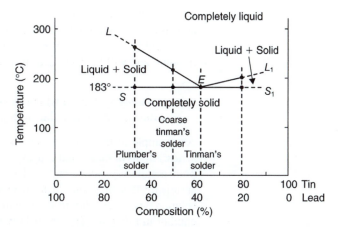

Figure 9.3 *Part of the lead–tin equilibrium diagram. The limited information obtained from the cooling curves mentioned enables us to construct only so much of the diagram.*

Similarly, given the composition of an alloy and its temperature at any instant, we can determine the state in which it exists. For example, an alloy containing 55% lead/45% tin at a temperature of 200°C will be in a pasty, part solid–part liquid state; whilst the same alloy at 250°C will be completely molten. Conversely, when cooled below 183°C, it will be completely solid.

9.3 Types of equilibrium diagram   Generally speaking, a useful alloy will only be formed when two metals are completely soluble in each other in the liquid state, i.e. they form a single homogeneous liquid in the crucible. Thus, attempts to form a lead–zinc alloy will fail because lead and zinc are only slightly soluble in each other as liquids. When melting a mixture of, say, equal parts of lead and zinc, we find a layer of zinc-rich liquid solution floats on top of a layer of lead-rich liquid solution – rather like oil on water. On casting this mixture, the two separate layers would form within the mould and solidify as such. Consequently, complete mutual solubility in the liquid state is a prerequisite to producing alloys by the traditional melting-casting route.

There are a number of different types of thermal equilibrium diagrams, but we need to deal with only three of them, namely those in which the characteristics of the diagram are governed by the extent to which one metal forms a solid solution with the other. The possibilities are that:

1   The two metals are completely soluble in each other in all proportions in the solid state.
2   The two metals are completely insoluble in each other in the solid state.
3   The two metals are partially soluble in each other in the solid state.

Strictly speaking, equilibrium diagrams indicate only microstructures which will be produced when alloys cool under equilibrium conditions and, in most cases, that means extremely slowly. Under industrial conditions, alloys often solidify and cool far too rapidly for equilibrium to be reached and, as a result, the final structure deviates considerably from that shown in the diagram. The coring of solid solutions mentioned in Chapter 8 is a case in point, and will be discussed further in the section which follows.

9.3.1 An alloy system in which the two metals are soluble in each other in all proportions in both liquid and solid states

An example of this type of system is afforded by the nickel–copper alloy series. Atoms of nickel and copper are approximately the same size and, since both metals crystallise in similar face-centred cubic patterns, it is not surprising that they should form mixed crystals of a substitutional solid-solution type when a liquid solution of the two metals solidifies. The resulting equilibrium diagram (Figure 9.4) will have been derived from a series of cooling curves as described earlier in this chapter, except that a pyrometer capable of withstanding high temperatures would be required for taking the temperature measurements.

The equilibrium diagram consists of only two lines:

- The upper, or *liquidus,* above which any point represents in composition and temperature an alloy in the completely molten state.
- The lower, or *solidus,* below which any point represents in composition and temperature an alloy in the completely solid state.

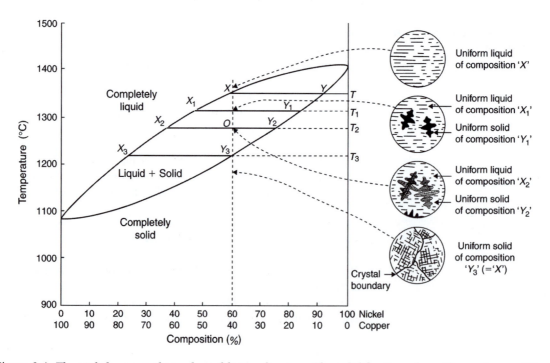

Figure 9.4 *The nickel–copper thermal equilibrium diagram. The solidification of an alloy under conditions of equilibrium (slow cooling) is illustrated.*

Any point between the two lines will represent in composition and temperature an alloy in the pasty or part solid–part liquid state. From the diagram, for an alloy in such a condition, we can read not only the compositions of the solid part and liquid part respectively, but also determine the relative proportions of the solid and liquid material.

Let us consider what happens when an alloy (X), containing 60% nickel and 40% copper, cools and solidifies *extremely slowly,* so that its structure is able to reach equilibrium at every stage of the process. Solidification will begin when the temperature falls to T (the vertical line representing the composition X and the horizontal line representing the temperature T intersect on the liquidus line). Now, it is a feature of equilibrium diagrams that, when a horizontal line representing some temperature cuts two adjacent phase-boundaries in this way, the compositions indicated by those two intersections can exist in equilibrium together. In this case, liquid solutions of composition X can exist in equilibrium with solid solutions of composition Y at the temperature T. Consequently, when solidification begins, crystal nuclei of composition Y begin to form.

Since the solid Y contains approximately 92% nickel/8% copper (as read from the diagram), it follows that the liquid which remains will be less rich in nickel (but correspondingly richer in copper) than the original 60% nickel/40% copper composition. In fact, as the temperature falls slowly, solidification continues; and the *composition of the liquid changes along the liquidus line, whilst the composition of the solid changes – by means of diffusion – along the solidus line.* Thus, by the time the temperature has fallen to T_1, the liquid solution has changed in composition to X_1, whilst the solid solution has changed in composition to Y_1. At a lower temperature – say T_2 – solidification will have progressed further, and the composition of the liquid will have changed to X_2, whilst the composition of the corresponding solid will have changed to Y_2.

Clearly, since the alloy is gradually solidifying so that the *proportions* of solid and liquid are continually changing, the *compositions* of solid and liquid must also change, because the overall composition of the alloy as a whole remains at 60% nickel/40% copper throughout the process. The relative weights of solid and liquid – as well as their compositions – can be obtained from the diagram, assuming that the alloy is cooling slowly enough for equilibrium to be attained by means of diffusion. Thus, at temperature T_2:

$$\text{weight of liquid (composition } X_2) \times OX_2$$
$$= \text{weight of solid (composition } Y_2) \times OY_2$$

This is commonly referred to as the *lever rule*. Engineers will appreciate that this is an apt title, since, in this particular case, it is as though moments had been taken about the point O (the overall composition of the alloy). We will now substitute actual values (read from the equilibrium diagram Figure 9.4) in the above expression. Then:

$$\text{weight of liquid (38\% nickel/62\% copper)} \times (60 - 38)$$
$$= \text{weight of solid (74\% nickel/26\% copper)} \times (74 - 60)$$

or

$$\frac{\text{weight of liquid (38\% nickel/62\% copper)}}{\text{weight of solid (74\% nickel/26\% copper)}} = \frac{(74-60)}{(60-38)} \quad \frac{14}{22}$$

Thus, assuming that the alloy is cooling slowly and is therefore in equilibrium, we can obtain the above information about it at the temperature T_2 (1280°C).

The solidification process will finish as the temperature falls to T_3. Here, the last trace of liquid (X_3) will have been absorbed into the solid solution which, due to diffusion, will now be of uniform composition Y_3.

Composition Y_3 is of course the same as X – the composition of the original liquid. Obviously, it cannot be otherwise if the solid Y_3 has become uniform throughout due to diffusion. 'Why go through all this complicated procedure to demonstrate an obvious point?', the reader may ask. Unfortunately – whether in engineering or in other branches of applied science – it is not often possible to make a straightforward application of a simple scientific principle. Influences of other variable factors usually have to be taken into account, such as the effects of friction in a machine, or of the pressure of wind in a civil engineering project.

In the earlier description of the solidification of the 60% nickel/40% copper alloy, we have assumed that diffusion has taken place completely,

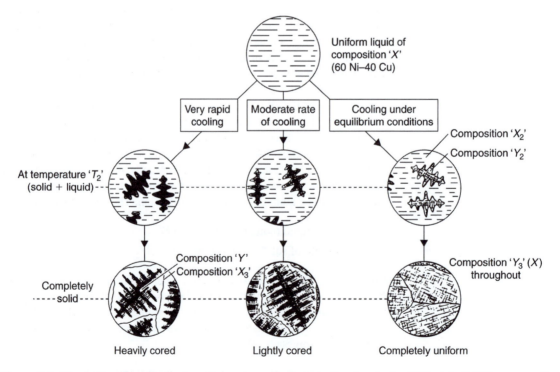

Figure 9.5 *Illustrating the effects of cooling-rate on the extent of coring in the 60% nickel/40% copper alloy dealt with above.*

resulting in the formation of a *uniform* solid solution. Under industrial conditions of relatively rapid cooling, this is rarely possible; there just isn't time for the atoms to 'jiggle' around as described in Chapter 8. Consequently, the composition of the solid solution always lags behind that indicated by the equilibrium diagram for some particular temperature, and this leads to some residual coring in the final solid (Figure 9.5).

If the rate of solidification has been very rapid, the cores of the crystal may be of a composition almost as rich in nickel as $Y;$ whilst the outer fringes of the crystals may be of a composition somewhere in the region of X_3. Slower rates of solidification will lead to progressively lesser degrees of coring, as, under these circumstances, the effects of diffusion make themselves felt. Alternatively, if this 60% nickel/40% copper alloy were annealed for some hours at a temperature just below T_3, i.e. just below the solidus temperature, any coring would be dissipated by diffusion.

In this section, we have been dealing with different modes of solidification of a 60% nickel/40% copper alloy. Since nickel and copper are soluble in each other in all proportions in the solid state, any other alloy composition of these two metals will solidify in a similar manner.

9.3.2 An alloy system in which the two metals are soluble in each other in all proportions in the liquid state, but completely insoluble in the solid state

In this case, the two metals form a single homogeneous liquid when they are melted together, but on solidification they separate again and form individual crystals of the two *pure* metals.

Cadmium and bismuth form alloys of this type. Both metals have low melting-points, but, whilst cadmium is a malleable metal used to some extent for electroplating, bismuth is so brittle as to be useless for engineering purposes. It should be noted that the name 'bismuth' is often used to describe a compound of the actual metal which is sometimes used in medicine.

Again, the equilibrium diagram (Figure 9.6) consists of only two boundaries: the liquidus *BEC* and the solidus *AED* – or, more properly, *BAEDC*. As in the previous case, any point above *BEC* represents in composition and temperature an alloy in the completely molten state; whilst any point below *AED* represents an alloy in the completely solid state. Between *BEC* and *AED,* any point will represent in composition and temperature an alloy in the part liquid–part solid state.

Let us consider a molten alloy of composition X, containing 80% cadmium and 20% bismuth. This will begin to solidify when the temperature falls to T (Figure 9.6). In this case, the appropriate 'temperature horizontal' through T cuts that part of the solidus *BA* which represents a composition of 100% cadmium. Consequently, nuclei of pure cadmium begin to crystallise. As a result, the remaining molten alloy is left less rich in cadmium and correspondingly richer in bismuth; so, as the temperature falls and cadmium continues to solidify, the liquid composition follows the liquidus line from X to X_1. This process continues, and by the time the temperature has fallen to T_2, the remaining liquid will be of composition X_2.

The crystallisation of pure cadmium continues in this manner until the temperature has fallen to 140°C (the final solidus temperature), when the remaining liquid will be of composition E (40% cadmium/60% bismuth). Applying the lever rule at this stage:

weight of pure cadmium $\times AO$

= weight of liquid (composition E) $\times OE$

or

$$\frac{\text{weight of pure cadmium}}{\text{weight of liquid (composition } E)} = \frac{OE}{AO} = \frac{(60-20)}{(20-0)} = \frac{40}{20} = 2$$

Thus, there will be twice as much solid cadmium by weight as there is liquid at this stage.

The two metals are now roughly in a state of equilibrium in the remaining liquid, which is represented in composition and temperature by the point E (the eutectic point). Until this instant, the liquid has been adjusting its composition by rejecting what are usually called *primary* crystals of cadmium. However, due to the momentum of solidification, a little too much cadmium solidifies, and this causes the composition of the liquid to swing

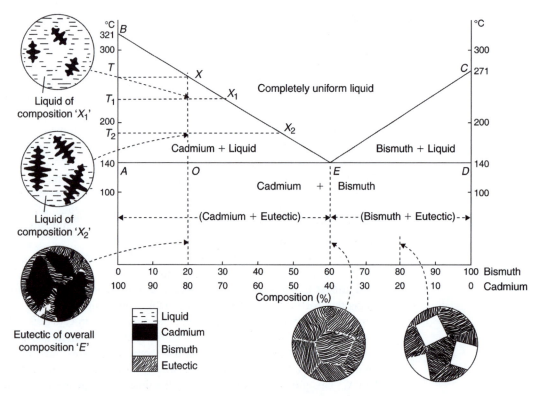

Figure 9.6 *The cadmium–bismuth thermal equilibrium diagram.*

back across point *E,* by depositing for the first time a thin layer of bismuth. Since this now upsets equilibrium in the other direction, a layer of cadmium is deposited, and so the liquid composition continues to swing to-and-fro across the eutectic point, by depositing alternate layers of each metal until the liquid is all used up. This see-saw type of solidification results in the laminated structure of eutectics and takes place while the temperature remains constant – in this case at the eutectic temperature of 140°C.

It was shown earlier that, just before solidification of the eutectic began:

$$\text{weight of pure cadmium} \ = \ 2 \times (\text{weight of liquid composition } E)$$

Since the liquid of composition *E* has changed to eutectic, it follows that the final structure will contain two parts by weight of primary cadmium to one part by weight of eutectic. By the same reasoning, an alloy containing 70% cadmium/30% bismuth will contain equal parts of primary cadmium and eutectic; whilst an alloy containing 40% cadmium/60% bismuth will consist entirely of eutectic.

Alloys containing more than 60% bismuth will begin to solidify by depositing primary crystals of bismuth, and the general procedure will be similar to that outlined earlier for the cadmium-rich alloy. Whatever the *overall* composition of the alloy, the eutectic it contains will always be of the same composition, i.e. 40% cadmium/60% bismuth. If the overall composition contains more than 40% cadmium, then some primary cadmium must deposit first; whilst if the overall composition has less than 40% cadmium, then some primary bismuth will deposit first.

It must be admitted here that *complete insolubility* in the solid state probably does not exist in metallic alloys – there is always some small amount of metal dissolved in the other. However, the solubility of solid cadmium and bismuth in each other is so small that their equilibrium system is used here to illustrate this case which serves as a useful introduction to the more general case which follows. Some older textbooks in fact use the lead–antimony system as an example of complete insolubility in the solid state whereas quite high mutual solubility of several per cent is now known to exist between these two metals.

9.3.3 An alloy system in which the two metals are soluble in each other in all proportions in the liquid state, but only partially soluble in each other in the solid state

This is a situation intermediate between the two previous cases of complete solid solubility in all proportions on the one hand and total insolubility on the other. As might be expected, very many alloy systems fall into this category.

In the early paragraphs of this chapter, part of the lead–tin thermal equilibrium diagram was used to give a general idea of a method by which the thermal equilibrium diagrams can be produced. We shall now explore the lead–tin system more fully, by reference to the complete equilibrium diagram (Figure 9.7). The diagram indicates that at 183°C lead will dissolve

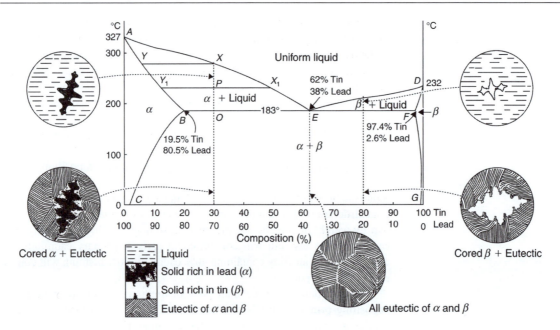

Figure 9.7 *The lead–tin thermal equilibrium diagram, showing the effects of rapid cooling on representative microstructures.*

a maximum of 19.5% tin in the solid state, giving a solid solution designated α (metallurgists use letters of the Greek alphabet to indicate different solid phases occurring in an alloy system); whilst, at the same temperature, tin will dissolve a maximum of 2.6% lead, forming a solid solution β.

An alloy whose composition falls between B and F will show a structure consisting of primary crystals of either α or β, and also some eutectic of α and β. The overall composition of the eutectic part of the structure will be given by E. In fact, an alloy containing precisely 62% tin and 38% lead will have the structure which is entirely eutectic, consisting of alternate layers of α and β.

Let us consider what happens when an alloy containing 70% lead and 30% tin solidifies and cools *slowly* to room temperature. Solidification will commence at X (at about 270°C), when nuclei of α (composition Y) begin to form. By the time the temperature has fallen to, say, 220°C, the α will have changed in composition to Y_1, due to diffusion, whilst the remaining liquid will be of composition X_1. By applying the lever rule we have that, at 220°C:

$$\text{weight of solid } \alpha \text{ (composition } Y_1) \times Y_1P$$
$$= \text{ weight of liquid (composition } X_1) \times PX_1$$

$$\frac{\text{weight of solid } \alpha \text{ (composition } Y_1)}{\text{weight of liquid (composition } X_1)} = \frac{PX_1}{Y_1P}$$

Similarly, when the temperature has fallen to 183°C, we have α (now of composition B) and some remaining liquid (of composition E) in a ratio given by:

$$\frac{\text{weight of solid } \alpha \text{ (composition } B)}{\text{weight of liquid (composition } E)} = \frac{QE}{BQ}$$

At a temperature just below 183°C, the remaining liquid solidifies as a eutectic, by depositing alternate layers of α (composition B) and β (composition F), the *overall* composition of this eutectic being given by E. Thus the structure, represented by a point just below Q, will consist of primary crystals of α of uniform composition B, surrounded by a eutectic mixture of α (composition B) and β (composition F).

In this diagram, we have two phase boundaries of a type not previously encountered; namely the *solvus* lines BC and FG. These boundary lines separate phase fields in which only solid phases exist, and therefore denote microstructural changes which occur *after an alloy is completely solid*. The slope of BC indicates that, as the temperature falls, the solubility of solid tin in solid lead will diminish from 19.5% at 183°C to about 2% at 0°C (point C); and similarly the slope of FG indicates that the solubility of solid lead in solid tin will fall from 2.6% at 183°C to less than 1% at 0°C (point G). Consequently, as our 70% lead/30% tin alloy cools slowly from 183°C to room temperature, the composition of any α in the structure alters along BC, whilst the composition of any β will alter along FG.

In practice, such an alloy will never cool slowly enough for equilibrium to be reached at each stage of the process, and some coring will inevitably occur, particularly in the crystals of primary solid solution. Accordingly, the sketches representing microstructures in Figure 9.7 assume that cooling has been fairly rapid, and that considerable coring has occurred as a result. Extremely slow cooling (as outlined earlier) or prolonged annealing eventually eliminates coring.

9.4 Precipitation from a solid solution

In the preceding section, the significance of the sloping solvus lines – BC and FG (Figure 9.7) – was mentioned. It has been my experience that many students find it difficult to accept that changes in solid solubility can occur *after* an alloy has become completely solid. I would refer them to the discussion of diffusion in Chapter 8 and how (Figure 8.7) atoms can move from one part of a crystal to another. However, this change in solid solubility as temperature is raised or lowered is a very important feature involved in the heat-treatment of alloys. Amongst other phenomena, it explains largely how the mechanical properties of some aluminium alloys can be changed by 'precipitation hardening' (see Section 17.7). This was still very much a mystery (then known as *age hardening*) during my student days – and not only to us students!

9.4.1 A liquid solution

Let us first consider a parallel case concerning the solubility of common salt (sodium chloride) in water. If some of the salt is put into water in a beaker, and stirred, much of the salt may dissolve but some solid may remain at the bottom of the beaker. We thus can have two phases in the beaker – a *saturated* solution, and some solid salt.

If we now gently warm the solution, more and more salt will dissolve, until only solution remains. At a higher temperature still, the solution would dissolve more salt, if it were available in the beaker. The solution is therefore said to be *unsaturated,* and only a single phase remains in the beaker – the unsaturated solution.

It is quite easy to plot a graph showing the variation in the solubility of the salt with a rise in temperature, as in Figure 9.8. It indicates that, as the temperature increases, so does the solubility of salt in water.

Suppose we add some salt to pure water, so that X (Figure 9.8) denotes the total percentage of salt present. After mixing the two together at, say, 10°C, we find that we still have a quantity of solid salt remaining. In fact, the solubility diagram tells us that Y% of salt has actually dissolved, giving a saturated solution at that temperature, and that $(X - Y)$% remains at the bottom of the beaker.

If we now warm the beaker slowly to 30°C, we find that more salt dissolves and the solubility diagram shows that the amount in solution has

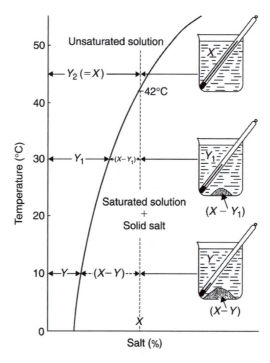

Figure 9.8 *The solubility curve for salt in water. Solubility increases as the temperature increases.*

increased to Y_1. At, say, 40°C, we would find that a very small quantity of solid salt remained, and at a slightly higher temperature (42°C) this would just dissolve. If the temperature were raised to, say, 45°C, the solution would be unsaturated – that is, it would dissolve more salt at that temperature, if solid salt were added to the beaker.

This process of solution is reversible and, if we allow the beaker to cool slowly, tiny crystals of salt will precipitate when the temperature has fallen a little below 42°C. These crystals will increase in size as the temperature falls. By the time 10°C has been reached, we shall again have an amount Y left in solution and $(X-Y)$ as solid crystals at the bottom of the beaker.

9.4.2 A solid solution

In the earlier case, we have been dealing with a *liquid* solution of salt in water, but exactly the same principles are involved if we consider instead a *solid* solution of, say, copper in aluminium. Naturally, in the case of a solid solution, the reversible process of solution and precipitation will take place much more slowly, since the individual atoms in a metallic structure are not able to move about as freely as the particles of salt and water in a beaker.

The aluminium-rich end of the aluminium–copper thermal equilibrium diagram is shown in Figure 9.9. The sloping phase boundary *AB* shows that the solubility of *solid* copper in *solid* aluminium increases from 0.2%

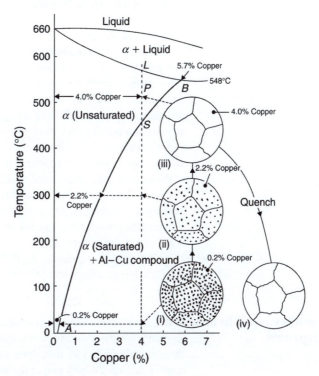

Figure 9.9 *The solubility curve for copper in aluminium. This is the aluminium-rich end of the aluminium–copper thermal equilibrium diagram.*

at 0°C (at *A*) to 5.7% at 548°C (at *B*). Any point to the left of *AB* will represent, in composition and temperature, an unsaturated solution (α) of copper in aluminium, whilst any point to the right of *AB* will represent, in composition and temperature, a structure consisting of saturated solid solution α, along with some excess aluminium–copper compound.

We will consider an alloy containing 4% copper, since this forms the basis of the well-known aluminium–copper alloy duralumin (see Section 17.7). If this has been allowed to cool very slowly to room temperature, its structure will have reached equilibrium and is represented by (i) in Figure 9.9. This consists of solid solution α, which at room temperature contains only about 0.2% of dissolved copper, the remainder of the 4% copper being present as particles of the aluminium–copper intermetallic compound scattered throughout the structure.

Suppose this alloy is now heated slowly. As the temperature rises, the solid aluminium–copper compound gradually dissolves in the solid solution α, by means of a process of diffusion. At, say, 300°C, the solid solution α will already have absorbed about 2.2% copper and, for this reason, there will be less of the intermetallic compound left in the structure (ii).

At about 460°C (point *S* in Figure 9.9), the solution of the intermetallic compound will be complete (iii), the whole of the 4% copper now being dissolved in the solid solution α. In practice, the alloy will be heated to about 500°C (point *P* in Figure 9.9), in order to ensure that all of the intermetallic compound has been absorbed by the solid solution α. Care must be taken not to heat the alloy above the point *L* in Figure 9.9, since at this point it would begin to melt.

The alloy is maintained at 500°C for a short time, so that its solid solution structure can become uniform in composition. It is then quickly removed from the furnace and immediately quenched in cold water. As a result of this treatment, *the rate of cooling will be so great that particles of the intermetallic compound will have no opportunity to be precipitated.* Therefore, at room temperature we shall have a uniform structure of a solid solution; though normally, with slow cooling, an equilibrium structure consisting of almost pure aluminium (composition *A* in Figure 9.9) along with particles of the intermetallic compound would be formed (Figure 9.9(i)). Quenching, however, has prevented equilibrium from being attained. Hence, the quenched structure is *not* an equilibrium structure and is in fact a *supersaturated* solid solution, since *A* contains much more dissolved copper than is normal at room temperature.

This treatment forms the basis of the first stage in the heat-treatment of duralumin-type alloys and will be dealt with in detail in Chapter 17.

9.5 Ternary equilibrium diagrams

In this chapter, we have been dealing only with equilibrium diagrams which represent *binary* systems, i.e. ones containing *two* metals. If an alloy contains *three* metals, this will introduce an additional variable quantity since the relative amounts of any two of the three metals can be altered independently. Temperature remains the other 'variable'. Thus, we have a system with a total of three variables. This can be represented graphically by a 'three-

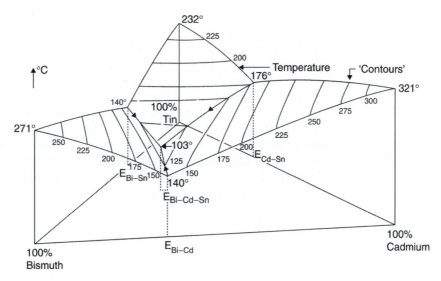

Figure 9.10 *The bismuth–cadmium–tin thermal equilibrium diagram. This is of course a ternary diagram. The three 'valleys' drain down to the ternary (or 'triple') eutectic point at 103°C. This alloy contains 53.9% bismuth, 25.9% tin and 20.3% cadmium. The 'temperature contours' are at 25°C intervals. The ternary eutectic would melt at a temperature just above that of boiling water.*

dimensional' or 'solid-type' diagram (Figure 9.10) which consists of a base in the form of an equilateral triangle, each point of the triangle representing 100% of one of the three metals (in this case, cadmium, tin and bismuth), whilst ordinates normal to this base represents temperature.

A 'solid' diagram of this type is not of much practical use since phase boundaries 'inside' the model are not visible. In my student days, research workers at the university used to construct models representing such systems by employing bits of coloured plastic-coated electrical wire to indicate phase boundaries. With a complex system, the resulting model was quite fantastic and resembled one of the more lurid examples of modern 'sculpture' – or possibly a parrot cage which had been designed by a committee.

However, for ternary, i.e. three-metal, alloys we can still draw useful *two-dimensional* diagrams by fixing the quantities of two of the metals so that only the quantity of the third is variable. Of course, we would then need a large number of separate sheets – each expressing two of the metals in different amounts – to cover a ternary system comprehensively. It is often useful to do this, as in the case of the diagram representing the structures of high-speed steel (Figure 13.1). Here, we have chosen a high-speed steel of standard composition and have indicated the effects of variations in the carbon content and the temperature on the structure of the alloy. The amounts of the other alloy additions, viz. tungsten, chromium and vanadium, are fixed at the values shown. The use of this diagram enables us to explain the basic principles of the heat-treatment of high-speed steel quite adequately. It is not a true equilibrium diagram, but is used like one, and is generally referred to either as *constitutional* or a *pseudo-binary* diagram.

It is of course possible to make alloys containing yet more than three metals. Thus, if we add lead to the bismuth–cadmium–tin alloy system described earlier, a *quaternary* alloy series will result. However, such a system could not be represented by an equilibrium diagram, even of the 'parrot's cage' type, simply because we have run out of dimensions – a point in space is represented by *three* co-ordinates in the Cartesian system.

In passing, it is interesting to note than an alloy from the quaternary system mentioned here and containing 50% bismuth/24% lead/14% tin/12% cadmium melts at 71°C. It was used in those frivolous unsophisticated days of my long-ago youth for the manufacture of tea spoons for practical jokers. Which is perhaps a note of light relief on which to end this serious and very important chapter.

10 Practical microscopy

10.1 Introduction

Examination of the microstructure of metals has been practised since it was developed by Professor Henry Sorby at Sheffield in the early 1860s and it is safe to say that of all the investigational tools available to him, the average materials scientist would be least likely to be without his microscope. With the aid of quite a modest instrument, a trained metallurgist can obtain an enormous amount of information from the microscopical examination of a metal or alloy. In addition to being able to find evidence of possible causes of failure of a material, he can often estimate its composition, as well as forecast what its mechanical properties are likely to be. Moreover, in the fields of pure metallurgical research, the microscope figures as the most frequently used piece of equipment.

Whilst it is recognised that nowadays industrial laboratories are likely to have at least some of the specimen preparation processes semi-automatic, it was felt useful to include the following to show the background to the processes and how the engineering student can acquire some skill in the preparation and examination of a micro-section, using a minimum of apparatus.

10.2 Selecting and mounting a specimen

Thought and care must be exercised in selecting a specimen from a mass of material, in order to ensure that the piece chosen is representative of the material as a whole. For example, free-cutting steels (see Section 7.5) contain a certain amount of slag (mainly manganese sulphide, MnS). This becomes elongated in the direction in which the material is rolled (Figure 10.1). If only the cross-section (A) were examined, the observer might be forgiven for assuming that the slag was present in free-cutting steel as more or less spherical globules instead of as elongated fibres. The latter fact could only be established if a longitudinal section (B) were examined in addition to the cross-section (A).

In some materials, both structure and composition may vary across the section, e.g. case-hardened steels have a very different structure in the surface layers from that which is present in the core of the material. The same may be said of steels which have been decarburised at the surface due to faulty heat-treatment. Frequently, it may be necessary to examine two or more specimens in order to obtain comprehensive information on the material.

A specimen approximately 20 mm in diameter or 20 mm^2 is convenient to handle. Smaller specimens are best mounted in one of the cold-setting plastic materials available for this purpose, since such a specimen may rock during the grinding process, giving rise to a bevelled surface. Moreover, mounting in plastic affords a convenient way of protecting the edge of the

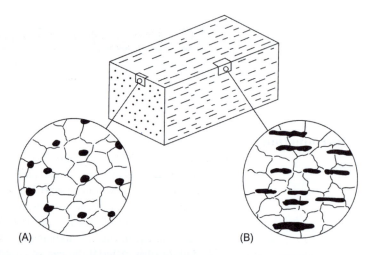

Figure 10.1 *At least two specimens are necessary to represent adequately the microstructure of a free-cutting steel.*

specimen in cases where investigation of the edge is necessary, as, for example, in examining a section through a carburised surface.

The cold-setting plastic materials, available under a variety of trade-names, consist of a white powder and a colourless liquid. When the powder is wetted with the liquid, copolymerisation (see Section 19.3) slowly takes place and a hard transparent solid is formed in about half an hour. A simple 'mould' (Figure 10.2) is all that is required to mount specimens in such a material. The specimen is placed face-down on a sheet of glass which has first been lightly smeared with Vaseline. The two L-shaped retaining pieces are then placed around the specimen, as shown, to give a mould of convenient size. The specimen is then covered with powder, which is in turn moistened with the liquid supplied. If a surplus of liquid is accidentally used, this can be absorbed by sprinkling a little more powder on the surface. In about 30 minutes, the mass will have hardened and the L-shaped members can be detached.

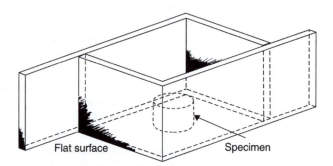

Figure 10.2 *A method of mounting a specimen in a cold-setting plastic material. No pressure is required.*

Specimens can also be mounted using thermoplastic or thermosetting materials in conjunction with a mould capable of being electrically heated and a suitable press. The advantage of using a thermosetting material, such as Bakelite, is that it is less likely to be dissolved by organic degreasing agents such as acetone or warm alcohol.

10.3 Grinding and polishing the specimen

It is first necessary to obtain a flat surface on the specimen. This is best achieved by using a file held flat on the workbench and then rubbing the specimen on the file. It is much easier for an unskilled operator to produce a single flat-surface on the specimen by using this technique, rather than by using a file in the orthodox manner. When the original hack-saw marks have been eliminated, the specimen should be rinsed in running water, to remove any coarse grit which may otherwise be carried over to the grinding papers.

10.3.1 Grinding

Grinding is then carried out by using emery papers of successively finer grades. These papers must be of the best quality, particularly in respect of uniformity of particle size. For successful wet-grinding, at least four grades of paper are required ('220', '320', '400' and '600', from coarse to fine), and these must be of the type with a waterproof base. Special grinding tables can be purchased, in which the standard 300 mm × 50 mm strips of grinding papers can be clamped. The surface of the papers is flushed by a current of water, which serves not only as a lubricant in grinding but also carries away coarse emery particles, which might otherwise scratch the surface of the specimen. If commercially produced grinding tables are not available – and certainly the prices of these simple pieces of apparatus seem to be unreasonably high – there is no reason why simple equipment should not be improvised, as indicated in Figure 10.3. Here, a sheet of 6 mm plate glass, about 250 mm × 100 mm, has a sheet of paper clamped to its surface by a pair of stout paper-clips. The paper should be folded round the edge of the glass plate, so that it will be held firmly. A suitable stream of water can be

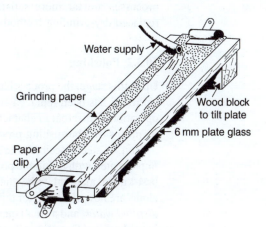

Figure 10.3 *A simple grinding table adapted from odds and ends.*

obtained by using a piece of rubber tubing attached to an ordinary tap, and the complete operation may be conducted in the laboratory sink. The glass plate is tilted at one end, so that the water flows fairly rapidly over the grinding paper.

In busy laboratories, rotating grinding tables are likely to be used. These are supplied with water which drips from a small reservoir above the rotating table.

The specimen is first ground on the '220' grade paper. Assuming that a stationary table is being used, this is achieved by rubbing it back and forth on the paper, in a direction which is roughly *at right angles* to the scratches left by the filing operation. In this way, it can easily be seen when the original deep scratches left by the file have been completely removed. If the specimen were ground so that the new scratches ran in the same direction as the old ones, it would be virtually impossible to see when the latter had been erased. With the primary grinding marks removed, the specimen is now washed free of '220' grit. Grinding is then continued on the '320' paper, again turning the specimen through 90° and grinding until the previous scratch marks have been erased. The process is repeated with the '440' and '600' papers.

If circumstances demand that dry-grinding be used, complete cleanliness must be maintained at all stages, in order to avoid the carrying-over of coarse grit to finer papers. After use, each paper should be shaken free of grit by smartly pulling it taut a number of times. Papers can be stored safely between the pages of a *glossy* magazine. Alternatively, a strip of each grade of paper can be permanently attached to its own polishing block. It is most important that the specimen be washed before passing from one grade of paper to the next, and particularly before transferring to the final polishing cloth.

Steels and the harder non-ferrous metals can be ground dry, provided that care is taken not to overheat them, since this may modify the microstructure. For the softer non-ferrous metals, such as aluminium alloys and bearing metals, the paper should be moistened with a lubricant such as paraffin. A lighter pressure can then be used, and there will be much less risk of particles of grit becoming embedded in the soft metal surface. Modern wet-grinding processes are far more satisfactory for all materials and have generally replaced dry-grinding methods.

10.3.2 Polishing

Up to this stage, the process has been one of grinding and each set of parallel 'furrows' has been replaced successively by a finer set. The final polishing operation is different in character, in that it removes the ridged surface layers by means of a burnishing process.

Iron or steel specimens are polished by means of a rotating cloth pad impregnated with a suitable polishing medium. 'Selvyt' cloth is probably the best-known material with which to cover the polishing wheel, though special cloths are now available for this purpose. The cloth is thoroughly wetted with *distilled water*, and a small quantity of the polishing powder is gently rubbed in with *clean* fingertips. Possibly the most popular polishing medium is alumina (aluminium oxide), generally sold under the name of 'Gamma

Alumina'. During the polishing process, water should be permitted to spot on to the pad, which should be run at a low speed until the operator has acquired the necessary manipulative technique. Light pressure should be used, since too heavy a pressure on the specimen may result in a torn polishing cloth. Moreover, the specimen is more likely to be scratched by grit particles embedded deep in the cloth if heavy pressure is applied.

A disadvantage of most of the water-lubricated polishing powders is that they tend to dry on the pad, which generally becomes hard and gritty as a result. If the pad is to be used only intermittently, it might be worthwhile to use one of the proprietary diamond-dust polishing compounds. In these materials, graded diamond particles are carried in a 'cream' base which is soluble both in water and the special polishing fluid, a few spots of which are applied to the pad in order to lubricate the work and lead to even spreading of the polishing compound. If properly treated, the pad remains in good condition until it wears out; so, although these diamond dust materials are more expensive than other polishing media, a saving may result in the long term, as polishing cloths will need to be changed less frequently.

Since non-ferrous metals and alloys are much softer than steels, they are best polished by hand, on a small piece of 'Selvyt' cloth wetted with 'Silvo'. During polishing, a circular sweep of the hand should be used, rather than the back-and-forth motion used in grinding.

When the surface appears free from scratches, it is cleaned thoroughly, dried and then examined under the microscope, using a magnification between 50 and 100. If reasonably free from scratches, the specimen can at this stage be examined for inclusions, such as those of manganese sulphide in steel, slag fibres in wrought iron or globules of lead in free-cutting brasses. Such inclusions would be less obvious were the specimen etched *before* this primary examination. Although the surface is made smooth by the polishing

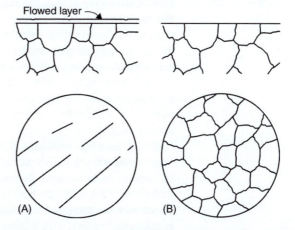

Figure 10.4 *The 'flowed layer' on the surface of a polished microsection. (A) In the polished state the structure is hidden by the lowed layer. (B) Etching removes the flowed layer, thus revealing the crystal structure beneath.*

operation, the structure still cannot be seen, because the nature of the polishing process is such that it leaves a 'flowed' or amorphous layer of metal on the surface of the specimen (Figure 10.4). In order that the structure can be seen, this flowed layer must be dissolved – or 'etched' away – by a suitable chemical reagent.

The most important points to be observed during the grinding and polishing processes are:

- Absolute cleanliness is necessary at each stage.
- Use very light pressure during both grinding and polishing.
- Deep scratches are often produced during the final stages of grinding. Do *not* attempt to remove these by prolonged polishing, as such scratches tend to be obliterated by the flowed layer, only to reappear on etching. Moreover, prolonged polishing of non-ferrous metals tends to produce a rippled surface. If deep scratches are formed, wash the specimen, and return to the last-but-one paper, remembering to grind in a direction at 90° to the scratches.
- Care must be taken not to overheat the specimen during preliminary filing or grinding. Hardened steels could be tempered by such treatment, particularly if a linishing machine is used.

10.4 Etching the specimen

Etching is generally the stage in preparing a microsection that the beginner finds most difficult. Often the first attempt at etching results in a badly stained or discoloured surface, and this is invariably due to inadequate cleaning and degreasing of the specimen before attempting to etch it.

The specimen should first be washed free of any adhering polishing compound. This can be rubbed from the *sides* of the specimen using the fingers, but great care must be taken in dealing with the polished face. The latter can be cleaned and degreased successfully by *very gently* smearing the surface with a *clean* fingertip dipped in grit-free soap solution, followed by thorough rinsing under the tap. Even now, traces of grease may still be present, as shown by the fact that a film of water will not flow evenly over the surface, but instead remains as isolated droplets. The last traces of grease are best removed by immersing the specimen for a minute or so in boiling alcohol, 'white industrial spirit' (on no account should alcohol be heated over a naked flame, as the vapour is highly inflammable – an electrically heated water-bath should be used, an electrical kettle with the lid removed is serviceable).

From this stage onwards, the specimen should not be touched by the fingers but handled with a pair of nickel tongs. It is lifted from the alcohol and cooled under the tap before being etched. Some thermoplastic mounting materials are dissolved by hot alcohol; in such cases, swabbing with a piece of cotton wool saturated with dilute caustic soda may degrease the surface effectively.

When the specimen is clean and free from grease, it is etched by plunging it into the etching solution, and agitating it vigorously for a few seconds. The specimen is then *very quickly* transferred to running water, in order to wash away the etchant as rapidly and evenly as possible. It is then examined with the naked eye, to see to what extent etching has taken place. If successfully etched, the highly polished surface will now appear dull, and, in the case of cast metals, individual crystals may be seen. A bright surface at this stage will usually indicate that further etching is necessary. The time required for etching varies with different alloys and etchants, and may be limited to a few seconds for a specimen of carbon steel etched with 2% 'nital', or extended to as long as 30 minutes for a stainless steel etched in a mixture of concentrated acids.

After being etched, the specimen is washed in running water, and then quickly immersed in boiling alcohol, where it should remain for a minute. On withdrawal from the alcohol, the specimen is shaken with a flick of the tongs, to remove surplus alcohol so that it will dry almost instantaneously.

Table 10.1 *Etching reagents*

Type of etchant	Composition	Characteristics and uses
2% 'nital' – for iron, steel and bearing metals	2 cm³ nitric acid, 98 cm³ alcohol ('white industrial methylated spirit')	The best general etching reagent for irons and steels, both in the normalised and heat-treated condition. For pure iron and wrought iron, the quantity of nitric acid may be raised to 5 cm³. Also suitable for most cast irons and some alloys, such as bearing metals.
Alkaline sodium picrate – for steels	2 g picric acid, 25 g sodium hydroxide, 100 cm³ water	The sodium hydroxide is dissolved in the water and the picric acid is then added. The whole is heated on a boiling water-bath for 30 minutes and the clear liquid is poured off. The specimen is etched for 5–15 minutes in the boiling solution. It is useful for distinguishing between ferrite and cementite; the latter is stained black, but ferrite is not attacked.
Ammonia/hydroxide peroxide – for copper, brasses and bronzes	50 cm³ ammonium hydroxide (0.880) 50 cm³ water. Before use, add 20–50 cm³ hydrogen peroxide (3%)	The best general etchant for copper, brasses and bronzes. Used for swabbing or immersion. Must be freshly made, as the hydrogen peroxide decomposes. (The 50% ammonium hydroxide solution can be stored, however.)
Acid ferric chloride – for copper alloys	10g ferric chloride, 30 cm³ hydrochloric acid, 120 cm³ water	Produces a very contrasty etch on brasses and bronzes. Use at full strength for nickel-rich copper alloys, but dilute one part with two parts of water for brasses and bronzes.
Dilute hydrofluoric acid – for aluminium and its alloys	0.5 cm³ hydrofluoric acid, 99.5 cm³ water	A good general etchant for aluminium and most of its alloys. The specimen is best swabbed with cotton wool soaked in etchant.

N.B. On no account should hydrofluoric acid or its fumes be allowed to come into contact with the eyes or skin. Care must be taken with all concentrated acids.

With specimens mounted in a plastic that is likely to be affected by boiling alcohol, it is better to spot a few drops of cold alcohol on the surface of the specimen. The surplus is then shaken off, and the specimen is held in a current of hot air from a domestic hairdryer. Unless the specimen is dried *evenly and quickly,* it will stain.

A summary of the more popular etching reagents which can be used for most metals and alloys is given in Table 10.1.

10.5 The metallurgical microscope

The reader may have used a microscope during his school days, but the chances are that this would be an instrument designed for biological work. Biological specimens can generally be prepared as thin, transparent slices, mounted between thin sheets of glass, so that illumination can be arranged simply by placing a source of light *behind* the specimen. Since metals are opaque substances, which must be illuminated by frontal lighting, the source of light must be *inside* the microscope tube itself with a metallurgical microscope (Figure 10.5). This is generally accomplished, as shown in

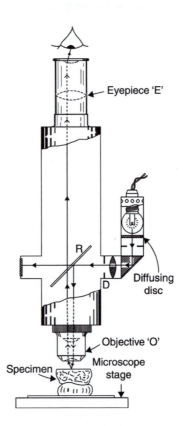

Figure 10.5 *A metallurgical microscope. This illustrates a very basic 'student's model'. More modern instruments are more streamlined and may be of inverted construction, i.e. the objective is at the upper end of the instrument so that the specimen is suspended above it. The basic method of illumination remains similar to that shown.*

Figure 10.5, by placing a small, thin plain-glass reflector R inside the tube. Since it is necessary for the returning light to pass through R, it is of unsilvered glass. This means that much of the total light available is lost, both by transmission through the glass when it first strikes the plate, and by reflection when the returning ray from the specimen strikes the plate again. Nevertheless, a small 6-volt bulb is generally sufficient as a source of illumination. The width of the beam is controlled by the iris diaphragm D. This should be closed until the width of the beam is just sufficient to cover the rear component of the objective lens O. Excess light, reflected within the microscope tube, would cause light-scatter and consequently 'glare' in the field of view, leading to a loss of contrast and definition in the image.

The optical system of the microscope consists essentially of two lenses, the objective O and the eyepiece E. The former is the more important and expensive of the two lenses, since it must resolve fine detail of the object under examination. Good-quality objectives, like camera lenses, must be of compound construction. However, at magnifications of $\times 1000$ or so, one is dealing with dimensions comparable with the wavelength of light itself, and further improvements in the magnification of the lens would produce no corresponding improvements in the 'sharpness' of the image. The magnification given by an objective depends upon its focal length – the shorter the focal length, the higher the magnification obtainable. The magnified image produced by the objective lens has a magnification of the, so termed, *tube length* (usually 200 mm for most instruments) divided by the focal length of the objective lens.

The eyepiece is so called because it is the lens nearest to the eye. Its purpose is to magnify the image formed by the objective. The magnification produced by the eyepiece is determined by its magnifying power (this is the least distance an image can be from the eye and be in focus, about 25 cm, divided by the focal length of the lens). Eyepieces are generally made in a number of powers, generally $\times 6$, $\times 8$, $\times 10$ and $\times 15$.

The overall magnification produced by the microscope is the product of the contributions made by each lens and is the linear magnification produced by the objective lens multiplied by the magnifying power of the eyepiece. The magnification of the system can thus be calculated from:

$$\text{magnification} = \frac{\text{tube-length (mm)} \times \text{power of eyepiece}}{\text{focal length of objective (mm)}}$$

Thus, for a microscope having a tube length of 200 mm, and using a 4 mm focal-length objective in conjunction with a $\times 8$ eyepiece:

$$\text{magnification} = \frac{200 \times 8}{4} = 400$$

It is not simply high magnification which is required of a microscope, we need to consider the detail we can see in the image. Thus, when working at high magnifications it can be useless to increase the size of the image by either increasing the tube length or using a higher-power eyepiece, as a point is reached where there is a falling-off in definition. A parallel example in photography is where a large enlargement fails to show any more *detail* than a smaller print but instead gives a rather blurred image; this is particularly obvious in digital photography where the number of pixels constituting the image is so obviously a limiting factor. The resolving power of a microscope is affected by the quality of the objective lens.

10.5.1 Using the microscope

There are two steps involved: mounting the specimen so that it can be viewed through the microscope and then focussing the microscope in order to see it.

The specimen must be mounted so that its surface is normal to the axis of the instrument. This is best achieved by fixing the specimen to a microscope slide, by means of a pellet of 'plasticine', using a mounting ring to ensure normality between the surface of the specimen and the axis of the parallel end-faces (Figure 10.6). Obviously, the mounting ring must have perfectly parallel end-faces. If the specimen is in a plastic mount, it can of course be placed directly on to the microscope stage, provided that the end-faces of the mount are completely parallel.

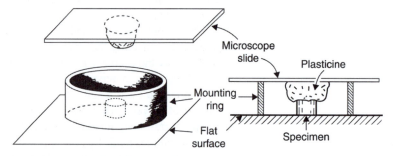

Figure 10.6 *Mounting a specimen so that its surface is normal to the axis of the microscope.*

The specimen is brought into focus by first using the coarse adjustment and then the fine adjustment. Lenses are designed to work at an optimum tube-length (usually 200 mm), and give the best results under these conditions. Hence, the tube carrying the eyepiece should be drawn out to the appropriate mark (a scale is generally engraved on the side of the tube). Slight adjustments in tube-length should then be made to suit the individual eye. Finally, the iris D in the illumination system should be closed to the point where the illumination begins to decrease. Glare due to internal reflection will then be at a minimum.

Invariably, the newcomer to the microscope selects the combination of objective and eyepiece which will give him the maximum magnification, but it is a mistake to assume that high magnifications in the region of ×1000 are necessarily the most useful. In fact, they may well give a misleading impression of the structure, by pinpointing some very localised feature rather than giving a general picture of the microstructure of the material. Directional properties in wrought structures, or dendritic formations in cast structures, are best examined using low powers of between ×30 and ×100. Even at ×30, a single crystal of a cast structure may completely fill the field of view. In a similar way, a more representative impression of the lunar landscape may be obtained by using a good pair of binoculars than by using the very high-powered system of an astronomical telescope. Hence, as a matter of routine, a low-powered objective should *always* be used in the initial examination of a microstructure, before it is examined at a high magnification.

10.5.2 The care of the microscope

Care should be taken never to touch the surface of optical glass with the fingers, since even the most careful cleaning may damage the surface coating (most high-quality lenses are 'bloomed' that is, coated with magnesium fluoride to increase light transmission). In normal use, dust particles may settle on a lens, and these are best removed by sweeping gently with a high-quality camel-hair brush.

If a lens becomes finger-marked, this is best dealt with by wiping *gently* with a piece of soft, well-washed linen moistened with the solvent xylol. Note that the operative word is *wipe,* not *rub.* Excess xylol should be avoided, as it may penetrate into the mount of the lens and soften the cement holding the glass components together so that they become detached. Such a mishap will involve an expensive re-assembly job!

High-power objectives of the oil-immersion type should be wiped free of cedar-wood oil before the latter has a chance to harden. If, due to careless neglect, the oil has hardened on a lens, then it will need to be removed with the minimum amount of xylol, but the use of the latter should be avoided whenever possible.

10.6 The electron microscope

As the reader will have gathered, the greater proportion of routine microscopic examinations of metals is carried out using magnifications in the region ×100. It is sometimes necessary, particularly in the field of research, to examine structures at much higher magnifications. Unfortunately, it is not practicable to use magnifications in excess of ×2000 with an ordinary optical microscope, since one is then dealing with an object size of roughly the same order as the wavelength of light itself and this leads to a loss of definition.

For high-power microscopy – between ×2000 and about ×200 000 – an electron microscope may be used. In this instrument, rays of light are replaced by beams of electrons. Since electrons are electrically charged particles, their path can be altered by electric and magnetic fields. Consequently, the 'lenses' in an electron microscope consist of a system of coils to focus the electron beam. The other important difference between an optical and an electron microscope is that, whereas in the former light is *reflected* from the surface of the specimen, in the electron microscope the electron beam *passes through* the specimen, rather in the manner of light rays in a biological microscope. Hence, thin foil specimens must be used, or, alternatively, a very thin replica of the etched metallic surface may be produced in a suitable plastic material. This replica is then examined by the electron microscope.

The limitations of the optical microscope have already been mentioned, but it should be noted that the range of the ordinary electron microscope too is limited. Whilst it is true that magnification obtained with this instrument can exceed a quarter of a million times, it should be appreciated that a magnification of something like 20 million times is required for us to be able to see an average-size atom. Magnifications of this order can now be obtained using a modern *field-ion* microscope.

11 Iron and steel

11.1 Introduction

Since the onset of the Industrial Revolution, the material wealth and power of a nation has depended largely upon its ability to make steel. During the nineteenth century, Britain was prominent among steel-producing nations, and, towards the end of the Victorian era, was manufacturing a great proportion of the world's steel. But exploitation of vast deposits of ore abroad changed the international situation so that by the middle of the twentieth century what we then referred to as the two Superpowers – the USA and the USSR – owed their material power largely to the presence of high-grade ore within, or very near to, their own vast territories. Consequently, they led the world in terms of the volume of steel produced annually. More recently, rapid technological developments in the Far East have meant that the People's Republic of China is currently the world's premier steel producer followed by the European Union, Japan, United States, Indonesia, Russia and South Korea. The UK now occupies twelfth position in the international steel producers' 'league table'.

Low-grade ores mined in Britain now account for only a small part of the input to British blast-furnaces and most of the ore used must be imported (as high-grade concentrates) from Canada, Brazil, Australia, Norway, North Africa, Venezuela and Sweden.

This chapter is about steel-making and the composition and structure of plain-carbon steels.

11.2 Smelting

Smelting of iron ore takes place in the blast-furnace (Figure 11.1). A modern blast-furnace is something like 60 m high and 7.5 m in diameter at the base, and may produce from 2000 to 10 000 tonnes of iron per day. Since a refractory lining lasts for several *years,* it is only at the end of this period that the blast-furnace is shut down; otherwise, it works a 365-day year. Processed ore, coke and limestone are charged to the furnace through the double-bell gas-trap system, whilst a blast of heated air is blown in through the tuyères near the hearth of the furnace. At intervals of several hours, the furnace team opens both the slag hole and the tap hole, in order to run off first the slag and then the molten iron. The holes are then plugged with clay.

The smelting operation involves two main reactions:

1 The chemical reduction of iron ore by carbon monoxide gas (CO) arising from the burning coke:

$$Fe_2O_3 + 3CO \rightarrow 2Fe + 3CO_2$$

Iron oxide ore   Iron   Carbon dioxide

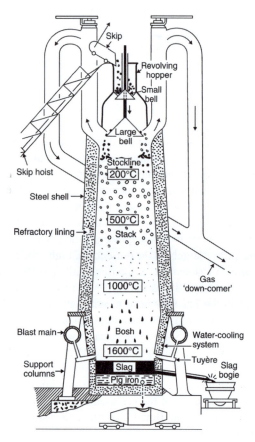

Figure 11.1 *A modern blast-furnace.*

Table 11.1 *Materials involved with a daily output of 2000 tonnes of pig iron*

Charge		Amount (tonnes)
Ore, say 50% iron		4 000
Limestone		800
Coke		1 800
Air		8 000
	Total	14 600

Products		Amount (tonnes)
Pig iron		2 000
Slag		1 600
Dust		200
Furnace gas		10 800
	Total	14 600

2   Lime (from limestone added with the furnace charge) combines with many of the impurities and also the otherwise infusible earthy waste (mainly silica SiO_2) in the ore to form a fluid slag which will run from the furnace:

The slag is broken up and used for road-making, or as a concrete aggregate. The molten iron is either cast into 'pigs', for subsequent use in an iron foundry, or transferred, still molten, to the steel-making plant. In the case of a large modern furnace, a daily output of 2000 tonnes of pig iron would involve the materials shown in Table 11.1.

One feature of Table 11.1 which may surprise the reader is the vast amount of furnace gas passing along the 'down-comer' each day. The gas contains a large amount of carbon monoxide, and therefore has a considerable calorific value. The secondary function of the blast-furnace is, in fact, to act as a large gas-producer. If the blast-furnace plant is part of an integrated steelworks, then much of this vast quantity of gas will be used for raising electric power; but its major function is to be burnt in the Cowper stoves and so provide heat which in turn will heat the air blast to the furnace. Two such stoves are required for each blast-furnace. One is being heated by the burning gas, whilst the other is heating the ingoing air.

In recent years, much research has been conducted into steel production by the 'direct reduction' of iron ore, but it seems that for some years yet the blast-furnace will survive as the major unit in iron production. The thermal efficiency of the blast-furnace is very high and this is improved further in modern management by the injection of oil or pulverised low-cost coal at the tuyères in order to reduce the amount of expensive metallurgical coke consumed. Be that as it may, we must not lose sight of the fact that the blast-furnace we have been considering briefly here is responsible for releasing some 6600 tonnes of 'greenhouse gas' (carbon dioxide) into the atmosphere each day of the year.

11.3 Steel-making

Until Henry Bessemer introduced his process for the mass-production of steel in 1856, all steel was made from wrought iron. Nowadays, wrought iron is no longer produced, except perhaps in small quantities for decorative purposes (though much of the ornamental 'wrought iron' work is in fact 'mild steel'). The Bessemer process, too, is obsolete, and, as far as steel production in Great Britain is concerned, has been followed into obscurity by the open-hearth process, though the latter is still used in a few countries abroad. In Britain, the bulk of steel is made either by one of the *basic oxygen processes* developed since 1952 or in the *electric-arc furnace*.

11.3.1 Basic oxygen steel-making (BOS)

The process of steel-making is mainly one involving oxidation of impurities present in the original charge, so that they form a slag which floats on the surface of the molten steel or are lost as fume. In the Bessemer process, impurities were removed from the charge of molten pig iron by blowing air through it. The impurities, mainly carbon, phosphorus, silicon and manganese, acted as fuel and so the range of compositions of pig iron was limited, because sufficient impurities were necessary in order that the charge did not 'blow cold' from lack of fuel. The oxidised impurities either volatilised or formed a slag on the surface of the charge.

Since the air blast contained only 20% of oxygen by volume, much valuable heat was carried away from the charge by the 80% nitrogen also present. Worse still, a small amount of this nitrogen dissolved in the charge, and, in the case of mild steel destined for deep-drawing operations, caused a deterioration in its mechanical properties. The new oxygen processes produce mild steel very low in nitrogen, so that its deep-drawing properties are superior to those of the old Bessemer steel. Improvements of this type are essential if mild steel is to survive the challenge of reinforced plastics, such as ABS, in the field of automobile bodywork.

The earliest of these oxygen processes was the L–D process, so called because it originated in the Austrian industrial towns of Linz and Donawit in 1952. It was made possible by the low-cost production of 'tonnage oxygen' and it is interesting to note that Bessemer had foreseen these possibilities almost a century earlier but of course did not have access to tonnage oxygen. Since 1952, a large number of variations of the original process have been developed but have become rationalised under the general heading of 'basic oxygen steel-making' or 'BOS'. Generally, the basic oxygen furnace (BOF) is a pear-shaped vessel of up to 400 tonnes capacity lined with basic refractories – magnesite bricks covered with a layer of dolomite. This lining must be basic to match the basic slag which is necessary for the removal of impurities from the charge. If the lining were chemically acid (silica bricks) it would be attacked by the basic slag and would quickly disintegrate.

In this process, no heat is carried away by useless nitrogen (as was the case in the old Bessemer process), so a charge containing up to 40% scrap can be used. This scrap is loaded to the converter first, followed by lime and molten pig iron (Figure 11.2). Oxygen is then blown at the surface of the molten charge from a water-cooled lance which is lowered through the mouth of the converter to within 0.5 m of the surface of the metal. At the end of the blow, this slag is run off first and any adjustments made to the carbon content of the charge which is then transferred to the ladle, preparatory to being cast as ingots, or, much more probably, fed to a continuous-casting unit which, in the UK (and the European Union) now deals with some 90% of the steel cast.

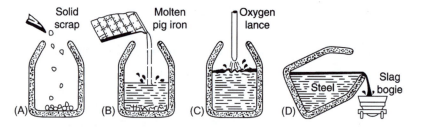

Figure 11.2 *Stages in the manufacture of steel by the basic oxygen process. Steel scrap is added first (A), followed by molten pig iron (B), Oxygen is blown through a lance (C). At the end of the 'blow', the slag is run off first (D), before 'teeming' the steel into a ladle.*

11.3.2 Electric-arc steel-making

Electric-arc steel-making is now the only alternative process in Britain to BOS, to which it is complementary rather than competitive. Originally, electric-arc furnaces were used for the manufacture of high-grade tool and alloy steels but are now widely employed both in the treatment of 'hot metal' and of process scrap as well as scrap from other sources. The high cost of electricity is largely offset by the fact that cheap scrap can be processed economically to produce high-quality steel.

Since electricity is a perfectly 'clean' fuel, no impurities are transmitted to the charge as was the case with producer gas used in the largely extinct open-hearth process. Moreover, the chemical conditions within the electric-

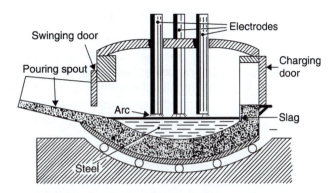

Figure 11.3 *The principles of the electric-arc furnace for steel-making. Modern furnaces have water-cooled panels built into the side walls to protect the refractory lining. Some furnaces lack the pouring spout and the charge is poured from a tap hole in the bottom of the furnace.*

arc furnace can be varied at will to favour successive removal of the various impurities present in the charge. Sulphur, which was virtually impossible to eliminate in either the Bessemer or open-hearth processes, can be effectively reduced to extremely low limits in the electric-arc process. The furnace (Figure 11.3) employs carbon electrodes which strike an arc onto the charge. Lime and mill scale are added in order to produce a slag which removes most of the carbon, silicon, manganese and phosphorus. This is run off and is often replaced by a slag containing lime and anthracite which effectively removes sulphur.

11.4 Composition of steel

Plain-carbon steels are those alloys of iron and carbon which contain up to 1.7% carbon. In practice, most ordinary steels also contain up to 1.0% manganese, which is left over from a deoxidisation process carried out at the end of the steel-making process. This excess of manganese dissolves in the solid steel, slightly increasing its strength and hardness. It also helps to reduce the sulphur content of the steel. Both sulphur and phosphorus are extremely harmful impurities which give rise to brittleness in steels. Consequently, most specifications allow no more than 0.05% of either of these elements, whilst specifications for higher-quality steels limit the amount of each element to 0.04% or less. In fact the quality, in respect of chemical composition, of mild steel is continually improving and it is common for specifications of steel used in gas and oil pipelines to demand sulphur contents as low as 0.002% with phosphorus at 0.015% maximum and carbon at 0.04%.

Although the remainder of this chapter and most of the next are devoted to the heat-treatment of carbon steels used for constructional and tool purposes, we must not lose sight of the fact that by far the greater quantity of steel produced is mild steel and low-carbon steel for structural work, none of which is heat-treated except for stress relief.

11.4.1 Cementite

At ordinary temperatures, most of the carbon in a steel which has not been heat-treated is chemically combined with some of the iron, forming an extremely hard compound known by chemists as iron carbide, though metallurgists know it as *cementite*. Since cementite is very hard, the hardness of ordinary carbon steel increases with the carbon content. Carbon steels can be classified into groups (see Table 12.3):

- dead-mild with up to 0.15% carbon,
- mild 0.15–0.25%,
- medium-carbon 0.25–0.60%,
- high-carbon with 0.60–1.50%.

11.5 The structure of plain-carbon steels

In Chapter 4, we saw that iron is what we call a polymorphic element; that is, an element which leads a sort of Jekyll-and-Hyde existence by appearing in more than one crystal form. Below 910°C, pure iron has a body-centred cubic crystal structure; but on heating the metal to a temperature above 910°C, its structure changes to one which is face-centred cubic. Now face-centred cubic iron will take quite a lot of carbon – up to 2.0% in fact – into solid solution, whereas body-centred cubic iron will dissolve scarcely any – a maximum of only 0.02%. Since the solid solubility of carbon in iron alters in this way, it follows that changes in the structure will also occur on heating or cooling through the polymorphic transformation temperature. Thus, it is the polymorphic transformation, and the structural changes which accompany it, which cause the thermal equilibrium diagram (Figure 11.4) to have a somewhat unusual shape as compared with those already dealt with in Chapter 9.

Any solid solution of carbon up to a maximum of 2.0% in face-centred cubic iron is called *austenite* (γ), whilst the very dilute solid solution formed when up to 0.02% carbon dissolved in body-centred cubic iron is called *ferrite* (α). For all practical purposes, we can regard ferrite as being more or less pure iron, since less than 0.02% carbon will have little effect on its properties.

Thus, in carbon steel at, say 1000°C, all of the carbon present is dissolved in the solid austenite. When this steel cools, the austenite changes to ferrite, which retains practically no carbon in solid solution. 'What happens to this carbon?', we may ask. The answer is that, assuming the cooling has taken place fairly slowly, the carbon will be precipitated as the hard compound cementite, referred to previously in Section 11.4.1.

11.5.1 A 0.4% carbon steel

By referring to Figure 11.4, let us consider what happens in the case of a steel containing 0.4% carbon as it solidifies and cools to room temperature. It will begin to solidify at a temperature of about 1500°C (Q_1) by forming dendrites of δ-iron (a body-centred cubic polymorph of iron) but as the temperature falls to 1493°C the δ-dendrites react with the remaining liquid to form crystals of a new phase – γ-iron containing 0.16% carbon (γ is the

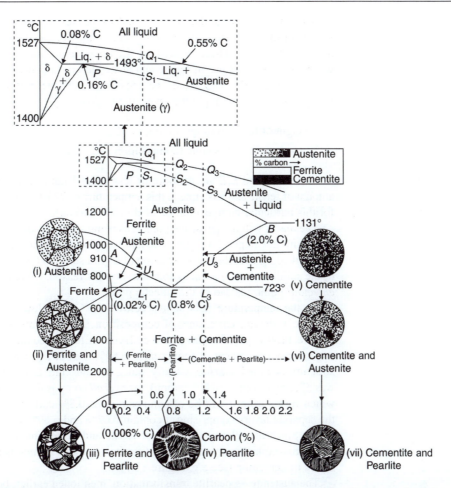

Figure 11.4 *The iron–carbon equilibrium diagram. The small dots in the diagrams depicting structures containing austenite do not represent visible particles of cementite – they are meant to indicate the concentration of carbon atoms dissolved in the austenite and in the real microstructures would of course be invisible. The inset shows the 'peritectic part' of the diagram in greater detail.*

phase we call austenite). This process of change in structure takes place by what is termed a *peritectic reaction.* As the temperature continues to fall, the remaining liquid solidifies as austenite, the composition of which changes along PS_1. The steel is completely solid at about 1450°C (S_1). The structure at this stage will be uniform austenite – there will be no coring of the dendrites because the diffusion of the interstitially dissolved carbon atoms is very rapid, particularly at high temperatures in the region of 1400°C. As this uniform austenite cools, nothing further will happen to its structure – except possibly grain growth – until it reaches the point U_1, which is known as the *upper critical point* for this particular steel. Here, austenite begins to change to ferrite, which will generally form as small new crystals at the grain boundaries of the austenite (Figure 11.4(ii)). Since ferrite contains very little carbon, it follows that at this stage the bulk of the carbon must remain behind in the shrinking crystals of austenite; and so the composition of the latter moves to the right. Thus, by the time the temperature has fallen to 723°C, we shall have a mixture of ferrite and austenite crystals of compositions C and E, respectively. The *overall* composition of the piece of steel is given by L_1, and so we can apply the lever rule to give:

weight of ferrite (composition C) $\times CL_1$

= weight of austenite (composition E) $\times L_1E$

Since CL_1 and L_1E are of more or less equal length, it follows that the amount of ferrite and austenite at this temperature of 723°C are roughly equal for this particular composition of steel (0.4% carbon).

The reader will recognise the point E as being similar to the eutectic points dealt with in Chapter 9 (Figures 9.6 and 9.7). In the present case, however, we are dealing with the transformation of a *solid* solution (austenite), instead of the solidification of a *liquid* solution. For this reason, we refer to E as the *eutectoid* point, instead of the eutectic point.

As the temperature falls to just below 723°C (the 'lower critical temperature'), the austenite, now of composition E, transforms to a eutectoid (Figure 11.4(iii)) by forming alternate layers of ferrite (composition C) and the compound cementite (containing 6.69% carbon). Such a laminated structure is called *pearlite*. Clearly, since the austenite at this temperature was of composition E (0.8% carbon), the *overall* composition of the eutectoid which forms from it will be of composition E (0.8% carbon), even though the separate layers comprising it contain 0.02% and 6.69% carbon, respectively. Since the relative densities of ferrite and cementite are roughly the same, this explains why the layers of ferrite are about 7 times as thick as the layers of cementite (see e.g. Figure 11.7).

The austenite → pearlite transformation, mentioned earlier, begins at the grain boundaries of the austenite (Figure 11.5). It is thought that carbon atoms congregate there in sufficient numbers to form cementite nuclei which grow inwards across the austenite grains. Since carbon atoms are removed from the austenite by this process, the adjacent layer of austenite is left very low in carbon and so it transforms to produce a layer of ferrite which grows inwards, following closely behind the cementite. Beyond the new ferrite layer, an increase in carbon atoms occurs so that further cementite nucleates and so on. In this way, the structure builds up as alternate layers of cementite and ferrite. In most cases, a eutectic or eutectoid in an alloy system is not given a separate name, since it is really a mixture of two phases. The iron–carbon system, however, is the most important of the alloy systems with which the metallurgist or engineer has to deal; so the eutectoid of ferrite and cementite referred to above is given the special name of *pearlite*. This name is derived from the fact that the etched surface of a high-carbon steel reflects an iridescent sheen like that from the surface of 'mother of pearl'. This is due to the diffraction of white light as it is 'unscrambled' by minute ridges (of cementite in the case of steel) protruding from the surface of the structure, into the colours of the spectrum.

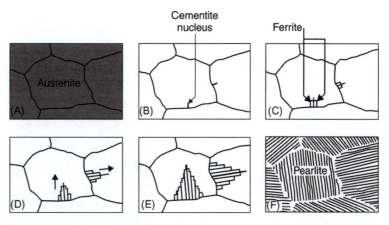

Figure 11.5 *The transformation of austenite to pearlite (in this case in a 0.8% carbon steel).*

These then are the main stages in the foregoing process of solidification and cooling of the 0.4% carbon steel:

1 Solidification is complete at S_1 and the structure consists of uniform austenite.
2 This austenite begins to transform to ferrite at U_1, the upper critical temperature of this steel (about 825°C).
3 At 723°C (the lower critical temperature of all steels), formation of primary ferrite ceases, and, as the austenite is now saturated with carbon, the eutectoid pearlite is produced as alternate layers of ferrite and cementite.
4 Below 723°C, there is no further significant change in the structure.

Figure 11.6 shows a series of micrographs of steels with carbon contents ranging from 0% to 0.8%.

11.5.2 A 0.8% carbon steel

A steel which contains exactly 0.8% carbon will begin to solidify at about 1490°C (Q_2) and be completely solid at approximately 1410°C (S_2). For a steel of this composition, the upper critical and lower critical temperatures coincide at E (723°C), so that no change in the uniformly austenitic structure occurs until a temperature slightly below 723°C is reached, when the austenite will transform to pearlite by precipitating alternate layers of ferrite and cementite. The final structure will be entirely pearlite (Figure 11.4(iv)).

11.5.3 A 1.2% carbon steel

Now let us consider the solidification and cooling of a steel containing, say, 1.2% carbon. This alloy will begin to solidify at approximately 1480°C (Q_3), by depositing dendrites of austenite and these will grow as the temperature

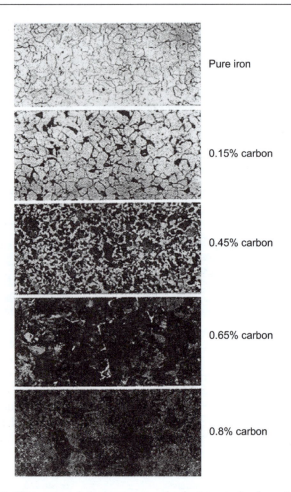

Pure iron

0.15% carbon

0.45% carbon

0.65% carbon

0.8% carbon

Figure 11.6 *This series of photomicrographs depicts steels of varying carbon contents, in the normalised condition. As the carbon content increases, so does the relative proportion of pearlite (dark), until with 0.8% carbon the structure is entirely pearlitic. The light areas consist of primary ferrite. The magnification (×80) is not high enough to reveal the laminated nature of the pearlite.*

falls, until at about 1350°C (S_3) the structure will be uniform solid austenite. No further change in the structure occurs until the steel reaches its upper critical temperature, at about 880°C (U_3). Then, needle-like crystals of cementite begin to form, mainly at the grain boundaries of the austenite (Figure 11.4(vi)). In this case, the remaining austenite becomes less rich in carbon, because the carbon-rich compound cementite has separated from it. This process continues, until at 723°C the remaining austenite contains only 0.8% carbon (E). This is, of course, the eutectoid composition; so, at a temperature just below 723°C, the remaining austenite transforms to pearlite, as in the previous two cases (Figure 11.7).

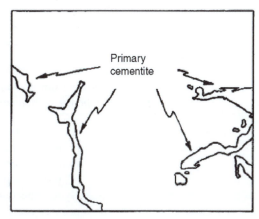

Figure 11.7 *A high-carbon tool steel (1.2% carbon) in the cast condition (×620). Since this steel contains more than 0.8% carbon, its structure shows some primary cementite (indicated on the 'key' diagram). The remainder of the structure consists of typical laminar pearlite, comprising layers of cementite sandwiched between layers of ferrite which have tended to join up and so form a continuous background or 'matrix'.*

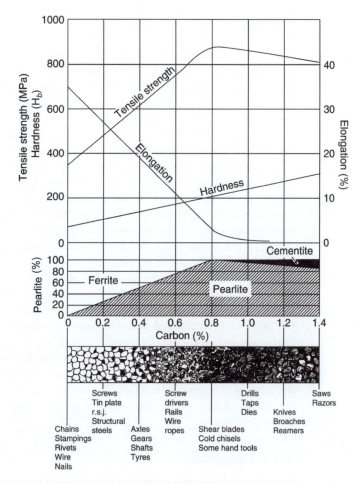

Figure 11.8 *A diagram showing the relationship between carbon content, mechanical properties and uses of plain-carbon steels which have been slowly cooled from above their upper critical temperatures.*

11.5.4 Hypo- and hyper-eutectoid steels

Thus, the structure of a carbon steel which has been allowed to cool fairly slowly from any temperature above its upper critical temperature will depend upon the carbon content as follows:

- Hypo-eutectoid steels, i.e. those containing less than 0.8% carbon – primary ferrite and pearlite (Figure 11.4(iii)).
- Eutectoid steels, containing exactly 0.8% carbon – completely pearlite (Figure 11.4(iv)).
- Hyper-eutectoid steels, i.e. those containing more than 0.8% carbon – primary cementite and pearlite (Figure 11.4(vii)).

Naturally, the proportion of primary ferrite to pearlite in a hypo-eutectoid steel, and also the proportion of primary cementite to pearlite in a hyper-eutectoid steel, will vary with carbon content, as indicated in Figure 11.8. This diagram summarises the structures, mechanical properties and uses of plain-carbon steels which have been allowed to cool slowly enough for equilibrium structures to be produced.

11.6 Heat-treatment of steel

The exploitation of the properties of iron–carbon alloys is both a tribute to Man's ingenuity – or probably blind chance (see Section 12.1) – and the great diversity of the properties of the elements, in this case the polymorphism of iron. Depending upon the composition, steel has a wide range of properties such as are found in no other engineering alloy. Steel can be a soft, ductile material suitable for a wide range of forming processes, or it can be the hardest and strongest metallurgical material in use. This enormous range of properties is controlled by varying both the carbon content and the programme of heat-treatment. Structural effects of the type obtained by the heat-treatment of steel would not be possible were it not for the natural phenomenon of polymorphism exhibited by the element iron. It is the transformation from a face-centred cubic structure to one which is basically body-centred cubic, occurring at 910°C when iron cools, that makes it possible to heat-treat these iron–carbon alloys.

There is not just one heat-treatment process, but *many* which can be applied to steels. In the processes we shall deal with in this chapter, the object of the treatment is to obtain a pearlitic type of structure; that is, one in which the steel has been allowed to reach structural equilibrium by employing a fairly slow rate of cooling following the heating process. In Chapter 12, we shall deal with those processes where quenching is employed to arrest the formation of pearlite and, as a result, increase the hardness and strength of the steel. Within certain limits, *the properties of a steel are independent of the rate at which it has been heated, but are dependent on the rate at which it was cooled.*

11.6.1 Normalising

The main purpose in normalising is to obtain a structure which is uniform throughout the work-piece, and which is free of any 'locked-up' stresses.

For example, a forging may lack uniformity in structure, because its outer layers have received much more deformation than the core. Thicker sections, which have received little or no working, will be coarse-grained, whilst thin sections, which have undergone a large amount of working, will be fine-grained. Moreover, those thin sections may have contracted faster than thicker sections and so set up residual stresses. If a forging were machined in this condition, its dimensions might well be unstable during subsequent heat-treatment.

Normalising is a relatively simple heat-treatment process. It involves heating a piece of steel to just above its upper critical temperature, allowing it to remain at that temperature for only long enough for it to attain a uniform temperature throughout, then withdrawing it from the furnace and allowing it to cool to room temperature in still air.

When, on heating, the work-piece reaches the lower critical temperature L (Figure 11.9), the pearlitic part of the structure changes to one of *fine-grained* austenite and, as the temperature rises, the remaining primary ferrite will be absorbed by the new austenite crystals until, at the upper critical temperature U, this process will be complete and the whole structure will be of uniformly fine-grained austenite. In practice, a temperature about 30°C above the upper critical temperature is used, to ensure that the whole structure has reached a temperature just above the upper critical. When the work-piece is withdrawn from the furnace and allowed to cool, the uniformly fine-grained austenite changes back to a structure which is of uniformly fine-grained ferrite and pearlite. Naturally, the grain size in thin sections may be a little smaller than that in heavy sections, because of the faster rate of cooling of thin sections.

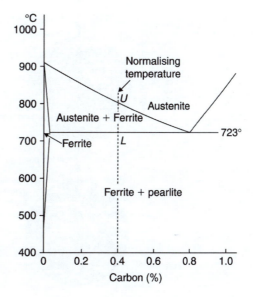

Figure 11.9 *The normalising temperature of a medium-carbon steel in relation to the equilibrium diagram.*

11.6.2 Annealing

A number of different heat-treatment processes are covered by the general description of annealing. These processes are applied to different steels of widely ranging carbon content. The three principal annealing processes are as follows.

1   *Annealing of castings*

Sand castings in steel commonly contain about 0.3% carbon so that a structure consisting of ferrite and pearlite is obtained. Such a casting, particularly if massive, will cool very slowly in the sand mould. Consequently, its grain size will be somewhat coarse, and it will suffer from brittleness because of the presence of what is known as a *Widmanstätten* structure. This consists of a directional plate-like formation of primary ferrite grains along certain crystal planes in the original austenite. Since fracture can easily pass along these ferrite plates, the whole structure is rendered brittle as a result.

The annealing process which is applied in order to refine such a structure (Figure 11.10) is fundamentally similar to that described earlier under 'normalising'; that is, the casting is heated to just above its upper

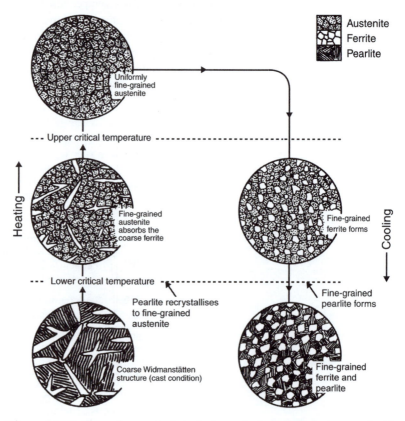

Figure 11.10 *The refinement of grain in a steel casting during a suitable annealing process.*

critical temperature, so that the coarse grain structure is replaced by one of the fine-grained austenite. It is held at this temperature for a sufficient time for the temperature of the casting to become uniform throughout and for the recrystallisation to fine-grained austenite to be complete. On cooling, this gives rise to a structure of fine-grained ferrite and pearlite. It is in the cooling stage where the two processes differ. Whereas air-cooling is employed in normalising, in this process the casting is allowed to cool within the furnace. This ensures complete removal of all casting stresses which might otherwise lead to distortion or cracking of the casting, without causing a substantial increase in grain size over that obtained by normalising. Whilst the tensile strength is not greatly improved by this treatment, both toughness and ductility are considerably increased, so that the casting becomes more resistant to mechanical shock.

An annealing temperature of 30–50°C above the upper critical temperature for the casting is commonly used. If the temperature range is exceeded, then the newly formed small crystals of austenite are likely to grow and result in a final structure almost as coarse-grained as the original cast structure. A prolonged holding time at the maximum temperature will also produce a coarse-grained structure. On the other hand, annealing at too low a temperature, i.e. below the upper critical temperature, or for too short a period, may mean that the original coarse as-cast structure does not recrystallise completely so that some coarse-grained Widmanstätten plates of ferrite are retained from the original as-cast structure.

2   *Spheroidisation annealing*

Although this may appear an onerous title, it is hoped that its meaning will become clearer in the following paragraphs. Essentially, it is an annealing process which is applied to high-carbon steels in order to improve their machinability and, in some cases, to make them amenable to cold-drawing.

It is an annealing process which is carried out *below* the lower critical temperature of the steel; consequently, no phase change is involved and we do not need to refer to the equilibrium diagram. The work-piece is held at a temperature between 650°C and 700°C for 24 hours or more. The pearlite, which of course is still present in the structure at this temperature, undergoes a physical change in pattern – due to a surface tension effect (surface tension causes water to form rounded droplets when on a greasy plate, the surface tension causing the surfaces to shrink to the minimum area possible – that of spheres) at the surface of the *cementite* layers within the pearlite. These layers break up into small plates, due to the tendency of the surface to shrink. This effect causes the plates to become gradually more spherical in form (Figure 11.11) – that is, they spheroidise, or 'ball up'. When this condition has been reached, the charge is generally allowed to cool in the furnace. Steel is more easily machined in this state, since stresses set up by the pressure on these cementite globules of the cutting edge cause minute chip cracks

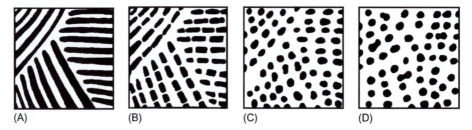

Figure 11.11 *The spheroidisation of pearlitic cementite during a subcritical annealing process. (A) The normal lamellar form of pearlite. (B) Cementite layers begin to break up (C) and ultimately form globules. (D) Finally, the original pattern of the pearlite is lost.*

to form in advance of the cutting edge. This is a standard method of improving machinability.

3   *Annealing of cold-worked steel*

Like the spheroidisation treatment described earlier, this also is a subcritical annealing process. It is employed almost entirely for the softening of cold-worked mild steels, in order that they may receive further cold-work. Such cold-worked materials must be heated to a temperature above the minimum which will cause recrystallisation to take place. Again, the equilibrium diagram is not involved, and the reader should *not* confuse this recrystallisation temperature with the lower critical temperature. The latter is at 723°C, whilst the recrystallisation temperature varies according to the amount of previous cold-work the material has received, but is usually about 550°C. Consequently, stress-relief annealing of mild steel usually involves heating the material at about 650°C for 1 hour. This causes the distorted ferrite crystals to recrystallise (Figure 11.12), so that the structure becomes soft again, and its capacity for cold-work is regained.

Since the cold-working of mild steel is usually confined to the finishing stages of a product, any annealing is generally carried out in a controlled atmosphere, in order to avoid oxidation of the surface of

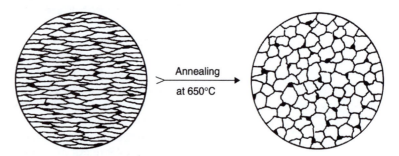

Figure 11.12 *Annealing of cold-worked mild steel causes recrystallisation of the distorted ferrite, so producing new ferrite crystals which can again be cold-worked.*

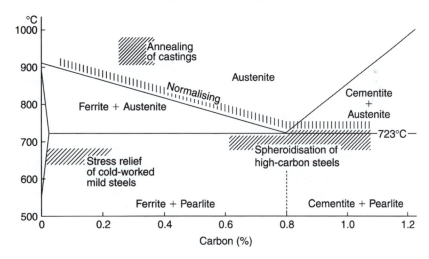

Figure 11.13 *Temperature ranges of various annealing and normalising treatments for carbon steels related to the carbon contents.*

the charge. The furnace used generally consists of an enclosed 'retort' through which an inert gas mixture passes whilst the charge is being heated. Such mixtures are based on 'burnt ammonia' or burnt town gas.

All of the foregoing heat-treatment processes produce a microstructure in the steel which is basically pearlitic. By this, metallurgists mean that the structure contains some pearlite (unless, of course, the steel is dead-mild). Thus, a hypo-eutectoid steel will contain ferrite and pearlite, a eutectoid steel only pearlite, and a hyper-eutectoid steel cementite and pearlite. Figure 11.13 summarises the temperature ranges at which these treatments are carried out, and also indicates the carbon contents of the steels most commonly involved in the respective processes.

11.7 Brittle fracture in steels

Fracture in metals generally takes place following a measurable amount of slip. Sometimes, however, fracture occurs with little or no such plastic extension. This is termed *cleavage* or *brittle* fracture. In a tensile specimen, after fracture the two specimen halves may be fitted back together so that the appearance is virtually that of the specimen before fracture. Brittle fracture is the main mode of failure of glass. If you drop a drinking glass and it breaks, it is possible to stick the broken bits together and end up with the same shape drinking glass.

In plain-carbon steels, the phenomenon is temperature dependent and a low-carbon steel which at ambient temperatures is tough and ductile will suddenly become extremely brittle at temperatures not far below 0°C. This type of failure was experienced in the welded 'liberty ships' manufactured during the Second World War for carrying supplies from the USA to Europe, particularly when these ships were used in the very cold North Atlantic.

Plastic flow depends upon the movement of dislocations during a finite time. As temperature decreases, the movement of these dislocations becomes more sluggish so that when a force is increased very rapidly it is possible for stress to increase so quickly that it cannot be relieved by slip. A momentary increase in stress to a value above the yield stress will produce fracture. Such failure will be precipitated from faults like sharp corners or arc-welded spots. In the liberty ships, a crack, once started, progressed right round the hull, whereas in a normal riveted hull the crack would have been halted at an adjacent rivet hole.

Metals with a face-centred cubic (FCC) structure maintain ductility at low temperatures, whilst some metals with structures other than FCC exhibit this type of brittleness. Body-centred cubic (BCC) ferrite is very susceptible to brittle fracture at low temperatures and the temperature at which brittleness suddenly increases is called the *transition temperature*. For applications involving near atmospheric temperatures, the transition point can be depressed to a safe limit by increasing the manganese content to about 1.3%. Where lower temperatures are involved, it is better to use a low-nickel steel.

12 The heat-treatment of plain-carbon steels

12.1 Introduction

Some of the simpler heat-treatments applied to steel were described in Chapter 11. In the main, they were processes in which the structure either remained or became basically pearlitic as a result of the treatment. Here we shall deal with those processes which are better known because of their wider use, namely *hardening* and *tempering*.

Almost any schoolboy knows that a piece of carbon steel can be hardened by heating it to redness and then plunging it into cold water. The Ancients knew this too and no one can tell us who first hardened steel. Presumably, such knowledge came by chance, as indeed did most knowledge in days before systematic research methods were instituted. One can imagine that sooner or later, some prehistoric metal-worker heating an iron tool among the glowing charcoal of his fire prior to forging it, and then, seeking to cool the tool quickly, would plunge it into water.

Although the fundamental *technology* of hardening steel has been well established for centuries, the *scientific principles* underlying the process were for long a subject of argument and conjecture. More than a century ago, Professor Henry Sorby of Sheffield began his examination of the microstructure of steel, but it was only during my undergraduate years that convincing explanations of the phenomenon were forthcoming.

12.2 Principles of hardening

If a piece of steel containing sufficient carbon is heated until its structure is austenitic – that is, until its temperature is above the upper critical temperature – and is then quenched, i.e. cooled quickly, it becomes considerably harder than it would be were it cooled slowly.

Generally, when a metallic alloy is quenched, there is a tendency to suppress any change in structure which might otherwise take place if the alloy were cooled slowly. In other words, it is possible to 'trap' or 'freeze in' a metallic structure which existed at a higher temperature, and so preserve it for examination at room temperature. Metallurgists often use this technique when plotting their equilibrium diagrams, and it is also used industrially as, for example, in the solution treatment of some aluminium alloys (see Section 17.7).

Clearly, things do not happen in this way when we quench a steel. Austenite, which is the phase present in a steel above its upper critical temperature, is a soft malleable material – which is why steel is generally shaped by hot-working processes. Yet when we quench austenite, instead of trapping the soft malleable structure, a very hard, brittle structure is produced, which is most unlike austenite. Under the microscope, this structure appears as a mass of uniform needle-shaped crystals, and is known as *martensite* (Figure 12.1A). Even at very high magnifications, no pearlite can be seen, so we must conclude that all of the cementite (which is one of the components of pearlite) is still dissolved in this martensitic structure. So far, this is what we would expect. However, investigations using X-ray methods, tell us that,

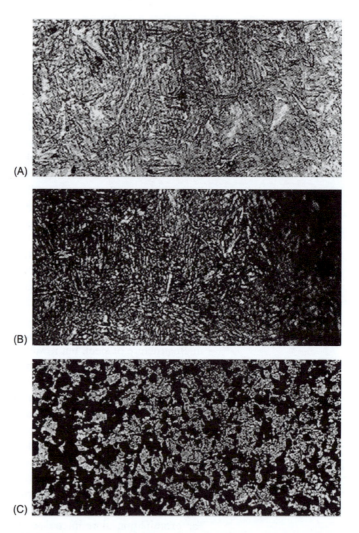

(A)

(B)

(C)

Figure 12.1 *Representative structures of quenched and tempered specimens of a 0.5% carbon steel: (A) water-quenched from 850°C – martensite which appears as an irregular mass of needle-shaped crystals, but what we see is a cross-section through roughly discus-shaped crystals (×700); (B) water-quenched from 850°C and tempered at 400°C – tempered martensite, the crystals of which have become darkened by precipitated particles of cementite (×700); (C) oil-quenched from 850°C – the slower cooling rate during quenching has allowed a mixture of bainite (dark) and martensite (light) to form (×100).*

although the rapid cooling has prevented the formation of pearlite, *it has not arrested the polymorphic change from face-centred cubic to body-centred cubic.*

Ferrite is fundamentally body-centred cubic iron which normally will dissolve no more than 0.006% carbon at room temperature (see Figure 11.4). Thus, the structure of martensite is one which is essentially ferrite very much super-saturated with carbon (assuming that the steel we are dealing with contains about 0.5% carbon). It is easy to imagine that this large amount of carbon remaining in *super-saturated* solid solution in the ferrite causes considerable distortion of the internal crystal structure of the latter. Such distortion will tend to prevent slip from taking place in the structure. Consequently, large forces can be applied, and no slip will be produced. In other words, the steel is now hard and strong.

In order to obtain the hard martensitic structure in a steel, it must be cooled quickly enough. The minimum cooling rate that will give a martensitic structure is termed the *critical cooling rate.* If the steel is cooled at a rate slower than this, then the structure will be less hard, because some of the carbon has had the opportunity to precipitate as cementite. Under the microscope, some dark patches will be visible among the martensitic needles (Figure 12.1C), these being due to the tiny precipitated particles of cementite, and the structure so produced is called *bainite,* after Dr. E. C. Bain – the American metallurgist who did much of the original research into the relationship between structure and rate of cooling of steels. Bainite is, of course, softer than martensite, but is tougher and more ductile. Even slower rates of cooling will give structures of fine pearlite.

12.2.1 TTT diagrams

The ultimate structure obtained in a plain-carbon steel is independent of the rate of heating, assuming that it is heated slowly enough to allow it to become completely austenitic before being quenched. It is the *rate* of subsequent cooling, however, which governs the resultant structure and hence the degree of hardness.

What are termed *TTT* (time–temperature–transformation) curves describe the relationship between the rate of cooling of a steel and its final microstructure and properties. A test-piece is heated to a temperature at which it is completely austenitic and then quickly transferred to a liquid bath at a measured temperature for a measured amount of time. After this time, the test-piece is quenched in cold water to convert the remaining austenite to martensite, then microscopic examination is used to determine the amount of martensite and hence the amount of austenite that had not changed in the liquid bath phase. By repeating such experiments for a number of different temperature liquid baths and a range of times, TTT curves can be constructed. However, they are as a result of experiments, for steels which had transformed *isothermally,* that is at a series of single fixed temperatures. Nevertheless, in practice we are generally more interested in transformations which occur on falling temperature gradients such as those that prevail during the water quenching or oil quenching of steels. For this reason, slightly modified

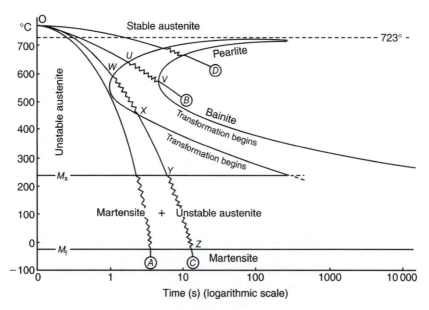

Figure 12.2 *Modified TTT curves for a 0.8% plain-carbon steel. Note that a logarithmic scale is used on the horizontal axis in order to compress the right-hand end of the diagram without at the same time cramping the important left-hand side.*

TTT curves are generally used and these are displaced a little to the right of the original TTT curves. Modified TTT curves are used here (Figure 12.2). It should be appreciated that a set of TTT curves relates to *one* particular steel of fixed composition.

A TTT diagram consists essentially of two C-shaped curves. The left-hand curve indicates the time interval which elapses at any particular temperature before a carbon steel (in this case, one containing 0.8% carbon) in its austenitic state begins to transform, whilst the right-hand curve shows the time which must elapse before this transformation is complete. The expected transformation product at that temperature is indicated in the diagram. The two parallel lines near the foot of the diagram are, strictly speaking, not part of the TTT curves but indicate the temperatures where austenite will start to transform to martensite (M_s) and where this transformation will finish (M_f).

It will be apparent that in order to obtain a completely martensitic structure, the steel, previously heated to point O to render it completely austenitic, must be cooled at a rate at least as rapid as that indicated by curve (A) (Figure 12.2). This represents the *critical cooling rate* for the steel mentioned earlier. Thus, curve (A) just grazes the nose of the 'transformation begins' curve so that the austenitic structure is retained right down to about 180°C (M_s) when this unstable austenite suddenly begins to change to martensite, this change being completed at about –40°C (M_f). Since quenching media are at a temperature higher than –40°C, some 'retained austenite' may be present in the quenched component. This retained austenite usually transforms to martensite during subsequent low-temperature tempering processes.

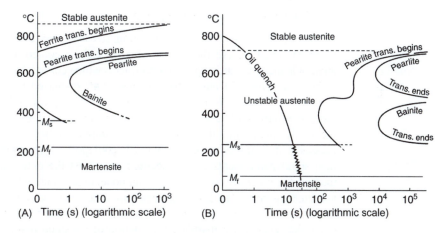

Figure 12.3 *(A) The TTT curves for a 0.35% carbon steel showing that it is virtually impossible to produce a completely martensitic structure by quenching, since, however rapid the cooling rate, ferrite separation inevitably begins as the 'ferrite transformation begins' curve is cut. (B) The TTT curves for an alloy steel containing 0.6% C, 0.6% Mn, 1.8% Ni, 0.6% Cr and 0.3% Mo. This indicates the effects of alloying in slowing transformation rates so that the TTT curves are displaced far to the right. This steel can be oil-quenched to give a martensitic structure.*

It follows then that a 0.8% carbon steel must be cooled very rapidly, i.e. from above 750°C to 120°C in little more than 1 second if it is to be completely martensitic (with possibly a little retained austenite as mentioned earlier).

The situation is even more difficult for carbon steels with either more or less than 0.8% carbon. For both hypo- and hyper-eutectoid steels, the TTT curves are displaced further to the left, making their critical cooling rates even faster. Fortunately, the presence of alloying elements slows down transformation rates considerably so that TTT curves are displaced to the right (Figure 12.3B), giving much lower critical cooling rates so that oil- or even air-quenching is possible to give a completely martensitic structure. Even so-called *plain-carbon steels* contain enough manganese, residual from deoxidation processes (see Section 11.3), to give lower critical cooling rates than equivalent pure iron–carbon alloys.

In Figure 12.2, curve (B) illustrates the result of quenching a plain 0.8% carbon steel in oil. Here, transformation begins at *U* and is completed at *V*, the resultant structure being bainite. Curve (*C*) indicates a rate of cooling intermediate between (*A*) and (*B*). Here, transformation to bainite begins at *W* but is interrupted at *X* and no further transformation takes place until the remaining austenite begins to change to martensite (at *Y*), this final transformation being complete at *Z*. Thus the resulting structure is a mixture of bainite and martensite.

The conditions prevailing during normalising are indicated by curve (*D*). Here, transformations to pearlite begin a few degrees below 723°C (the lower-critical temperature) and is complete a few degrees lower still.

12.2.2 Factors affecting cooling rates

In practice, factors such as the composition, size and shape of the component to be hardened govern the rate at which it can be cooled. Generally, no attempt is made to harden plain-carbon steels which contain less than 0.35% carbon, since the TTT curves for such a steel are displaced so far to the left (Figure 12.3A) that it is impossible to cool the steel rapidly enough to avoid the precipitation of large amounts of soft ferrite as the cooling curve inevitably cuts far into the nose of the 'transformation begins' curve. Large masses of steel of heavy section obviously cool more slowly when quenched than small components of thin-section; so, whilst the outer skin may be martensitic, the inner core of a large component may contain bainite or even pearlite (Figure 12.4). More important still, articles of heavy-section will be more liable to suffer from *quench-cracking*. This is due to the fact that the outer skin changes to martensite a fraction of a second before layers just beneath the surface, which are still austenitic. Since sudden *expansion* takes place at the instant when face-centred cubic austenite changes to body-centred cubic martensite, considerable stress will be set up between the skin and the layers beneath it, and, as the skin is now hard and brittle due to martensite formation, cracks may develop in it.

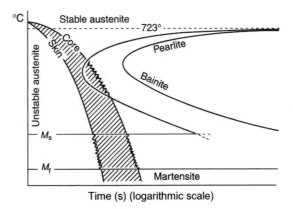

Figure 12.4 *TTT curves for a thick component.*

Design also affects the susceptibility of a component to quench-cracking. Sharp variations in cross-section, and the presence of sharp angles, grooves and notches are all likely to increase the possibility of quench-cracking by causing uneven rates of cooling throughout the component.

12.2.3 Quenching media

The rate at which a quenched component cools is governed by the quenching medium and the amount of agitation it receives during quenching. The media shown in Table 12.1 are commonly used, and are arranged in order of the quenching speeds shown in the table.

Table 12.1 *Order of quenching speeds for different media*

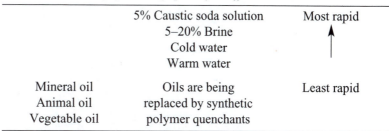

	5% Caustic soda solution	Most rapid
	5–20% Brine	
	Cold water	
	Warm water	
Mineral oil	Oils are being	Least rapid
Animal oil	replaced by synthetic	
Vegetable oil	polymer quenchants	

The very drastic quenching resulting from the use of caustic soda solution or brine is used only when extreme hardness is required in components of relatively simple shape. For more complex shapes, it would probably be better to use a low-alloy steel, which has a much lower critical cooling rate and can therefore be hardened by quenching in oil. Mineral oils used for quenching are derived from petroleum, whilst vegetable oils include those from linseed and cottonseed. Animal oils are obtained from the blubber of seals and whales, though near extinction of the latter has led to the long overdue banning of whaling by most civilised countries. Synthetic polymer quenchants are now being developed to replace oils – which is good news for those gentle giants of the ocean and, incidentally, for the heat-treatment operatives since less fumes and offensive smells are generated by the 'synthetics'. These new quenchants consist of synthetic polyalkane glycols which can be mixed with water in varying proportions to give different quenching rates.

12.3 The hardening process

To harden a hypo-eutectoid steel component, it must be heated to a temperature of 30–50°C above its upper critical temperature, and then quenched in some medium which will produce in it the required rate of cooling. The medium used will depend upon the composition of the steel, the size of the component and the ultimate properties required in it. Symmetrically shaped components, such as axles, are best quenched 'end-on', and all components should be violently agitated in the medium during the quenching operation.

The procedure in hardening a hyper-eutectoid steel is slightly different. Here, a quenching temperature about 30°C above the lower critical temperature is generally used. In a hyper-eutectoid steel, primary cementite is present, and, on cooling from above the upper critical temperature, this primary cementite tends to precipitate as long, brittle needles along the grain boundaries of the austenite. This type of structure would be very unsatisfactory, so its formation is prevented by continuing to forge the steel whilst the primary cementite is being deposited – that is, between the upper and lower critical temperatures. In this way, the primary cementite is broken down into globules during the final stages of shaping the steel. During the subsequent heat-treatment, it must never be heated much more than 30°C above the lower critical temperature, or there will be a tendency for primary cementite to be absorbed by the austenite, and then precipitated again as long brittle needles on cooling. When a hyper-eutectoid steel has been correctly hardened, its structure should consist of small, near spherical globules of very hard cementite (Figure 12.5) in a matrix of hard, strong martensite.

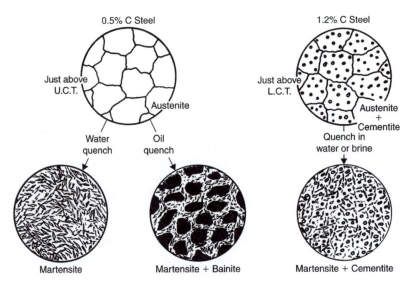

Figure 12.5 *Typical microstructures produced when quenching both medium-carbon and tool steels in their appropriate media.*

12.4 Tempering

A fully hardened carbon steel is relatively brittle, and the presence of quenching stresses makes its use in this condition inadvisable unless extreme hardness is required. For these reasons, it is usual to reheat, or 'temper', the quenched component, so that stresses are relieved, and, at the same time, brittleness and extreme hardness are reduced.

As we have seen, the martensitic structure in hardened steel consists essentially of ferrite which is heavily super-saturated with carbon. By heating such a structure to a high enough temperature, we enable it to begin to return to equilibrium, by precipitating carbon in the form of tiny particles of cementite.

On heating the component up to 200°C, no change in the microstructure occurs, though quenching stresses are relieved to some extent whilst hardness may even increase slightly as any retained austenite transforms to martensite. At about 230°C, tiny particles of cementite are precipitated from the martensite, though these are so small that they are difficult to see with an ordinary microscope. Generally, the microstructure appears somewhat darker, but still retains the shape of the original martensite needles. This type of structure persists as the temperature is increased to about 400°C (Figure 12.1B) with more and more tiny cementite particles being precipitated, and the steel becoming progressively tougher – though softer than the original martensite. The structure so produced was commonly known as *troostite*.

Tempering at temperatures above 400°C causes the cementite particles to coalesce (or fuse together) to such an extent that they can be seen clearly at magnifications of about ×500. At the same time, more cementite is precipitated. The structure, which is relatively granular in appearance, was known as *sorbite*. It must be emphasised that there is no fundamental difference between troostite and sorbite, since both are formed by precipitation of cementite from martensite; and there is no definite temperature where troostite formation ceases and formation of sorbite begins. Naturally, sorbite is softer and tougher than troostite, because still more carbon has been precipitated from the original martensite structure.

The names 'troostite' and 'sorbite' are long since obsolete and should not be used. The modern metallurgist describes these structures as 'tempered martensite', mentioning the temperature used during the tempering process.

Generally speaking, low temperatures (200–300°C) are used for tempering various types of high-carbon steel tools where hardness is the prime consideration, higher temperatures (400–600°C) being used for tempering stress-bearing medium-carbon constructional steels where strength, toughness and general reliability are more important.

12.4.1 Tempering colours

Furnaces used for tempering are usually of the batch type, in which the charge is carried in a wire basket through which hot air circulates. By this method,

Table 12.2 *Tempering colours for carbon steels*

Temperature (°C)	Colour	Types of components
220	Pale yellow	Scrapers, hack-saws, light turning-tools
230	Straw	Hammer faces, screwing-dies for brass, planing- and slotting-tools, razor blades
240	Dark straw	Shear blades, milling cutters, drills, boring-cutters, reamers, rock-drills
250	Light brown	Penknife blades, taps, metal shears, punches, dies, woodworking tools for hardwood
260	Purplish-brown	Plane blades, stone-cutting tools, punches, reamers, twist-drills for wood
270	Purple	Axes, augers, gimlets, surgical tools, press-tools
280	Deeper purple	Cold chisels (for steel and cast iron), chisels for wood, plane-cutters for softwood
290	Bright blue	Cold chisels (for wrought iron), screwdrivers
300	Darker blue	Wood-saws, springs

Table 12.3 *Heat-treatments and typical uses of plain-carbon steels*

Type of steel	Carbon (%)	Heat-treatment			Typical uses
		Hardening temperature (°C)	Quenching medium	Tempering temperature (°C)	
Dead-mild	Up to 0.15	These do not respond to heat-treatment, because of their low carbon content			Nails, chains, rivets, motor-car bodies
Mild	0.15–0.25	These do not respond to heat-treatment, because of their low carbon content			Structural steels (RSJ), screws, tinplate, drop-forgings, stampings, shafting, free-cutting steels
Medium-carbon	0.25–035	880–850*	Oil or water, depends upon type of work	Temper as required	Couplings, crankshafts, washers, steering arms, lugs, weldless steel tubes
	0.35–0.45	870–830*			Crankshafts, rotor shafts, crank pins, axles, gears, forgings of many types
	0.45–0.60	850–800*	Oil, water or brine, depends upon type of tool		Hand-tools, pliers, screwdrivers, gears, die-blocks, rails, laminated springs, wire ropes
High-carbon tool	0.60–0.75	820–800*	Water or brine. Tools should not be allowed to cool below 100°C before tempering	275–300	Hammers, dies, chisels, miners tools, boilermakers' tools, set-screws
	0.75–0.90	800–820		240–250	Cold-chisels, blacksmiths' tools, cold-shear blades, heavy screwing dies, mining drills
	0.90–1.05	780–800		230–250	Hot-shear blades, taps, reamers, threading and trimming dies, mill-picks
	1.05–1.20	760–780		230–250	Taps, reamers, drills, punches, blanking-tools, large turning tools
	1.20–1.35	760–780		240–250	Lathe tools, small cold-chisels, cutters, drills, pincers, shear blades
	1.35–1.50	760–780		200–230	Razors, wood-cutting tools, drills, surgical instruments, slotting-tools, small taps

* *The higher temperature for the lower carbon-content.*

the necessarily accurate temperature can easily be maintained. The traditional method of treating tools is to 'temper by colour', and this still provides an accurate and reliable method of dealing with plain-carbon steels. After the tool has been quenched, its surface is first cleaned to expose bright metal. The tool is then slowly heated until the thin oxide layer which forms on the surface attains the correct colour (Table 12.2). It should be noted that this technique only applies to plain-carbon steels, since some of the alloy steels, particularly those containing chromium, do not oxidise readily.

12.4.2 Applications of heat-treated plain-carbon steels

A summary of typical heat-treatment programmes, and uses of the complete range of plain-carbon steels, is given in Table 12.3.

12.5 Isothermal heat-treatments

The risk of cracking and/or distortion during the rather drastic water-quenching of carbon steels has already been mentioned (see Section 12.2) and such difficulties may be overcome in the case of suitably dimensioned work-pieces by both *martempering* and *austempering* (these two processes are stuck with these clumsy titles, which unfortunately do not accurately describe the principles involved). These processes are known as *isothermal heat-treatments*.

12.5.1 Martempering

The principles of martempering are indicated in Figure 12.6A. Here, a carbon steel component has been quenched into a bath (either of hot oil or molten low-melting point alloy) held at a temperature just above M_s. The component is allowed to remain there for a time sufficient for the whole component to have *reached a uniform temperature throughout*. It is then removed from the bath and allowed to cool *very slowly* in warm air. Note that in Figure 12.6A, the cooling curve is foreshortened by the use of the logarithmic scale. Thus several minutes will elapse before the steel, at the quench-bath temperature, would begin to transform, giving ample time for uniformity of temperature to be attained in the work-piece. Since, under these conditions, both skin and core of the component pass through the M_s and M_f lines almost simultaneously, there is little chance of stresses being set up which may induce either distortion or cracking in the hard martensitic structure which results.

12.5.2 Austempering

Austempering (Figure 12.6B) is a means of obtaining a tempered type of structure without the necessity of the preliminary drastic water quench which is involved in traditional methods of heat-treatment. As with martempering, the work-piece is quenched from the austenitic state into a bath held at some suitable temperature above M_s, but in this case it remains there long enough for transformation to occur to completion, between B and E on the TTT curves. This yields a structure of bainite which will be similar in properties

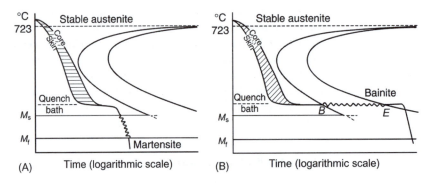

Figure 12.6 *The isothermal treatment of (A) martempering and (B) austempering.*

to those structures of traditionally tempered martensite. The rate at which the work-piece is finally cooled is not important since transformation is already complete at *E*.

12.5.3 Limitations of martempering and austempering

Although austempering and martempering would seem to provide enormous advantages in that risks of distortion and cracking of work-pieces are largely removed, there is one very obvious drawback to the wide application of such processes, namely that their use is limited generally to components of thin-section since the *whole* of the cross-section of the work-piece must be capable of being cooled rapidly enough to miss the nose of the 'transformation begins' curve of the TTT diagram appropriate to the composition of the steel being used. Thin-sectioned components which are austempered include steel toe-caps of industrial boots, whilst garden spades and forks were similarly heat-treated long before the fundamental theory of the process was investigated by Davenport and Bain in the 1930s.

Similarly, the *patenting* of high-tensile steel wire was achieved by winding it through an austenitising furnace (at 970°C), followed by a bath of molten lead (at 500°C) where the structure transformed directly from austenite to bainite. In this condition, the wire can be further hardened by cold-drawing.

12.6 Hardenability

In order to harden a piece of steel completely, it must be cooled quickly from its austenitic state. The rate of cooling must be as great as, or greater than, the 'critical rate', otherwise the section will not be completely martensitic. Clearly, this will not be possible for a work-piece of very heavy section; for, whilst the outer skin will cool at a speed greater than the critical rate, the core will not. Consequently, whilst the outer shell may be of hard martensite, the core may be of bainite, or even fine pearlite.

This phenomenon is generally referred to as the *mass effect* of heat-treatment, and plain-carbon steel is said to have a 'shallow depth of hardening', or, alternatively, 'a poor hardenability'. The term *hardenability* is used as a measure of the depth of martensitic hardening introduced by

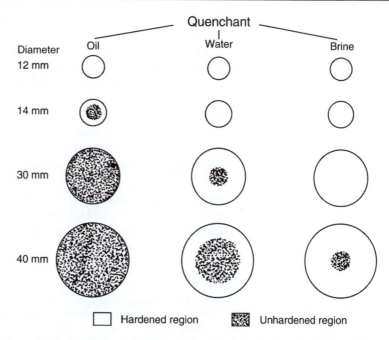

Figure 12.7 *Specimens of different diameter have been quenched in different media and the depth of hardening assessed in each case. A low-alloy steel containing 0.25% C, 0.6% Mn, 0.2% Ni and 0.2% Mo was used.*

quenching, the smaller the depth the poorer the hardenability. Whilst a rod of, say, 12 mm diameter in plain-carbon steel can be water-quenched and have a martensitic structure throughout, one of, say, 30 mm diameter will have a core of bainite and only the outer shell will be martensitic. The influence of section thickness and quenching medium used on the structure produced in a low-alloy steel is illustrated in Figure 12.7.

For some applications, this variation in structure across the section will not matter, since the outer skin will be very hard and the core reasonably tough; but in cases where a component is to be quenched and then tempered for use as a stress-bearing member, it is essential that the structure obtained by quenching is uniform throughout.

12.6.1 Ruling section

Fortunately, the addition of alloying elements to a steel reduces its critical rate so that such a steel can be oil-quenched in thin-sections, or, alternatively, much heavier sections can be water-quenched. Thus, an alloy steel generally has a much greater depth of hardening, i.e. better hardenability, than has a plain-carbon steel of similar carbon content.

However, it would be a mistake to assume that *any* alloy steel in *any* thickness will harden right through when oil-quenched from above its upper critical temperature. The low-alloy steel used as an example in Figure 12.7 illustrates this point. This steel has a critical cooling rate just a little lower than that of a plain-carbon steel; but, as the diagram shows, sections much over 14 mm diameter will not harden completely unless quenched in brine, and the maximum diameter which can harden completely – even with this drastic treatment – is only about 30 mm.

In order to prevent the misuse of steels by those who imagine that an alloy steel can be hardened to almost any depth, both the British Standards Institution and manufacturers now specify limiting ruling sections for each particular composition of steel. The *limiting ruling section* is the *maximum diameter* which can be heat-treated (under conditions of quenching and tempering suggested by the manufacturer) for the stated mechanical properties to be obtained. As an example, Table 12.4 shows a set of limiting ruling sections for a low-alloy steel (530M40), along with a manufacturer's suggested heat-treatment, and the corresponding BS specifications in respect of tensile strength. Formulae are given by British Standards for deriving the equivalent diameters of rectangular and other bars, so that ruling section information can be applied to other cross-sections.

Table 12.4 *Limiting ruling sections*

Limiting ruling section (mm)	Suggested heat-treatment	Tensile strength (MPa)
100	Oil harden from 850°C	695
62.5	and temper at 650°C	770
29		850

12.7 The Jominy test

This test (Figure 12.8) is of considerable value in assessing the hardenability of a steel.

A standard test-piece (Figure 12.8B) is heated to above the upper critical temperature of the steel, i.e. until it becomes completely austenitic. It is then quickly transferred from the furnace and dropped into position in the frame of the apparatus shown in Figure 12.8A. Here it is quenched at one end only, by a standard jet of water at 25°C; thus, different rates of cooling are obtained along the length of the test-piece. When the test-piece has cooled, a 'flat' approximately 0.4 mm deep is ground along the length of the bar and hardness determined every millimetre along the length from the quenched end. The results are then plotted (Figure 12.9).

These curves show that a low nickel–chromium steel hardens to a greater depth than a plain-carbon steel of similar carbon content, whilst a chromium–molybdenum steel hardens to an even greater depth.

Whilst the Jominy test gives a good indication as to how deeply a steel will harden, there is no simple mathematical relationship between the results of this test and the ruling section of a steel. Using the results of the Jominy test as a basic, it is often more satisfactory to find the ruling section by trial and error.

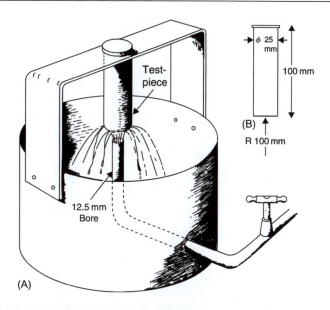

Figure 12.8 *The Jominy end-quench test. (A) The apparatus: the end of the water pipe is 12 mm below the bottom of the test-piece, but the 'free height' of the water jet is 63 mm. (B) A typical test-piece.*

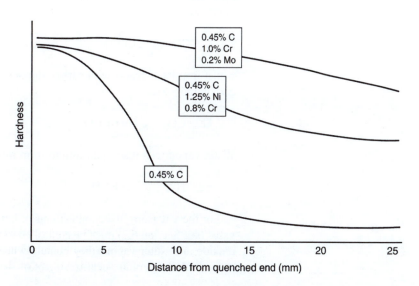

Figure 12.9 *Hardness as a function of depth for various steels of similar carbon content, as shown by the Jominy test.*

12.8 Hardenability

The factors affecting the hardenability of a steel are:

- The quench rate.
- *Grain size:* pearlite starts to grow at grain boundaries and so a coarse grain structure, because it has less grain boundaries, will have greater hardenability.
- *Composition of the steel:* alloying elements affect the hardenability (see Chapter 13).

12.9 Heat-treatment furnaces

It was mentioned earlier (Section 12.4) that hardened steel is generally tempered in some form of furnace in which hot air is circulated over the charge. Since the temperatures involved are relatively low, little or no change in composition occurs at the surface of the steel. Hardening temperatures, however, are much higher so that both decarburisation and oxidation of the surface can occur unless preventative measures are taken. Baths of electrically heated molten salt are sometimes used and these not only maintain an accurate quenching temperature, due to the high heat capacity of the salt, but also provide protection of the surface from decarburisation and oxidation.

When furnace atmospheres derived from hydrocarbons (see Section 6.7) are used, both the carbon dioxide and water vapour produced can act as decarburisers and oxidisers. Thus, carbon (in the surface of the steel) will combine with carbon dioxide (in the furnace atmosphere) to form carbon monoxide:

$$C + CO_2 \rightleftharpoons 2CO$$

whilst iron itself reacts with carbon dioxide to form iron oxide (scale) and carbon monoxide:

$$Fe + CO_2 \rightleftharpoons FeO + CO$$

Water vapour can react with iron to form iron oxide and hydrogen:

$$Fe + H_2O \rightleftharpoons FeO + H_2$$

For these reasons, the compositions of both exothermic and endothermic gases (see Section 6.7) must be controlled to reduce the quantities of carbon dioxide and water vapour they contain when used as furnace atmospheres in the high-temperature heat-treatment of steels.

13 Alloy steels

13.1 Introduction

The first deliberate attempt to develop an alloy steel was made by Sir Robert Hadfield during the 1880s, but most of these materials are products of the twentieth century. Possibly, the greatest advance in this field was made in 1900, when Taylor and White, two engineers with the Bethlehem Steel Company in the USA, introduced the first high-speed steel and so helped to increase the momentum of the Industrial Revolution and, in particular, what has since been known as *mass-production*.

This chapter is about 'alloy steels', but there is a problem in defining what is meant by the term. So-called *plain-carbon steels* contain up to 1.0% manganese, which is the residue of that added to deoxidise and desulphurise the steel just before casting. Consequently, a steel was not originally classified as an 'alloy steel' unless it contained more than 2% manganese and/or other elements (nickel, chromium, molybdenum, vanadium, tungsten, etc.), with each in amounts of at least between 0.1% and 0.5%. Now that steels of very high purity are commonly produced commercially, alloy additions of as little as 0.005% can significantly influence mechanical properties and a range of 'micro-alloyed' or 'high-strength low-alloy' (HSLA) steel has been introduced. This renders the above definition of an 'alloy steel' somewhat obsolete.

Although there are alloy steels with special properties, such as the stainless steels and heat-resisting steels, the main purpose of alloying is to improve the existing properties of carbon steels, making them more adaptable and easier to heat-treat successfully. In fact, one of the most important and useful effects of alloying was mentioned in Chapter 12 (Section 12.6) – the improvement in 'hardenability'. Thus, an alloy steel can be successfully hardened by quenching in oil, or even in an air blast, with less risk of distortion or cracking of the component than is associated with water-quenching. Moreover, suitable alloy steels containing as little as 0.2% carbon can be hardened successfully because of the considerable slowing down of transformation rates imparted by the alloying elements (see Figure 12.3B).

13.1.1 Alloying elements

Alloying elements can be divided into two main groups:

1   Those which strengthen and toughen the steel by dissolving in the ferrite. These elements are used mainly in constructional steels, and include nickel, manganese, small amounts of chromium and even smaller amounts of molybdenum.
2   Alloying elements which combine chemically with some of the carbon in the steel, to form carbides which are much harder than iron carbide

(cementite). These elements are used mainly in tool steels, die steels and the like. They include chromium, tungsten, molybdenum and vanadium.

Other alloying elements which are added in small amounts and for special purposes include titanium, niobium, aluminium, copper, boron and silicon. Even sulphur, normally regarded as the steel-maker's greatest enemy, is utilised in free-cutting and 'bright-drawn' steels (see Section 7.5).

13.1.2 Alloy steels

Alloy steels may be classified into three main groups:

1   Constructional steels which are generally used for machine parts highly stressed in tension or compression.
2   Tool steels which require great hardness and, in some cases, resistance to softening by heat.
3   Special steels, e.g. stainless steels and heat-resisting steels.

13.2 Constructional steels

Whilst the 'nickel-chrome' steels are the best known in this group, other alloy steels containing the elements just nickel or just chromium are also important.

13.2.1 Nickel steels

Nickel increases the strength of a steel by dissolving in the ferrite. Its main effect, however, is to increase toughness by limiting grain growth during heat-treatment processes. For this reason, up to 5.0% nickel is present in some of the better quality steels used for case-hardening. Table 13.1 indicates the uses of such steels.

Unfortunately, nickel does not combine chemically with carbon, and, worse still, tends to make iron carbide (cementite) decompose and so release free graphite. Consequently, nickel steels are always low-carbon steels, or, alternatively, medium-carbon steels with very small amounts of nickel. However, because of their shortcomings in respect of carbide instability, they have been almost entirely replaced in recent years by other low-alloy steels.

13.2.2 Chromium steels

When chromium is added to a steel, some of it dissolves in the ferrite (which is strengthened as a result), but the remainder forms chromium carbide. Since chromium carbide is harder than ordinary iron carbide (cementite), the hardness of the steel is increased. Because chromium forms stable carbides, these steels may contain 1.0%, or even more, of carbon. The main disadvantage of chromium as an alloying element is that, unlike nickel, it increases grain growth during heat-treatment. Thus, unless care is taken to limit both the temperature and the time of such treatment, brittleness may arise from the coarse grain produced.

As indicated by the uses mentioned in Table 13.2, these low chromium steels are important because of their increased hardness and wear resistance.

Table 13.1 *Nickel steels*

BS specification	Composition (%)	Typical mechanical properties			Heat-treatment	Uses
		Yield point (MPa)	Tensile strength (MPa)	Izod (J)		
	0.4 C 1.0 Mn 1.0 Ni	500	700	96	Oil-quench from 850°C, temper between 550°C and and 650°C	Crankshafts, axles, other parts in the motor-car industry and in general engineering
	0.12 C 0.45 Mn 3.0 Ni	510	775	86	After carburising: refine grain by an oil-quench from 860°C, then harden by a water-quench from 770°C	A case-hardening steel: crown-wheels, differential pinions, cam-shafts

Table 13.2 *Chromium steels*

BS specification	Composition (%)	Typical mechanical properties			Heat-treatment	Uses
		Yield point (MPa)	Tensile strength (MPa)	Izod (J)		
530M40	0.45 C 0.9 Mn 1.0 Cr	880	990	40	Oil-quench from 860°C, temper at 550–700°C	Agricultural machine parts, machine-tool components, parts for concrete and tar mixers, excavator teeth, automobile axles, connecting rods and steering arms, spanners
535A99	1.0 C 0.45 Mn 1.4 Cr	850			Oil-quench from 810°C, temper at 150°C	Ball- and roller-bearings, roller- and ball-races, cams, small rolls

13.2.3 Nickel–chromium steels

In the foregoing sections dealing with nickel and chromium, it was seen that in some respect the two metals have opposite effects on the properties of a steel. Thus, whilst nickel is a grain-refiner, chromium tends to cause grain-growth. On the other hand, whilst chromium is a carbide stabiliser, nickel tends to cause carbides to break down, releasing graphite. Fortunately, the beneficial effects of one metal are stronger than the adverse effects of the other, and so it is advantageous to add these two metals together to form a steel. Generally, about two parts of nickel to one of chromium is found to be the best proportion.

In other respects, the two metals, as it were, work together, and so the hardenability is increased to the extent that, with 4.25% nickel and 1.25% chromium, an *air-hardening* steel is produced; that is, one which can be 'quenched' in an air blast, thus making cracking or distortion even less likely than if the steel were oil-quenched. However, for air-hardening, a ruling section of 62.5 mm diameter must be observed, and for greater diameters than this the steel must be oil-quenched if the stated properties are to be obtained.

Unfortunately, these straight nickel–chromium steels suffer from a defect known as 'temper brittleness' (described in Section 13.2.4) and for this reason straight nickel–chromium steels have been almost entirely replaced by nickel–chromium–molybdenum steels.

13.2.4 Nickel–chromium–molybdenum steels

As mentioned earlier, a severe drawback in the use of straight nickel–chromium steels is that they suffer from a defect known as *temper brittleness*. This is shown by a very serious decrease in toughness (as indicated by a low Izod or Charpy impact value) when a quenched steel is subsequently tempered in the range 250–580°C. Further, if such a steel is tempered at 650°C, it must be cooled quickly through the 'dangerous range' by quenching it in oil, following the tempering process. Even with this disastrous reduction in impact toughness, the tensile strength and percentage elongation may not be seriously affected. Consequently, a tensile test alone would not reveal the shortcomings of such a steel, and the importance of impact testing in cases like this is obvious. Fortunately, temper brittleness can be largely eliminated by adding about 0.3% molybdenum to the steel, thus establishing the well-known range of 'nickel–chrome–moly' steels. Table 13.3 shows typical properties and uses of such steels.

13.2.5 Manganese steels

Most steels contain some manganese remaining from the deoxidisation and desulphurisation processes, but it is only when the manganese content exceeds 1.0% that it is regarded as an alloying element. Manganese increases the strength and toughness of a steel, but less effectively than does nickel. Like all elements, it increases the depth of hardening. Consequently, low-manganese steels are used as substitutes for other, more expensive, low-alloy steels.

Manganese is a metal with a structure somewhat similar to that of austenite at ordinary temperatures; therefore, when added to a steel in sufficient quantities, it tends to stabilise the FCC (austenitic) structure of iron at lower temperatures than is normal for austenite. In fact, if 12.0% manganese is added to a steel containing 1.0% carbon, the structure remains austenitic even

Table 13.3 *Nickel–chromium–molybdenum steels*

BS specification	Composition (%)	Typical mechanical properties			Heat-treatment	Uses
		Yield point (MPa)	Tensile strength (MPa)	Izod (J)		
817M40	0.4 C 0.55 Mn 1.5 Ni 1.1 Cr 0.3 Mo	990	1010	72	Oil-quench from 840°C, temper at 600°C	Differential shafts, crankshafts and other high-stressed parts. (If tempered at 200°C, it is suitable for machine-tool and automobile gears)
826M40	0.4 C 0.65 Mn 2.5 Ni 0.65 Cr 0.55 Mo	850	1000	45	Oil-quench from 830°C, temper at 600°C	Thin-sections where maximum shock resistance and ductility are required, e.g. connecting rods, inlet-valves, cylinder-studs, valve-rockers
835M30	0.3 C 0.5 Mn 4.25 Ni 1.25 Cr 0.3 Mo	1450	1700	37	Oil-quench from 830°C, temper at 150–200°C	An air-hardening steel for aeroengine connecting rods, valve mechanisms, gears, differential shafts, etc. and other highly stressed parts

after the steel has been *slowly cooled* to room temperature. The curious – and useful – fact about this steel is that, if the surface suffers any sort of mechanical disturbance it immediately becomes extremely hard. Some suggest that this is due to spontaneous martensitic formation but others think it is simply due to work-hardening. Whatever the reasons, the result is a soft but tough austenitic core with a hard wear-resistant shell which is useful in conditions where both mechanical shock and severe abrasion prevail, as in dredging, earth-moving and rock-crushing equipment. A further point of interest regarding this steel is that it was one of the very first alloy steels to be developed – by Sir Robert Hadfield in 1882 – though little use was made of it until the early days of the twentieth century.

Table 13.4 shows examples of manganese steels, their properties and their uses.

13.2.6 Boron steels

Boron is not a metal and is generally familiar in the form of its compound 'borax'. The pure element is a hard grey solid with a melting point

Table 13.4 *Manganese steels*

BS specification	Composition (%)	Typical mechanical properties				Heat-treatment	Uses
		Yield point (MPa)	Tensile strength (MPa)	Izod (J)	Brinell		
150M36	0.35 C 1.5 Mn	510	710	71		Oil-quench from 850°C, temper at 600°C	Automobile and general engineering, as a cheaper substitute for the more expensive nickel-chromium steels
605M36	0.36 C 1.5 Mn	1000	1130	72		Oil-quench from 850°C, temper at 600°C	
– (Hadfield steel)	1.2 C 12.5 Mn				Case-550 Core-200	Finish by quenching from 1050°C to keep carbides in solution – quenching does *not* harden the steel, however	Rock-crushing equipment, buckets, heel-plates and bucket-lips for dredging equipment, earth-moving equipment, trackway crossings and points

of 2300°C. In recent years, it has been developed as an alloying element in some steels.

Extremely small amounts – 0.0005–0.005% – added to fully deoxidised steels are effective in reducing the austenite →ferrite + pearlite transformation rates in those steels containing between 0.2% and 0.5% carbon, i.e. the TTT curves are displaced appreciably to the right so that these low-carbon steels can then be effectively hardened. Moreover, in some low-alloy steels, the amounts of other expensive elements like nickel, chromium and molybdenum can be reduced by as much as a half if small amounts of boron are added. Low-carbon, manganese steels containing boron are used for high-tensile bolts, thread-rolled wood screws and in automobiles where extra strength is required, e.g. in sills, chassis areas and rear cross members. Table 13.5 gives the composition, properties and uses of typical boron steels.

13.2.7 Maraging steels

Maraging steels (Table 13.6) are a group of very high-strength alloys which have been used in aerospace projects, such as the Lunar Rover Vehicle, but in general engineering have found a wide variety of uses, such as the flexible drive shafts for helicopters, barrels for rapid-firing guns, die-casting dies and extrusion rams.

On examining the composition of such alloys (Table 13.6), one notices that the amount of carbon present is very small. Indeed, carbon plays no part in developing the high strength and is only residual from the manufacturing

Table 13.5 *Boron steels*

Typical composition (%)						Typical mechanical properties			Uses
C	Mn	Ni	Cr	Mo	B	Yield point (MPa)	Tensile strength (MPa)	Percentage elongation	
0.2	1.5				0.0005	690	790	18	Some structural steels
0.17	0.6			0.6	0.003	690	860	18	High-tensile constructional steels
0.45	0.8	0.3	0.4	0.12	0.002	700	1000	15	Automobile engineering

Table 13.6 *Typical maraging 'steels'*

Typical composition (%)						Heat-treatment	Typical mechanical properties			
Ni	Co	Mo	Ti	Al	C		0.2% proof stress (MPa)	Tensile strength (MPa)	Percentage elongation	Impact, Charpy (J)
18	8.5	3	0.2	0.1	0.01	Solution treated at 820°C for 1 hour, air-cooled and age hardened at 480°C for 3 hours	1430	1565	9	52
18	9	3	0.5	0.1	0.01	Solution treated at 820°C for 1 hour, air-cooled and age hardened at 480°C for 3 hours	1930	1965	7.5	21
17.5	12.5	3.75	1.6	0.15	0.01	Solution treated at 820°C for 1 hour, air-cooled and age hardened at 480°C for 12 hours	2390	2460	8	11

process, so that these alloys should be thought of as high-strength alloys rather than steels, in the sense that steels normally depend for their properties on the presence of carbon.

Cobalt and nickel are essential constituents of maraging 'steels'. If such an alloy is solution treated at 820°C to absorb precipitated intermetallic compounds, uniform austenite is formed. On cooling in air, an iron–nickel variety of martensite is produced due to the retardation of transformation rates caused by the large amounts of alloying elements present. This form of 'martensite', however, is softer and tougher than ordinary martensite based on the presence of carbon. If the alloy is 'age-hardened' (see Section 17.7) at 480°C for 3 hours or more, coherent precipitates of intermetallic compounds ($TiNi_3$, $MoNi_3$ or $AlNi_3$) are formed, making slip along crystal planes more difficult so that high tensile strengths up to 2400 MPa result. The main function of cobalt seems to be in providing more lattice positions where the coherent precipitates are able to form.

These alloys combine considerable toughness with high strength when heat-treated and are far superior – but also very much more expensive – than conventional alloy constructional steels. Heat-treatment, however, is relatively uncomplicated since there can be no decarburisation and no water quench is required. They are also very suitable for surface hardening by nitriding (see Section 14.5).

13.3 Tool and die steels

The primary requirement of a tool or die steel is that it shall have considerable hardness and wear-resistance, combined with reasonable mechanical strength and toughness. A plain high-carbon tool steel possesses these properties, but unfortunately its cutting edge softens easily on becoming over-heated during a high-speed cutting process. Similarly, dies which are to be used for hot-forging or extrusion operations cannot be made from plain-carbon steel, which, in the heat-treated state, begins to soften if heated to about 220°C. Consequently, tool steels which work at high speeds, or die steels which work at high temperatures, are generally alloy steels containing one or more of those elements which form very hard carbides – chromium, tungsten, molybdenum or vanadium. Of these elements, tungsten and molybdenum also cause the steel, once hardened, to develop a resistance to tempering influences, whether from contact with a hot work-piece or from frictional heat. Thus, either tungsten or molybdenum is present in all high-speed steels, and in most high-temperature die steels. Table 13.7 shows examples of tool and die steels.

13.3.1 Die steels

As mentioned earlier, these materials will contain at least one of the four metals which form hard carbides; whilst hot-working dies will in any case contain either tungsten or molybdenum to provide resistance to tempering and, hence, the necessary strength and hardness at high temperatures. The heat-treatment of these steels resembles that for high-speed steels, which will be described in Section 13.3.2.

13.3.2 High-speed steel

High-speed steel, as we know it, was first shown to an amazed public at the Paris Exposition of 1900. A tool was exhibited cutting at a speed of some 0.3 m/second, with its tip heated to redness. Soon after this, it was found that the maximum cutting efficiency was attained with a composition of 18%

tungsten, 4% chromium, 1% vanadium and 0.75% carbon, and this remains possibly the best-known general-purpose high-speed steel to this day.

Since high-speed steel is a complex alloy, containing at least five different elements, it cannot be represented by an ordinary equilibrium diagram (see Section 9.5); however, by grouping all the alloying elements together under

Table 13.7 *Tool and die steels (other than high-speed steels)*

BS EN ISO 4957 specification	Composition (%)	Hardness (VPN)	Heat-treatment	Uses
	0.6 C 0.65 Mn 0.65 Cr	700	Oil-quench from 830°C. Temper: (i) for cold-working tools at 200–300°C, (ii) for hot-working tools at 400–600°C.	Blacksmith's and boiler-maker's tools and chisels, swages, builders', masons' and miners' tools, chuck- and vice-jaws, hot-stamping and forging dies.
BD3	2.1 C 12.5 Cr	850	Heat slowly to 800°C, then raise to 980°C and oil-quench. Temper between 150°C and 400°C for 30–60 minutes.	Blanking punches, dies, shear blades for hard thin materials, dies for moulding ceramics and other abrasive powders, master-gauges, thread-rolling dies.
BH11	0.35 C 5.0 Cr 1.25 Mo 0.3 V 1.0 Si	Tempered at 550°C: 600; Tempered at 650°C: 375	Preheat to 850°C, then heat to 1000°C. Soak for 10–30 minutes, and air-harden. Temper at 550–650°C for 2 hours.	Hot-forging dies for steel and copper alloys where excessive temperatures are not encountered, extrusion dies for aluminium alloys, pressure and gravity dies for casting aluminium.
BD2A	1.6 C 13.0 Cr 0.8 Mo 0.5 V	Tempered at 200°C: 800; Tempered at 400°C: 700	Preheat to 850°C, and then heat to 1000°C. Soak for 15–45 minutes, and quench in oil or air. Temper at 200–400°C for 30–60 minutes.	Fine press-tools, deep drawing and forming dies for sheet metal, wire-drawing dies, blanking dies, punches and shear blades for hard metals.
BH12	0.35 C 1.0 Si 5.0 Cr 1.5 Mo 0.45 V 1.35 W		Preheat to 800°C. Soak, and then heat quickly to 1020°C. Air-quench, and then temper for 90 minutes at 540–620°C.	Extrusion dies for aluminium and copper alloys, hot-forming, piercing- and heading-tools, brass-forging dies.
BH21	0.35 C 2.85 Cr 0.35 V 10.0 W		Preheat to 850°C, and then heat rapidly to 1200°C. Oil-quench (or air-quench thin sections). Temper at 600–700°C for 2–3 hours.	Hot-forging dies and punches for making bolts, rivets, etc. where tools reach high temperatures. Hot-forging dies, extrusion dies and die-casting dies for copper alloys, pressure die-casting dies for aluminium alloys.

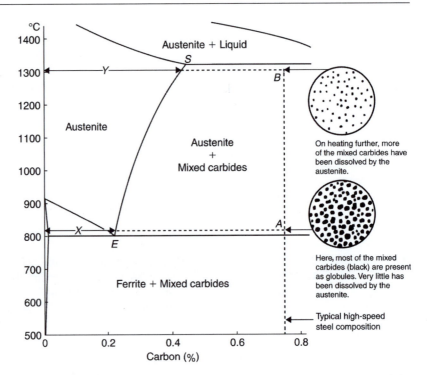

Figure 13.1 *A modified equilibrium diagram for high-speed steel. If the 'typical high-speed steel composition' is heated to A, only an amount of mixed carbides equivalent to X is dissolved by the austenite. Hence, on quenching, the steel would be soft and would not resist temperature influence. The maximum amount of mixed carbides which can be safely dissolved (without beginning to melt the tool) is shown by Y, and this involves heating the tool to 1300°C – just short of its melting point.*

the title 'complex carbides', a simplified two-dimensional diagram (Figure 13.1) can be used to explain the heat-treatment of this material.

Hot-forging dies for steel and copper alloys where excessive temperatures are not encountered, extrusion dies for aluminium alloys, pressure and gravity dies for casting aluminium.

Fine press-tools, deep drawing and forming dies for sheet metal, wire-drawing dies, blanking dies, punches and shear blades for hard metals.

Extrusion dies for aluminium and copper alloys, hot-forming, piercing- and heading-tools, brass-forging dies.

Hot-forging dies and punches for making bolts, rivets, etc. where tools reach high temperatures, hot-forging dies, extrusion dies and die-casting dies for copper alloys, pressure die-casting dies for aluminium alloys.

It will be seen that this diagram still resembles the ordinary iron–carbon diagram in general shape. The main difference is that the lower critical temperature has been raised (alloying elements usually raise or lower this temperature), and the eutectoid point E is now at only 0.25% carbon (instead of 0.8%). All alloying elements cause a shift of the eutectoid point to the left, hence alloy steels generally contain less carbon than the equivalent plain-carbon steels.

In the normalised condition, a typical high-speed steel contains massive globules of carbide in a matrix (or background) of ferrite. If this is now heated to just above the lower critical temperature (A in Figure 13.1), the ferrite will change to austenite and begin to dissolve the carbide globules. If the steel were quenched from this point, in the manner of a plain-carbon tool steel, the resultant structure would lack hardness, since only 0.25% carbon, or thereabouts, would be dissolved in the martensite so produced. Moreover, it would not resist tempering influences, as little tungsten would be dissolved in the martensite and it is the presence of *dissolved* tungsten which provides resistance to tempering. It is therefore necessary to ensure that the maximum amount of tungsten carbide is dissolved in the austenite before the steel is quenched.

The slope of the boundary line *ES* shows that, as the temperature rises, the amount of carbides dissolved in the austenite increases to a maximum at *S,* where the steel begins to melt (approximately 1320°C). Hence, to make sure that the maximum amount of carbide is dissolved before the steel is quenched, a high quenching temperature in the region of 1300°C is necessary. Since this is just short of the temperature at which melting begins, grain growth will proceed rather quickly. For this reason, a special heat-treatment furnace must be used. This consists of a lower chamber, usually heated by gas, and running at the quenching temperature of 1300°C, and above this a preheater chamber, maintained at about 850°C by the exhaust gases which have already circulated around the high-temperature chamber. The tool is first preheated to 850°C, and then transferred to the high-temperature compartment, where it will reach the quenching temperature in a few minutes. In this way, the time of contact between tool and high-temperature conditions is reduced below that which would be necessary were the tool not preheated. At such temperatures, decarburisation of the tool surface would be serious, so a controlled non-oxidising atmosphere is generally used in the furnace chamber.

As soon as it has reached the quenching-temperature, the tool is quenched in oil or in an air-blast (depending upon its size and composition). The resultant structure contains not only some martensite but also some soft austenite, because the high alloy content considerably reduces the rate of transformation. Hence the steel is heated to about 550°C to promote trans-formation of this austenite to martensite. This process is known as *secondary hardening* and gives an increase in hardness from about 700 to over 800 VPN. 'Super high-speed' steels contain up to 12% cobalt and are harder than the ordinary tungsten types.

Since molybdenum is now cheaper than tungsten, many modern highspeed steels contain large amounts of molybdenum to replace much of the tungsten. These molybdenum-type steels are reputed to be more difficult to heat-treat successfully and, whilst they are widely used in the USA, they are less popular in Britain. Table 13.8 gives examples.

Table 13.8 *High-speed steels*

BS 4659 specification	Composition (%)	Heat-treatment		Hardness (VPN)	Uses
		Quench in oil or air from (°C)	Secondary heat-treatment		
BT21	0.65 C 4.0 Cr 14.0 W 0.5 V	1300	Double temper at 565°C for 1 hour. (The double temper treatment gives extra hardness as more austenite transforms to martensite.)	860	General shop practice – all-round work, moderate duties, also for blanking-tools and shear blades.
BT1	0.8 C 4.5 Cr 18.0 W 1.2V	1310		890	Lathe-, planer- and shaping-tools, millers and gear-cutters, reamers, broaches, taps, dies, drills, hacksaws, roller-bearings for gas turbines.
BT6	0.8 C 4.25 Cr 20.0 W 1.5 V 0.5 Mo 12.0 Cr	1320		950	Lathe-, planer- and shaper-tools, milling cutters, drills for very hard materials. So-called super high-speed steel, which has maximum hardness and toughness.
BM1	0.8 C 3.75 Cr 1.6 W 1.25 V 9.0 Mo	1230		900	A general-purpose molybdenum-type high-speed steel for drills, taps, reamers, cutters. Susceptible to decarburisation during heat-treatment, which therefore requires careful control.

13.4 Stainless steels

Although Michael Faraday had attempted to produce stainless steel as long ago as 1822, it was not until 1912 that Brearley discovered the rust-resisting properties of high-chromium steel.

Chromium imparts the 'stainless' properties to these steels by coating the surface with a thin but extremely dense film of chromium oxide, which effectively protects the surface from further attack. Ordinary steel, on the other hand, becomes coated with a loose, porous layer of rust, through which the atmosphere can pass and cause further corrosion. Thus, ordinary steel rusts quickly, the top flakes of rust being pushed off by new layers forming beneath.

Much corrosion in metals is of the 'electrolytic' type (see Section 26.5). Readers will be familiar with the working of a simple cell, in which a copper and zinc plate are immersed in dilute sulphuric acid (called the *electrolyte*). As soon as the plates are connected, a current flows, and the zinc plate dissolves ('corrodes') rapidly.

In many alloys containing crystals of two different compositions, corrosion of one type of crystal will occur in this manner when the surface of the alloy is coated with an electrolyte – which, incidentally, may be rainwater. In stainless steels, however, the structure is a uniform solid solution. Since all the crystals within a piece of the alloy are of the same composition, electrolytic action cannot take place.

13.4.1 Types of stainless steels

There are two main types of stainless steel:

1   The straight chromium alloys, which contain 13% or more of chromium. These steels, provided they contain sufficient carbon, can be heat-treated to give a hard martensitic structure. Stainless cutlery steel is of this type. Some of these steels, however, contain little or no carbon, and are pressed and deep-drawn to produce such articles as domestic kitchen sinks, refrigerator parts, beer-barrels and tableware.
2   The '18/8' chromium/nickel steels are austenitic, even after being cooled slowly to room temperature (Figure 13.2). This type of steel cannot be hardened (except of course by cold-work). Much of it is used in chemical plant, where acid-resisting properties are required, whilst the cheaper grades are widely used in tableware and kitchen equipment.

Although these austenitic stainless steels cannot be hardened by heat-treatment, they are usually 'finished' by quenching from 1050°C. The purpose of this treatment is to prevent the precipitation of particles of chromium carbide, which would occur if the steel were allowed to cool slowly to room temperature. The precipitation of chromium carbide particles would draw out chromium from the surrounding structure, leaving it almost free of chromium (Figure 13.3) so that rusting would occur in that region. Such corrosion would be due to a combination of electrolytic action and direct attack. Because of the risk of precipitation of chromium carbide, these steels are unsuitable for welding, and suffer from a defect known as *weld-decay*.

13.4.2 Weld-decay

During welding, some regions of the metal near to the weld will be maintained between 650°C and 800°C long enough for chromium carbide to precipitate there (Figure 13.4). Subsequently, corrosion will occur in this area near to the weld. The fault may be largely overcome by adding about 1% of either titanium or niobium. These metals have a great affinity for carbon, which therefore combines with them in preference to chromium. Thus chromium is not drawn out of the structure, which, as a result, remains uniform.

13.4.3 Stainless steels and their uses

Table 13.9 gives examples of stainless steels, their composition, heat-treatment and uses.

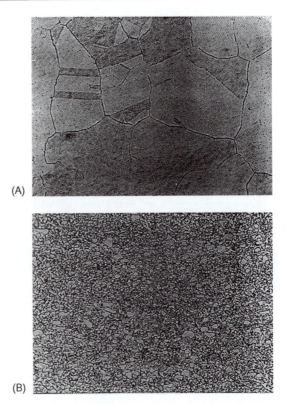

(A)

(B)

Figure 13.2 *(A) Austenitic 18/8 stainless steel, furnace-cooled from 1100°C. Note the annealing 'twins' (left-hand side of the picture) which are common in FCC structures which have been cold-worked and then annealed. The grain boundaries appear thick because some chromium carbide is precipitated there (see also Figure 13.3) due to the slow cooling process. (B) High-speed steel (18% tungsten, 4% chromium, 1% vanadium) annealed at 900°C. Rounded particles of tungsten carbide (light) in a ferrite-type matrix.*

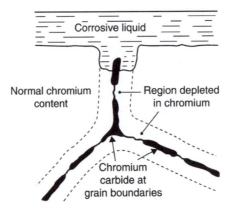

Figure 13.3 *The effect of carbide precipitation on the resistance to corrosion.*

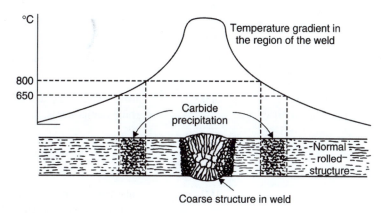

Figure 13.4 *Microstructural changes during welding which lead to subsequent corrosion ('weld-decay') in some stainless steels.*

Table 13.9 *Stainless steels*

BS 970 specification	Composition (%)	Heat-treatment	Uses
403S17	0.04 C 0.45 Mn 14.0 Cr	Non-hardening, except by cold-work.	'Stainless iron' – domestic articles such as forks and spoons. Can be pressed, drawn and spun.
420S45	0.3 C 0.5 Mn 13.0 Cr	Oil- or water-quench (or air-cool) from 960°C. Temper (for cutting) 150–180°C, temper (for springs) 400–450°C.	Specially for cutlery and sharp-edged tools, springs, circlips.
302S25	0.1 C 0.8 Mn 8.5 Ni 18.0 Cr	Non-hardening except by cold-work. (Cool quickly from 1050°C, to keep carbides dissolved.)	Particularly suitable for domestic and decorative purposes.
347S17	0.05 C 0.8 Mn 10.0 Ni 18.0 Cr 1.0 Nb		Weld-decay proofed by the presence of Nb. Used in welded plant where corrosive conditions are severe, e.g. nitric-acid plant.

Table 13.10 *Heat-resisting steels*

Composition (%)	Heat-treatment	Maximum working temperature (°C)	Uses
0.4 C, 0.2 Si, 1.4 Mn, 10.01 Cr, 36.0 Ni		600	Steam-turbine blades and other fittings.
0.1 C, 0.7 Mn, 12.0 Cr, 2.5 N, 1.8 Mo, 0.35 V	Hardened from 1050°C and tempered at 650°C.	600	Turbine blades and discs, bolts, some gas-turbine components.
0.35 C, 1.5 Si, 21.0 Cr, 7.0 Ni, 4.0 W		950	Resists a high concentration of sulphurous gases.
0.15 C, 1.5 Si, 25.0 Cr, 19.0 Ni		1100	Heat-treatment pots and muffles, aircraft-engine manifolds, boiler and super-heater parts.
0.35 C, 0.6 Si, 28.0 Cr		1150	Furnace parts, automatic stokers, retorts. Resistant to sulphurous gases.

13.5 Heat-resisting steels

The main requirements of a steel to be used at high temperatures are:

- It must resist oxidation and also attack by other gases in the working atmosphere.
- It must be strong enough at the working temperature.

Resistance to oxidation is effected by adding chromium and sometimes a small amount of silicon to steels. Both of these elements coat the surface with a tenacious layer of oxide, which protects the metal beneath from further attack. Nickel toughens the alloy by restricting grain-growth, but increased strength at high temperatures is achieved by adding small amounts of tungsten, titanium or niobium. These form small particles of carbide, which raise the limiting creep stress at the working temperature. Such steels are used for exhaust valves of internal-combustion engines, conveyor chains and other furnace parts, racks for enamelling stoves, annealing-boxes, rotors for steam and gas turbines and retorts.

Table 13.10 gives the composition, maximum working temperature and uses of some examples of heat-resisting steels.

13.6 Magnet alloys

Magnetic fields are generated by the spin of electrons within the orbits of atoms. Most elements produce only extremely weak magnetic fields because the electrons are in pairs in their orbits and, since they spin in opposite directions, the magnetic fields so produced cancel each other. However, iron, nickel, cobalt and the rare earth metal gadolinium are strongly magnetic, or *ferromagnetic*.

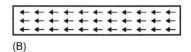

(A)   (B)

Figure 13.5 *The principles of ferromagnetism. (A) Random orientation of domain field before magnetisation. (B) Alignment of domain fields to produce a resultant field following magnetisation.*

In ferromagnetic metals, unpaired electrons produce resultant magnetic fields. Even so, the metal generally shows no resultant field in the unmagnetised condition. This is because the material has individual atoms grouped in blocks, termed *domains,* within which they are all pointing in the same direction. However, in the unmagnetised condition, the domains are orientated at random and the magnetic fields cancel. The process of magnetisation permits domains to grow that are in the direction of the field at the expense of the other domains and so a strong resultant field is produced in one direction (Figure 13.5).

13.6.1 Magnetic hysteresis

Suppose a piece of magnetic material is placed in a solenoid through which an electric current is passing. This current will produce a magnetising field H in the solenoid and a 'magnetic flux' B will be *induced* in the magnetic material. If the current is progressively increased, H will increase and so will B until a point P is reached (Figure 13.6A) where the magnet becomes saturated so that any further increase in H will not produce any further increase in B. If the current in the solenoid is now reduced, H will of course decrease, reaching zero when the current has reached zero (point O). B on the other hand will decrease less rapidly and, when H has reached zero, B will not be zero but the value R. That is, a *magnetic flux density R* remains in the magnet in the absence of a magnetising field. This value R is termed the *remanence* – or residual magnetism.

If the current in the solenoid is now increased in a reverse direction, a magnetising field $(-H)$ is generated in a reverse direction and RUQ represents the change in the induced magnetic flux in the magnet. The value UO is the strength of the reverse magnetic field $(-H)$ which is required to *demagnetise* the magnet completely. It is called the *coercive force* of the material and represents its *resistance to demagnetisation* by opposing magnetic fields. If the magnetising/demagnetising cycle is completed by again reversing the field H from S, then a *hysteresis loop* is formed. The *area enclosed by a loop* is a measure of the energy dissipated in the material during the magnetisation cycle, the bigger the area the greater the energy dissipated.

An alloy suitable for use as a *permanent magnet* will require a high remanence (B_{rem}) and/or a high coercive force (H_c). The ultimate standard by which a permanent magnet material is judged is the maximum product of B and H that is possible, this being the largest area rectangle that can be drawn in the upper-left quadrant of the hysteresis loop and represented in

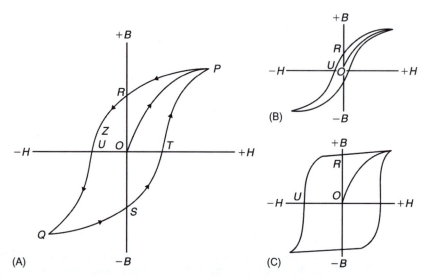

Figure 13.6 *Magnetic hysteresis: (A) the derivation of a hysteresis 'loop', (B) typical hysteresis loop for a magnetically 'soft' alloy, (C) typical hysteresis loop for a magnetically 'hard' (permanent magnet) alloy.*

Figure 13.6A by the point Z. This value BH_{max} – the product of B and H at Z – is a measure of the energy needed to demagnetise a material and so the larger the value the greater the energy that is needed for demagnetisation.

13.6.2 Soft and hard magnetic materials

Magnetic materials used in engineering fall into two groups:

1   Magnetically *soft* materials are easily demagnetised. They thus have very *low* values of B_{rem} and H_c, so producing a very 'narrow' hysteresis loop (Figure 13.6B). These materials are used in transformer cores and dynamo pole-pieces where it is necessary for the induced magnetic field to fluctuate with the applied magnetising field and little energy dissipated in magnetising it. Pure soft iron and iron–silicon alloys (4% silicon) containing no carbon are generally used, but a 'metallic glass' (see Section 22.7) containing 80% iron and 20% boron is much 'softer' magnetically and gives up to 25% savings on 'heat loss', i.e. energy dissipation, in power transmission transformers.

2   Magnetically *hard* materials are those which possess very *high* values of B_{rem} and H_c so that a very high degree of magnetism is retained in the absence of a magnetising field. These are thus used for permanent magnets. The earliest of these materials included tungsten steels and cobalt steels (up to 35% cobalt), both of which are now obsolete. Alnico (16% Ni, 12% Co, 9% Al, 5% Cu, 1% Ti, the balance Fe) and Hycol 3 (34% Co, 15% Ni, 7% Al, 5% Ti, 4% Cu, balance Fe) are representative of cast and sintered alloys with very high remanence and coercive force and are now used in the electrical, communications and general

Table 13.11 *Permanent magnet materials*

Material	Trade names	Chemical composition (%)	B_{rem} (T)	H_c (kA/m)	BH_{max} (kJ/m³)
Fe–Co–Ni–Al alloy	Alni	25 Ni, 4 Cu, 13 Al, 0–0.5 Ti, remaining Fe	0.56	46	10
	Alnico	16–20 Ni, 12–14 Co, 6 Cu, 9–10 Al, remaining Fe	0.72	45	13.5
	Alcomax 3	12–14 Ni, 22–26 Co, 3 Cu, 7.8–8.5 Al, 0–3 Nb, 0–1 Ti, remaining Fe	1.26	52	43
	Columax	13–14 Ni, 24–25 Co, 3 Cu, 7.8–8.5 Al, 0.5–3 Nb, remaining Fe	1.35	59	59.5
Ferrites	Feroba 1	13.5–14.5 BaO (+ SrO), remaining Fe_2O_3	0.22	135	8
Steels	6% tungsten	0.7 C, 1 Cr, 6 W, remaining Fe	1.05	5.2	2.4
	6% chromium	1.0 C, 6 Cr, remaining Fe	0.95	5.2	2.4
	9% cobalt	1.0 C, 9 Cr, 1.5 Mo, 9 Co, remaining Fe	0.79	12.7	4.0
	35% cobalt	0.9 C, 6 Cr, 35 Co, remaining Fe	0.90	20	7.5
Precipitation-hardening alloys	Comalloy	12–14 Co, 15–20 Mo, remaining Fe	1.0	20	10
	Vicalloy 1	52–53 Co, 10–12 V, remaining Fe	0.9	24	8

Table 13.12 *The more important effects of the main alloying elements*

Element	Chemical symbol	Principal effects when added to steel
Manganese	Mn	Acts as a deoxidiser and a desulphuriser. Stabilises carbon.
Nickel	Ni	Toughens steel by refining grain. Strengthens ferrite. Causes cementite to decompose – hence used by itself only in low-carbon steel.
Chromium	Cr	Stabilises carbides, and forms hard chromium carbide – hence increases hardness of steel. Promotes grain growth, and so causes brittleness. Increases resistance to corrosion.
Molybdenum	Mo	Reduces 'temper brittleness' in nickel–chromium steels. Stabilises carbides. Improves high-temperature strength.
Vanadium	V	Stabilises carbides. Raises softening temperature of hardened steels (as in high-speed steels).
Tungsten	W	Forms very hard stable carbides. Raises the softening temperature, and renders transformations very sluggish (in high-speed steels). Reduces grain growth. Raises the limited creep stress at high temperatures.

engineering industries. A number of sintered ceramic materials are also important. These are based mainly on mixtures of barium oxide and iron oxide. Such a substance is Feroba, used extensively in loudspeakers, door catches and computer devices. In powder form, it can be incorporated in a rubber matrix for use on display boards. Table 13.11 gives examples of such magnetically hard materials.

13.7 The principal effects of the main alloying elements

The most important effects of the main alloying elements added to steels are summarised in Table 13.12. Remember that *all* of these elements increase the depth of hardening of a steel, so an alloy steel has a bigger 'ruling section' than a plain-carbon steel. This is because alloying elements *slow down* the austenite → martensite transformation rates, so it is possible to oil-harden or, in some cases, air-harden a suitable steel (see Figure 12.3B).

14 The surface hardening of steels

14.1 Introduction

Many metal components require a combination of mechanical properties which at first sight seems impossible to attain. Thus, bearing metals (see Section 18.6) must be both hard and, at the same time, ductile, whilst many steel components, like cams and gears, need to be strong and shock-resistant, yet also hard and wear-resistant. In ordinary carbon steels, these two different sets of properties are found only in materials of different carbon content, e.g. a steel with about 0.1% carbon will be tough, whilst one with 0.9% carbon will be very hard when suitably heat-treated.

The problem can be overcome in two different ways:

1 By employing a tough low-carbon steel, and altering the composition of its surface, either by case-hardening or by nitriding.
2 By using a steel of uniform composition throughout, but containing at least 0.4% carbon, and heat-treating the surface differently from the core, as in flame- and induction-hardening.

In the first case, it is the hardening material which is localised, whilst in the second case, it is the heat-treatment which is localised. In this chapter, we look at both.

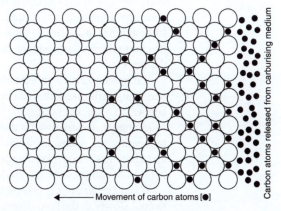

Figure 14.1 *An impression of the penetration by carbon atoms into the lattice structure of FCC iron (austenite).*

14.2 Case-hardening

This process makes use of the fact that carbon will dissolve in appreciable amounts in *solid* iron, provided that the latter is in the face-centred cubic crystal form. This is due to the fact that carbon dissolves interstitially in iron (see Section 8.3); the carbon atoms are small enough to infiltrate between the larger iron atoms (Figure 14.1), so solid iron can absorb carbon in much the same way that water is soaked up by a sponge.

Since only face-centred cubic iron will dissolve carbon in this way, it follows that steel must be carburised at a temperature *above* the upper critical temperature. As it is generally low-carbon steel which is carburised, this involves using a temperature in the region of 900–950°C. Thus, carburising consists of surrounding mild-steel components with some carbon-rich material, and heating them above their upper critical temperature for long enough to produce a carbon-rich surface layer of sufficient depth.

Solid, liquid and gaseous carburising materials are used, and the output quantity required largely governs the method used. Carburising is capable of producing a wide range of case depths, the main disadvantage being the distortion that can occur as a result of the thermal gradients produced by cooling from the austenitic state.

14.2.1 Carburising in solid media

So-called *pack-carburising* is probably the process with which the reader is most likely to be familiar. Components to be treated are packed into steel boxes, along with the carburising material, so that a space of roughly 50 mm exists between them. Lids are then fixed on the boxes, which are then slowly heated to the carburising temperature (900–950°C). They are then maintained at this temperature for up to 6 hours (much longer periods are sometimes necessary when deep cases are to be produced), according to the depth of case required (Figure 14.2).

Carburising mixtures vary in composition, but consist essentially of some carbon-rich material, such as charcoal or charred leather, along with an energiser which may account for about 40% of the total. This energiser is generally a mixture of sodium carbonate ('soda ash') and barium carbonate. Its function is to accelerate the solution of carbon by taking part in a chemical reaction which causes single carbon atoms to be released at the surface of the steel.

If it is necessary to prevent any parts of the surface of the component from becoming carburised, this can be achieved by electroplating these areas with copper, to a thickness of 0.07–0.10 mm, since carbon does not dissolve in solid copper. In small-scale treatment, the same objective can be achieved by coating the necessary areas of the components with a paste of fireclay and ignited asbestos mixed with water. This is allowed to dry on the surface, before the components are loaded into the carburising box.

When carburising is complete, the charge is either quenched or allowed to cool slowly in the box, depending on the subsequent heat-treatment it will receive.

Pack carburising can be used with low-carbon and low-alloy carburising steels to produce case depths of up to 2.5 mm with a case hardness of 45–65 HRC (scale C on the Rockwell hardness scale).

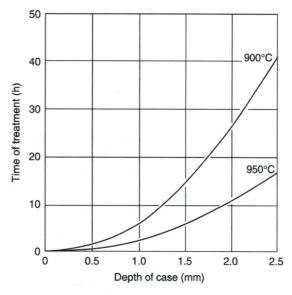

Figure 14.2 *The relationship between time of treatment, temperature and depth of case in a carburising process using solid media (0.15% plain-carbon steel).*

14.2.2 Carburising in liquid media

Liquid-carburising – or cyanide hardening, as it is usually called – is carried out in baths of molten salt which contain 20–50% sodium cyanide, together with as much as 40% sodium carbonate, and varying quantities of sodium or barium chloride. The cyanide-rich mixture is heated in iron pots to a temperature of 870–950°C, and the work, which is carried in wire baskets, is immersed for periods of about 5 minutes upwards, according to the depth of case required. The process is particularly suitable for producing shallow cases of 0.1–0.25 mm.

Carburising takes place due to the decomposition of sodium cyanide at the surface of the steel. Atoms of both carbon and nitrogen are released, so cyanide hardening is due to the absorption of nitrogen as well as carbon.

The main advantages of cyanide hardening are:

* The temperature of a liquid salt bath is uniform throughout, and can be controlled accurately by pyrometers.
* The basket of work can be quenched directly from the bath.
* The surface of the work remains clean.

Many readers will be aware of the fact that all cyanides are extremely poisonous chemicals. However, since *sodium cyanide is one of the most deadly poisonous materials* in common use industrially, it might be well to stress the following points, which should be observed by the reader should he find himself involved in the use of cyanides:

- Every pot should be fitted with an efficient fume-extraction system.
- The consumption of food by operators whilst working in a shop containing cyanide should be *absolutely forbidden*.
- Cyanide-rich salts should never be allowed to come into contact with an open wound.
- Advice should be sought before disposing of any waste hardening salts. They should *never* be tipped into canals or rivers.

Cyaniding can be used with low-carbon and low-alloy carburising steels and is used to produce case depths of the order 0.1–0.25 mm with a case hardness of 50–60 HRC.

14.2.3 Carburising by gaseous media

Gas-carburising is carried out in both continuous and batch-type furnaces. Whichever is used, the components are heated at about 900°C for 3 hours or more in an atmosphere containing gases which will deposit carbon atoms at the surface of the components. The gases generally used are the hydro-carbons methane ('natural gas') and propane (a by-product of petroleum production). These should be of high purity, otherwise oily soot may be deposited on the work-pieces. The hydrocarbon is usually mixed with a 'carrier' gas (generally a mixture of nitrogen, hydrogen and carbon mon-oxide) which allows better gas circulation and hence greater uniformity of treatment.

Gas-carburising is now used for most large-scale treatment, particularly for the mass-production of thin cases. Its main advantages as compared with other methods of carburising are:

- The surface of the work is clean after the treatment.
- The necessary plant is more compact for a given output.
- The carbon content of the surface layers can be more accurately controlled by this method.

Gas carburising can be used with low-carbon and low-alloy carburising steels to give case depths of the order of 0.07–3 mm with a case hardness of 45–85 HRC.

14.3 Heat-treatment after carburising

If the carburising treatment has been successful, the core will still have a low carbon content (0.1–0.3% carbon), whilst the case should have a maximum carbon content of 0.8% carbon (the eutectoid composition) (Figure 14.3). Unfortunately, prolonged heating in the austenitic range causes the formation of coarse grain, and further heat-treatment is desirable if optimum properties are to be obtained.

The most common method of producing a fine-grained structure in steel is by normalising it. This involves heating the steel to just above its upper critical temperature, followed by cooling it in air (see Section 11.5). The need for such treatment poses a problem here, since core and case are of

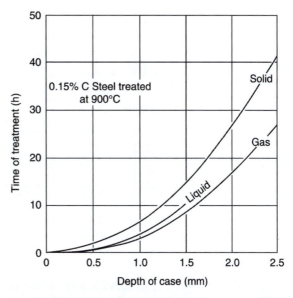

Figure 14.3 *The relationship between time of treatment and depth of case produced when carburising in solid, liquid and gaseous media (0.15% plain-carbon steel).*

widely different carbon contents, and therefore have different upper critical temperatures. Thus, if the best mechanical properties are to be obtained in both core and case, a double heat-treatment is necessary.

1 *Refining the core*
The component is first heat-treated to refine the grain of the core, and so toughen it. This is done by heating the component to a temperature just above the upper critical temperature for the core (point *A* in Figure 14.4), so that the coarse ferrite/pearlite will be replaced by fine-grained austenite. The component is then generally water-quenched, so that a mixture of fine-grained ferrite and a little martensite is produced. The temperature of this treatment is well above the upper critical temperature for the case (723°C), so at this stage the case will be of coarse-grained martensite (because the steel was quenched). Further heat-treatment is therefore necessary to refine the grain of the case.

2 *Refining the case*
The component is now heated to 760°C (point *B* in Figure 14.4), so that the structure of the case changes to fine-grained austenite. Quenching then gives a hard case of fine-grained martensite and, at the same time, any brittle martensite present in the core as a result of the *first* quenching process will be tempered to some extent by the second heating operation (point *C* in Figure 14.4). Finally, the component is tempered at 200°C to relieve any quenching-stresses present in the case.

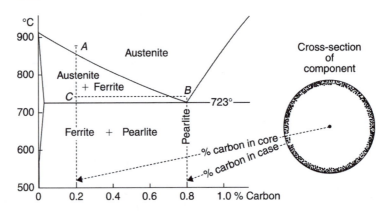

Figure 14.4 *Heat-treatment of a carburised component in relation to the equilibrium diagram.*

The above heat-treatment processes can be regarded to some extent as the council of perfection, and the needs of economy often demand that such treatments may be replaced by a single operation. Often the work may be 'pot-quenched'; that is, quenched direct from the carburising process, followed by a low-temperature tempering process to relieve any quenching stresses.

Alternatively, the work may be cooled slowly from the carburising temperature, to give maximum ductility to the core. It is then reheated to 760°C and water-quenched. This treatment leaves the core quite soft, but hardens the case, which will be fine-grained, due to the low quenching temperature.

14.4 Case-hardening steels

Plain-carbon and low-alloy steels are used for case-hardening, but in either type, the carbon content should not be more than 0.2% if a really tough core is to be obtained. Manganese may be present in amounts up to 1.4%, since it stabilises cementite and increases the depth of hardening. Unfortunately, it is also liable to increase the tendency of a steel to crack during quenching.

Alloy steels used for case-hardening contain up to 4.0% nickel, since this increases the strength of the core and retards grain-growth during the carburising process. This often means that the core-refining heat-treatment can be omitted. Chromium is sometimes added to increase hardness and wear-resistance of the case, but it must be present only in small quantities, as it tends to promote grain growth (see Section 13.2).

Table 14.1 gives the composition and uses of some examples of case-hardening steels.

14.5 Nitriding

Nitriding and case-hardening have one factor in common – both processes involve heating the steel for a considerable time in the hardening-medium, but, whilst in case-hardening the medium contains carbon, in nitriding it contains gaseous nitrogen. Special steels – 'Nitralloy' steels – are necessary for the nitriding process, since hardening depends upon the formation of very

Table 14.1 *Case-hardening steels*

Composition (%)					Characteristics and uses
C	Mn	Ni	Cr	Mo	
0.15	0.7				Machine parts requiring a hard surface and a tough core, e.g. gears, shafts, cams.
0.15	1.3				A carbon–manganese steel giving high surface hardness where severe shock is unlikely.
0.13	0.5	3.25	0.85		High surface hardness combined with core toughness – high-duty gears, worm gears, crown wheels, clutch gears.
0.17	0.5	1.75		0.25	High hardness and severe shock resistance – automobile parts (steering works, overhead valve mechanisms).
0.15	0.4	4.0	1.2	0.2	Best combination of surface hardness, core strength and shock resistance – crown wheels, bevel pins, intricate sections which need to be air-hardened (see Section 13.2).

hard compounds of nitrogen and such metals as aluminium, chromium and vanadium present in the steel. Ordinary plain-carbon steels cannot be nitrided, since any compounds of iron and nitrogen which form will diffuse into the core, so that the increase in hardness of the surface is lost. The hard compounds formed by aluminium, chromium and vanadium, however, remain near to the surface and so provide an extremely hard skin.

Nitriding is carried out at the relatively low temperature of 500°C. Consequently, it is made the final operation in the manufacture of the component, all machining and core heat-treatments having been carried out previously. The work is maintained at 500°C for between 40 and 100 hours, according to the depth of case required (Figure 14.5), though treatment for 90 hours is general. The treatment takes place in a gas-tight chamber through which ammonia gas is allowed to circulate. Some of the ammonia decomposes, releasing single atoms of nitrogen, which are at once absorbed by the surface of the steel.

$$NH_3 = 3H + N \text{ (atom)}$$

Ordinary 'atmospheric' nitrogen is not suitable since it exists in the form of molecules (N_2) which would not be absorbed by the steel.

Nitralloy steels containing aluminium are hardest, since aluminium forms very hard compounds with nitrogen. Unfortunately, aluminium tends to affect the core-strength adversely, and is replaced by chromium, vanadium and molybdenum in those Nitralloy steels in which high strength and toughness of core are important. Composition and uses of some nitriding steels are given in Table 14.2.

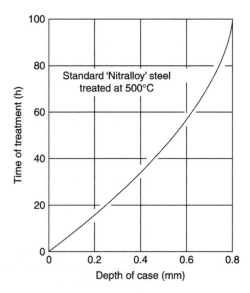

Figure 14.5 *The relationship between time of treatment and depth of case produced in the nitriding process.*

Table 14.2 *Nitriding steels*

Composition (%)					Typical mechanical properties		Characteristics and uses
C	Cr	Mo	V	Al	Tensile strength (MPa)	VPN	
0.5	1.5	0.2		1.1	1200	1075	Where maximum surface hardness, coupled with high core-strength, is essential.
0.2	1.5	0.2		1.1	600	1075	For maximum surface hardness, combined with ease of machining before hardening.
0.4	3.0	1.0	0.2		1400	875	Ball-races, etc. where high core-strength is necessary.
0.3	3.0	0.4			1000	875	Aero-crankshafts, air-screw shafts, aerocylinders, crank-pins and journals.

14.5.1 Heat treatment

Prior to being nitrided, the work-pieces are heat-treated to produce the required properties in the core. Since greater scope is possible in this heat-treatment than is feasible in that associated with case-hardening, Nitralloy steels often have higher carbon contents, allowing high core-strengths to be developed. The normal sequence of operations will be:

1   Oil-quenching from 850°C to 900°C, followed by tempering at between 600°C and 700°C.
2   Rough machining, followed by a stabilising anneal at 550°C for 5 hours, to remove internal stresses.
3   Finish-machining, followed by nitriding.

Any areas of the surface which are required soft are protected by coating with solder or pure tin, by nickel plating, or by painting with a mixture of whiting and sodium silicate.

14.5.2 Advantages and disadvantages of nitriding

Advantages of nitriding over case-hardening are:

- Since no quenching is required *after* nitriding, cracking or distortion is unlikely, and components can be machine-finished before treatment.
- An extremely high surface hardness of up to 1150 VPN is attainable with the aluminium-type Nitralloy steels.
- Resistance to corrosion is good, if the nitrided surface is left unpolished.
- Hardness is retained up to 500°C, whereas a case-hardened component begins to soften at about 200°C.
- The process is clean and simple to operate.
- It is cheap if large numbers of components are treated.

Disadvantages of nitriding as compared with case-hardening are:

- The initial outlay for nitriding plant is higher than that associated with solid- or liquid-medium carburising, so nitriding is only economical when large numbers of components are to be treated.
- If a nitrided component is accidentally overheated, the loss of surface hardness is permanent, unless the component can be nitrided again. A case-hardened component would need only to be heat-treated, assuming that it has not been so grossly overheated as to decarburise it.

14.5.3 Carbonitriding

Carbonitriding is a surface-hardening process which makes use of a mixture of hydrocarbons and ammonia. It is therefore a gas treatment, and is sometimes known as *dry-cyaniding* – a reference to the fact that a mixed carbide–nitride case is produced, as in ordinary liquid-bath cyanide processes (see Section 14.2.2).

Furnaces used for carbonitriding are generally of the continuous type, as the work is nearly always directly quenched in oil from the carbonitriding atmosphere. If 'stopping off' is necessary for any areas required soft, then good-quality copper-plating is recommended.

Carbonitriding is an ideal process for hardening small components where great resistance to wear is necessary. Case depths are less than those produced by gas carburising, typically being up to about 0.7 mm with a case hardness of 50–70 HRC.

14.6 Ionitriding

More correctly termed *ion-nitriding,* this process is also known as *plasma nitriding* and *ion implantation.* The work load has made the cathode in a sealed chamber containing nitrogen under near-vacuum conditions. Under a potential difference approaching 1000 V (D.C.), the low-pressure nitrogen ionises, i.e. the nitrogen atoms lose outer-shell electrons and so become *positively charged* ions. They are thus attracted, and so accelerated, towards the negatively charged cathode, i.e. the work load. They strike this at a very high velocity and so penetrate the surface, the kinetic energy lost on impact being converted into heat so that the surface temperature of the work load is raised to the nitriding temperature (400–600°C). The work load is surrounded by a glow of ionised nitrogen and treatment time is between 10 minutes and 30 hours, depending upon the type of steel and the depth of case required. Maximum hardness is achieved with Nitralloy-type steels (see Table 14.2) – the higher the alloy content, the thinner and harder the case.

The process is used to nitride components weighing several tonnes down to the tiny balls of ball-point pens. Automobile parts, hot- and cold-working dies and tools are now ion-nitrided.

14.7 Flame-hardening

In this process, the work-piece is of uniform composition throughout and the surface hardening occurs because the surface layers receive extra-heat treatment as compared with the core material.

The surface is heated to a temperature above its upper critical temperature, by means of a travelling oxyacetylene torch (Figure 14.6), and is immediately quenched by a jet of water issuing from a supply built into the torch-assembly.

Symmetrical components, such as gears and spindles, can be conveniently treated by being spun between centres so that the whole circumference is thus treated.

Only steels with a sufficiently high carbon content – at least 4% – can be hardened effectively in this way. Alloy steels containing up to 4.0% nickel

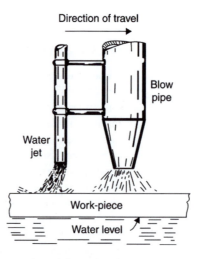

Figure 14.6 *The principles of flame-hardening.*

and 1.0% chromium respond well to such treatment. Before being hardened, the components are generally normalised, so that the final structure consists of a martensitic case some 4 mm deep, and a tough ferrite–pearlite core. Core and case are usually separated by a layer of bainite, which helps to prevent the hard case from cracking away from the core material. Should a final tempering process be necessary, this can also be carried out by flame-heating, though furnace treatment is also possible since such low-temperature treatment will have no effect on the core, particularly if it has been normalised.

Flame hardening can produce case depths of up to 0.8 mm with a case hardness of 55–65 HRC.

14.8 Induction-hardening

This process is similar in principle to flame-hardening, except that the component is usually held stationary whilst the whole circumference is heated simultaneously by means of an induction-coil. This coil carries a high-frequency current, which induces eddy currents in the surface of the component, thus raising its temperature. The depth to which heating occurs varies inversely as the square root of the frequency, so that the higher the frequency used, the shallower the depth of heating. Typical frequencies used are:

3000 Hz for depths of 3–6 mm

9600 Hz for depths of 2–3 mm

As soon as the surface of the component has reached the necessary temperature, the current is switched off and the surface simultaneously quenched by pressure jets of water, which pass through holes in the induction-block (Figure 14.7).

The process lends itself to mechanisation, so that selected regions of a symmetrical component can be hardened, whilst others are left soft. As in

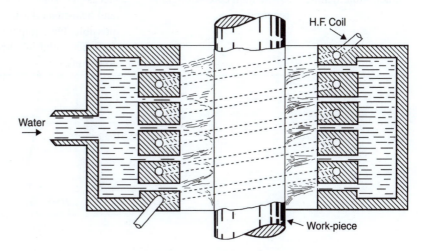

Figure 14.7 *The principles of induction-hardening.*

flame-hardening, the induction process makes use of the existing carbon content – which consequently must be at least 0.4% – whilst in case-hardening, nitriding and carbonitriding, an alteration in the composition of the surface layers takes place.

14.9 Summary of surface-hardening processes

The characteristics and uses of the processes dealt with in this chapter are summarised in Table 14.3.

Table 14.3 *Summary of surface-hardening processes*

Process	Type of work	Characteristics
Case-hardening (solid and gas)	Gears, king-pins, ball- and roller-bearings, rocker-arms, gauges.	A wide variety of low-carbon and low-alloy steels can be treated. Local soft-surfaces are easily retained. Gas carburisation is a rapid process.
Case-hardening (liquid cyanide)	Used mainly for light cases.	The case tends to be of poorer quality, but thin cases can be produced quickly.
Nitriding	Crankshafts, cam-shafts, gears requiring high core-strength.	A very high surface hardness, combined with a high core-strength when required. Surface will withstand tempering influences up to 500°C. Less suitable than other methods if surface has to withstand very high pressure, e.g. gear-teeth.
Carbonitriding	Particularly useful for treating small components.	Safe, clean and easy to operate, applicable to mass-producing methods.
Ionitriding	Crankshafts and many other components in various industries.	A high degree of control and uniformity is possible.
Flame- and induction-hardening	Tappets, cam-shafts, gears where high core-strength is required.	Particularly useful where high core-strength is necessary, since a high-carbon steel can be used and heat-treated accordingly. Rapid output possible, but equipment often needs to be designed for a particular job, hence suitable mainly for long runs.

15 Cast iron

15.1 Introduction

The Victorian era may well be remembered by the cast-iron monstrosities which it produced. Street lamps, domestic fireplaces and railings were typical cast-iron products of that period. Most of these relics are gone – the railings fell victim of the need for steel during the Second World War – but many an industrial town still boasts an ornamental drinking-fountain in its local park, or a cast-iron clock presiding over the public conveniences (of similar period) in the town square, whilst it seems that the once despised blackleaded cast-iron fireplaces of my childhood are now regarded as being 'eminently collectable'.

During the nineteenth century, much cast iron was also used for engineering purposes. Today, the whole production of cast iron is directed towards these purposes, and, as in other fields of metallurgical technology, considerable progress has been made during the twentieth century. Special high-duty and alloy compositions have made cast iron an extremely important engineering material, which is suitable for the manufacture of crankshafts, connecting rods and axles – components which were formerly made from forged steel.

Ordinary cast iron is similar in composition to the crude pig iron produced by the blast-furnace. The pig iron is generally melted in a cupola, any necessary adjustments in composition being made during the melting process. At present, the high cost of metallurgical coke, coupled with the desire to produce high-grade material, has led the foundryman to look for other methods of melting cast iron; consequently, line-frequency induction furnaces are being used on an increasing scale.

The following features make cast iron an important material:

- It is a cheap metallurgical substance, since it is produced by simple adjustments to the composition of ordinary pig iron.
- Mechanical rigidity and strength under compression are good.
- It machines with ease when a suitable composition is selected.
- Good fluidity in the molten state leads to the production of good casting-impressions.
- High-duty cast irons can be produced by further treatment of irons of suitable composition, e.g. spheroidal-graphite irons are strong, whilst malleable irons are tough.

15.2 Composition of cast irons

Ordinary cast irons contain the following elements: carbon 3.0–4.0%, silicon 1.0–3.0%, manganese 0.5–1.0%, sulphur up to 0.1%, phosphorus up to 1.0%.

1   *Carbon*

Carbon may be present in the structure either as flakes of graphite or as a network of hard, brittle, iron carbide (or cementite). If a cast iron contains much of this brittle cementite, its mechanical properties will be poor, and for most engineering purposes it is desirable for the carbon to be present as small flakes of graphite. Cementite is a silvery-white compound and, if an iron containing much cementite is broken, the fractured surface will be silvery white, because the piece of iron breaks along the brittle cementite networks. Such an iron is termed a *white iron*. Conversely, if an iron contains much graphite, its fractured surface will be grey, due to the presence of graphite flakes in the structure and this iron would be described as a *grey iron*. Under the microscope, these graphite flakes appear as two-dimensional strands but what we see is a cross-section through a micro-constituent shaped something like a breakfast cereal 'cornflake' (Figure 15.1).

Figure 15.1 *A three-dimensional impression of a cluster – or 'cell' – of graphite flakes in a grey iron. A micro-section of a high-silicon iron would show them as 'star-fish' shapes (see Figure 15.2C).*

2   *Silicon*

Silicon to some extent governs the form in which carbon is present in cast iron. It causes the cementite to be unstable, so that it decomposes, thus releasing free graphite. Therefore, a high-silicon iron tends to be a grey iron, while a low-silicon iron tends to be a white iron (Figure 15.2).

3   *Sulphur*

Sulphur has the opposite effect on the structure to that given by silicon; that is, it tends to stabilise cementite, and so helps to produce a white iron. However, sulphur causes excessive brittleness in cast iron (as it does in steel), and it is therefore always kept to the minimum amount which is economically possible. No self-respecting foundryman would think of altering the structure of an iron by adding sulphur to a cupola charge, any more than he would think of diluting his whisky with dirty washing-up water; the desired microstructure for an iron is obtained by adjusting the silicon content.

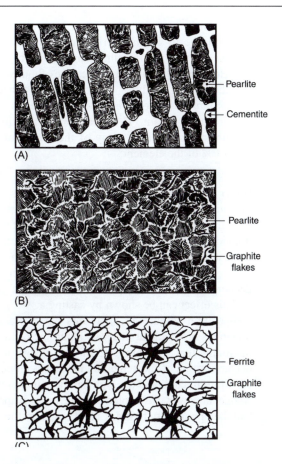

Figure 15.2 *The effects of silicon content on the structure of cast iron. The higher the silicon content, the more unstable the cementite becomes, until even the pearlitic cementite decomposes (C). Magnifications approximately ×100. (This illustration shows that a careful sketch of a microstructure can often reveal more than would a photomicrograph. Patience, assisted by a very sharp pencil, necessary and some microstructures are more conducive to being sketched than others.) (A) White cast iron (low silicon) – primary cementite network in a matrix of pearlite. (B) Fine grey iron (medium silicon) – small graphite flakes in a matrix of pearlite. (C) Coarse grey iron (high silicon) – large graphite flakes in a matrix of ferrite.*

During the melting of cast iron in a cupola, some silicon is inevitably burnt away, whilst some sulphur will be absorbed from the coke. Both factors tend to make the iron 'whiter', so the foundryman begins with a charge richer in silicon than that with which he expects to finish.

4   *Manganese*
Manganese toughens and strengthens an iron, partly because it neutralises much of the unwelcome sulphur by forming a slag with it, and partly because some of the manganese dissolves in the ferrite.

5   *Phosphorus*

Phosphorus forms a brittle compound with some of the iron; it is therefore kept to a minimum amount in most engineering cast irons. However, like silicon, it increases fluidity, and considerably improves the casting qualities of irons which are to be cast in thin sections, assuming that components are involved in which mechanical properties are unimportant. Thus, cast-iron water pipes contained up to 0.8% phosphorus, whilst many of the ornamental castings contained up to 1.0% of the element.

15.3 The influence of cooling rate on the properties of a cast iron

When the presence of silicon in an iron tends to make cementite unstable, the latter does not break up or decompose instantaneously; this process of decomposition requires time. Consequently, if such an iron is cooled so that it solidifies rapidly, the carbon may well be 'trapped' in the form of hard cementite, and so give rise to a white iron. On the other hand, if this iron is allowed to cool and solidify slowly, the cementite has more opportunity to decompose and form graphite, so producing a grey iron.

This effect can be shown by casting a 'wedge-bar' in an iron of suitable composition (Figure 15.3). If this bar is fractured, and hardness determinations are made at intervals along the centre line of the section, it will be found that the thin end of the wedge has cooled so quickly that decomposition of the cementite has not been possible. This is indicated by the white fracture and the high hardness in that region. The thick end of the wedge, however, has cooled slowly, and is graphitic, because cementite has had the opportunity to break up. Thus, here the structure is softer.

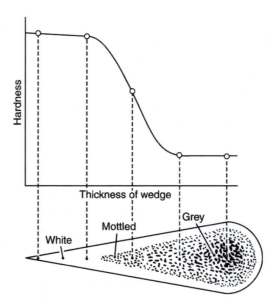

Figure 15.3 *The effect of sectional thickness on the depth of chilling of a grey iron.*

If the reader has had experience in machining cast iron, he will know that such a casting has a hard surface skin, but that, once this is removed, the material beneath is easy to machine. This hard skin consists largely of cementite which has been prevented from decomposing by the chilling action of the mould. Iron beneath the surface has cooled more slowly, so that cementite has decomposed, releasing graphite.

To summarise: the engineer requires a cast iron in which carbon is present in the form of small flakes of graphite. The form in which the carbon is present depends on:

- The silicon content of the iron.
- The rate at which the iron solidifies and cools, which in turn depends upon the cross-sectional thickness of the casting.

Thus, the foundryman must strike a balance between the silicon content of the iron and the rate at which it cools.

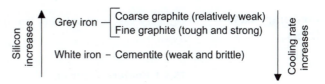

When casting *thin* sections, it will be necessary for the foundryman to choose an iron which has a coarser grey fracture in the pig form than that which is required in the finished casting. Thus, the iron must have a higher silicon content than that used for the manufacture of castings of heavier section, which will consequently cool more slowly.

Sometimes it is necessary to have a hard-wearing surface of white iron at some point in a casting which otherwise requires a tough grey iron structure. This can be achieved by incorporating 'chills' at appropriate points in the sand mould (Figure 15.4). The 'chill' usually consists of a metal block, which will cause the molten iron in that region to cool so quickly that a layer of hard cementite is retained adjacent to the chill.

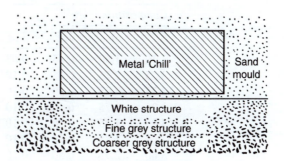

Figure 15.4 *The use of 'chills' in iron-founding.*

15.4 'Growth' in cast irons

An engineering cast iron contains some cementite as a constituent of the pearlitic areas of the structure. If such an iron is heated for a prolonged period in the region of 700°C or above, this cementite decomposes to form graphite and iron. Since the graphite and iron so formed occupy more space than did the original cementite, the volume of the heated region increases, and this expansion leads to warping of the casting, and the formation of cracks at the surface. Hot gases penetrate these cracks, gradually oxidising both the graphite and the iron, so that the surface ultimately disintegrates. Fire bars in an ordinary domestic grate often break up in this way.

The best way to prevent such 'growth' in cast irons which are to be used at high temperatures is to ensure that they contain *no* cementite in the first place. This can be achieved by using a high-silicon-content iron, in which decomposition of all the cementite will take place during the actual solidification of the casting. Thus, 'Silal' contains 5.0% silicon, with relatively low carbon, so that the latter is present in the finished casting entirely as graphite. Unfortunately, Silal is rather brittle, so, when the cost is justified, the alloy cast iron 'Nicosilal' (see Table 15.4) may be used.

15.5 Ordinary cast irons

Ordinary cast irons fall into two main groups: engineering irons and fluid irons.

15.5.1 Engineering irons

Engineering irons must possess reasonable strength and toughness, generally coupled with good machinability. The silicon content of such an iron will be chosen in accordance with the cross-sectional thickness of the casting to be produced. It may be as much as 2.5% for castings of thin section, but as low as 1.2% for bulky castings of heavy section. This relationship between silicon content and sectional thickness of a casting is illustrated by the three irons specified for light, medium and heavy machine castings in Table 15.1. Amounts of sulphur and phosphorus generally are kept low, since both elements cause brittleness, though some castings contain as much as 1.0% phosphorus, to give fluidity to the molten iron.

Table 15.1 *Composition and uses of some ordinary cast irons*

Composition (%)					Uses
C	Si	Mn	S	P	
3.30	1.90	0.65	0.08	0.15	Motor brake-drums
3.25	2.25	0.65	0.10	0.15	Motor cylinders and pistons
3.25	2.25	0.50	0.10	0.35	Light machine-castings
3.25	1.75	0.50	0.10	0.35	Medium machine-castings
3.25	1.25	0.50	0.10	0.35	Heavy machine-castings
3.60	1.75	0.50	0.10	0.80	Water pipes
3.60	2.75	0.50	0.10	0.90	Low-strength ornamental castings of yesteryear

Table 15.2 *Typical properties of engineering grey irons in accordance with BS EN 1561. It will be noted that the Grade Number indicates the minimum acceptable tensile strength*

Grade	Minimum tensile strength (MPa)	Brinell hardness at thickness of section (mm)				
		2.5	5	10	20	80
100	100	210				150
150	150	260				
180	180					
200	200	280				120
220	220					
250	250		280			145
300	300			280		165
350	350				280	185

The principal engineering grey irons are covered by BS 1452 which deals with eight different grades, each with its minimum acceptable tensile strength (Table 15.2).

Grades 100 and 150 are fluid irons, possibly containing phosphorus, and which will only be used where low strength is acceptable, e.g. domestic pipes and gutters. Grade 180 is a high-silicon iron in which most of the carbon is present as coarse graphite flakes in a matrix mainly of ferrite. At the other extreme, grade 350 is a low-silicon iron containing fine graphite flakes in a pearlitic matrix. The properties of flake graphite (grey) iron depend on the amount and form of graphite and on the matrix structure. This structure may be controlled by a variety of factors, such as production conditions, chemical composition, rate of solidification and rate of cooling *after* solidification, as well as complexity of design. BS 1452 therefore specifies only tensile strength.

15.5.2 Fluid irons

Fluid irons, in which mechanical strength is of secondary importance, were at one time widely used in the manufacture of railings, lamp-posts and fireplaces. High fluidity was necessary in order that the iron should fill intricate mould impressions, and this was achieved by using a high silicon content of 2.5–3.5%, as well as a high phosphorus content of up to 1.5%. Cast iron has been replaced by other materials for the purposes mentioned earlier. As a small boy, I was walking along the local High Street when a heavy lorry careered across the road, crashing into the ornamental cast iron 'skirting' of a lamp standard on the pavement in front of me. The skirting – presumably of brittle high-phosphorus iron – fragmented on impact, sending a hail of 'shrapnel' across the pavement. Shop windows were broken and some pedestrians received nasty wounds from the razor-sharp shards. Much scared, I scuttled home unscathed! The local authority soon began to remove the cast iron skirtings, ultimately replacing them by concrete supports. The menace of HGVs had begun even that early in the twentieth century.

15.6 High-duty cast irons

As described earlier for the carbon in a grey iron (Section 15.2), what we see under the microscope as a thread-like inclusion with sharp-pointed ends is actually a cross-section through a graphite flake in shape very like a 'cornflake' with a very sharp-edged rim. Graphite has no appreciable strength and so these flakes have the same effect on mechanical properties as would have cavities in the structure. The sharp edges act as stress-raisers within the structure; that is, they give rise to an increase in local stress in the same way that a sharp-cornered key-way tends to weaken a shaft. Hence, both the mechanical strength and the toughness of a cast iron can be improved by some treatment which disperses the graphite flakes, or, better still, which alters their shape to spherical globules.

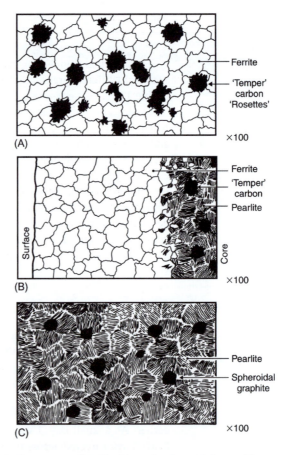

Figure 15.5 *Structures of malleable and spheroidal-graphite cast irons. The structure of a whiteheart malleable iron (B) is rarely uniform throughout, and often contains some carbon in the core. This has had insufficient time to diffuse outwards to the skin, and so to be lost. (A) Blackheart malleable iron – 'Rosettes' of temper carbon in a matrix of ferrite. (B) Whiteheart malleable iron – which, however, often has a black heart. (C) Spheroidal-graphite iron – nodules of graphite in a pearlite matrix.*

15.6.1 Spheroidal-graphite (SG) cast iron

Spheroidal-graphite (SG) cast iron, also known as *nodular iron,* or (in the USA) as *ductile iron,* contains its graphite in the form of rounded globules (Figure 15.5C). The sharp-edged, stress-raising flakes are thus eliminated and the structure is made more continuous. Graphite is made to deposit in globular form by adding small amounts of either of the metals cerium or magnesium to the molten iron, just before casting. Magnesium is the more widely used, and is generally added as a magnesium–nickel alloy, in amounts to give a residual magnesium-content of 0.1% in the iron. Such an iron may have a tensile strength as much as 775 MPa.

15.6.2 Compacted-graphite (CG) cast iron

Compacted-graphite (CG) cast irons are characterised by graphite structures – and consequentially mechanical properties – intermediate between those of ordinary grey irons and those of SG irons. Before being cast, the molten iron is first desulphurised and then treated with an alloy containing magnesium, cerium and titanium, so that traces of these elements remain in the casting. Titanium prevents the graphite from being completely spherical, as in SG iron. Instead, the flakes are short and stubby and have *rounded* edges. These irons are used in vehicle brake parts, gear pumps, eccentric gears and fluid and air cylinders.

15.7 Malleable cast irons

These are irons of such a composition as will give, in the ordinary cast form, a white (cementite) structure. However, they subsequently receive heat-treatment, the object of which is either to convert the cementite into small spherical particles of carbon (the 'black-heart' process), or, alternatively, to remove the carbon completely from the structure (the 'white-heart' process) (Figure 15.5). In either process, the silicon content of the iron is usually less than 1.0%, in order that the iron shall be 'white' in the cast condition. When the cementite has either been replaced by carbon or removed completely, a product which is both malleable and ductile is the result.

15.7.1 Blackheart malleable iron

In this process, the low-silicon white-iron castings are heated at about 900°C in a continuous-type furnace through which an oxygen-free atmosphere circulates. A moving hearth carries the castings slowly through the heating-zone, so that the total heating time is about 48 hours. This prolonged annealing causes the cementite to break down; but, instead of coarse graphite flakes, the carbon deposits as small 'rosettes' of 'temper carbon'. A fractured surface appears dark, because of the presence of this carbon, hence the term *blackheart.* Since the structure now consists entirely of temper carbon and ferrite, it is soft and ductile. Blackheart malleable castings are widely used in the automobile industries, because of their combination of castability, shock-resistance and good machinability. Typical components include brake-shoes, pedals, levers, wheel-hubs, axle-housings and door hinges. Table 15.3 shows the properties of some blackheart malleable irons.

Table 15.3 *Properties of malleable cast irons*

BS grade	Minimum tensile strength (MPa)	Mininimum 0.2% proof stress (MPa)	Minimum percentage elongation
Blackheart			
B32-10	320	190	10
B35-12	350	200	12
Whiteheart			
W38-12	400	210	8
W40-05	420	230	4
W45-07	480	280	4
Pearlitic			
P45-06	450	270	6
P50-06	500	300	5
P55-04	550	340	4
P60-03	600	390	3

15.7.2 Whiteheart malleable iron

Castings for this are also produced from a low-silicon white iron; but, in this process, the castings are heated at about 1000°C for up to 100 hours whilst in contact with some oxidising material such as haematite ore.

During heating, carbon at the surface of the castings is oxidised by contact with the haematite ore, and is lost as carbon dioxide gas. Carbon then diffuses outwards from the core – rather like a carburising process in reverse – and is in turn lost by oxidation. After this treatment is complete, *thin* sections will consist entirely of ferrite, and, on fracture, will give a steely-white appearance, hence the name 'whiteheart'. Thick sections may contain some particles of temper carbon at the core, because heating has not been sufficiently prolonged to allow all of the carbon to diffuse outwards from these thick sections (Figure 15.5B).

The whiteheart process is particularly suitable for the manufacture of thin sections which require high ductility. Pipe-fittings, parts for agricultural machinery, switchgear equipment and fittings for bicycle and motorcycle frames are typical of whiteheart malleable castings. Table 15.3 shows the properties of some whiteheart malleable irons.

15.7.3 Pearlitic malleable iron

This is similar in its initial composition to the blackheart material. The castings are malleabilised, either fully or partially, at about 950°C to cause the breakdown of the bulk of the cementite, and are then reheated to about 900°C. This reheating process causes some of the carbon to dissolve in austenite which forms above the lower critical temperature, and, on cooling, a background of pearlite is produced, the mechanism of the process being something like that of normalising in steels. Table 15.3 shows the properties of some pearlitic malleable irons.

15.8 Alloy cast irons

Generally speaking, the effects which alloying elements have on the properties of cast iron are similar to the effects which the same elements have on steel. Alloying elements used are:

1   *Nickel*
    Nickel, like silicon, has a graphitising effect on cementite, and so tends to produce a grey iron. At the same time, nickel has a grain-refining effect, which helps to prevent the formation of coarse grain in those heavy sections which cool slowly. It also toughens thin sections, which might otherwise be liable to crack.

2   *Chromium*
    Chromium is a carbide stabiliser and forms chromium carbide, which is harder than ordinary cementite. It is therefore used in wear-resistant irons. Since chromium forms very stable carbides, irons which contain chromium are less susceptible to 'growth'.

3   *Molybdenum*
    Molybdenum increases the hardness of thick sections, and also improves toughness.

4   *Vanadium*
    Vanadium increases both strength and hardness; but, more important still, promotes heat-resistance in cast irons by stabilising carbides so that they do not decompose on heating.

5   *Copper*
    Copper dissolves in iron in only very small amounts and has little effect on mechanical properties. It is added mainly to improve resistance to rusting.

Alloying elements can therefore be used to improve the mechanical properties of an iron, by:

- Refining the grain size.
- Stabilising hard carbides.
- In some cases, producing cast irons with a martensitic or austenitic structure.

Some typical alloy cast irons are shown in Table 15.4.

Table 15.4 *Properties and uses of some alloy cast irons*

Name or type of iron	Composition (%)				Typical mechanical properties		Uses
	C	Si	Mn	Others	Tensile strength (MPa)	Brinell	
Chromidium	3.2	2.1	0.8	Cr–0.3	270	230	Cylinder-blocks, brake-drums, clutch-casings, etc.
Ni-tensyl	2.8	1.4		Ni–1.5	350	230	An 'inoculated' cast iron.
Wear-resistant iron	3.6	2.8	0.6	V–0.2			Piston-rings for aero, automobile and diesel engines. Possesses wear-resistance and long life.
Ni-Cr-Mo iron	3.1	2.1	0.8	Ni–0.5 Cr–0.9 Mo–0.9	360	300	Hard, strong and tough – used for automobile crankshafts.
Heat-resistant iron	3.4	2.0	0.6	Ni–0.35 Cr–0.65 Cu–1.25	270	220	Good resistance to wear and to heat cracks – used for brake-drums and clutch-plates.
Ni-hard	3.3	1.1	0.5	Ni–4.5 Cr–1.5		600	A 'martensitic' iron, due to the presence of nickel and chromium – used to resist severe abrasion – chute plates in coke plant.
Ni-resist	2.9	2.1	1.0	Ni–15.0 Cr–2.0 Cu–6.0	210	130	Plant handling salt water – an austenitic iron.
Nicosilal	2.0	5.0	1.0	Ni–18.0 Cr–5.0	255	330	An austenitic corrosion- and heat-resistant iron.
Silal	2.5	5.0			165		'Growth' resistant at high temperatures.

15.9 Which iron?

The choice between ordinary grey irons, spheroidal-graphite iron and malleable iron is, as with most materials problems, governed by both economic and technical considerations. Grey iron is, of course, cheapest, and also the easiest in which to produce sound casings. For 'high-duty' purposes, the choice between SG iron and malleable iron may be less easy, and a number of factors may have to be considered in arriving at a decision, but in recent years malleable irons have lost ground to other forms of high-duty iron.

16 Copper and its alloys

16.1 Introduction

In my early history lessons we were told that the Stone Age was followed some 3500 years ago by the Bronze Age, but it seems very likely that, even earlier, metallic copper was used in Egypt some 1500 years before bronze – a copper–tin alloy. Since copper is more easily corroded than bronze, only limited evidence of this early use of copper remains. Nevertheless, we can regard copper as the most ancient metal of any engineering significance. Brass – a copper–zinc alloy – was, in classical times, made by smelting together ores of copper and zinc. The extraction of metallic zinc was not possible at that time, in fact not until the sixteenth century. The Romans made their brass by smelting copper along with the zinc ore calamine (zinc carbonate). The Romans used brass for coinage and ornaments, and bronze for fittings on important buildings, e.g. bronze pillars, tiles and doors. In ancient times, the world's output of copper – mainly for bronze manufacture – was ultimately outstripped by that of iron and, during the twentieth century, also by aluminium.

In former days, the bulk of the world's requirements of copper was smelted in Swansea, from ore mined in Cornwall, Wales or Spain. Later, deposits of ore were discovered in the Americas and Australia, and shipped to Swansea to be smelted, but it was subsequently realised that it would be far more economical to smelt the ore at the mine. Thus, Britain ceased to be the centre of the copper industry. Today, Chile, along with the USA, China and Australia, are the leading producers of copper.

16.2 The extraction of copper

Copper is extracted almost entirely from ores based on copper pyrites (a mineral in which copper is chemically combined with iron and sulphur). The metallurgy of the process is rather complex, but is essentially as follows:

1 The ore is 'concentrated'; i.e. it is treated by 'wet' processes to remove as much as possible of the earthy waste, or 'gangue'.
2 The concentrate is then heated in a current of air, to burn away much of the sulphur. At the same time, other impurities, such as iron and silicon, oxidise to form a slag which floats on top of the purified molten copper sulphide (called *matte*).
3 The molten matte is separated from the slag, and treated in a Pierce–Smith converter, the operation of which resembles to some extent that of the furnace used in steel-making by the 'oxygen process' (see Section 11.2). Some of the copper sulphide is oxidised, and the copper oxide

thus formed reacts chemically with the remainder of the sulphide, producing crude copper.

The crude copper is then refined by either of the following:

* Remelting it in a furnace, so that the impurities are oxidised, and are lost as a slag.
* Electrolysis, in which an ingot of impure copper is used as the anode, whilst a thin sheet of pure copper serves as the cathode. During electrolysis, the anode gradually dissolves, and high-purity copper is deposited on the cathode. 'Cathode copper' so formed is 99.97% pure.

16.3 Properties of copper

The most important physical property of copper is its *very high electrical conductivity*. In this respect, it is second only to silver; if we take the electrical conductivity of silver to be 100 units, then that of pure copper reaches 97, followed by gold at 71 and aluminium at 58. Consequently, the greater part of the world's production of metallic copper is used in the electrical/electronic industries.

Much of the copper used for electrical purposes is of very high purity, as the presence of impurities reduces the electrical conductivity, often very seriously. Thus, the introduction of only 0.04% phosphorus will reduce the electrical conductivity by almost 25%. Other elements have less effect, e.g. 1.0% cadmium, added to copper used for telephone-wires in order to strengthen them, has little effect on the conductivity.

The thermal conductivity and corrosion-resistance of copper are also high, making it a useful material for the manufacture of radiators, boilers and other heating equipment. Since copper is also very malleable and ductile, it can be rolled, drawn, deep-drawn, and forged with ease.

In recent years, the cost of copper production has risen steeply; so for many purposes – electrical and otherwise – it has been replaced by aluminium, even though the electrical and thermal conductivities of the latter (see Section 17.3) are inferior to those of copper.

16.4 Coppers and alloys

Commercial grades of 'pure' copper are available in a number of forms. Table 16.1 shows properties of such coppers.

1 *Oxygen-free high-conductivity (OFHC) copper*
   This is derived from the electrolytically refined variety. The cathodes are melted, cast, rolled and then drawn to wire or strip for electrical purposes. This grade is usually 99.97% pure and is of the highest electrical conductivity. It is widely used for electronic components.

2 *'Tough-pitch' copper*
   This is a fire-refined variety which contains small amounts of copper oxide as the main impurity. Since this oxide is present in the microstructure as tiny globules which have little effect on the mechanical properties but reduce the electrical conductivity by about 10%, it is suitable for purposes where maximum electrical conductivity is not

required. It is unsuitable where gas welding processes are involved because reactions between the oxide globules and hydrogen in the welding gas cause extreme brittleness:

$$Cu_2O + H_2 = 2Cu + H_2O \text{ (steam)}$$

Steam is insoluble in solid copper and forms fissures at the crystal boundaries.

3   *Deoxidised copper*
This is made from tough-pitch grade copper by treating it with a small amount of phosphorus just before casting, in order to remove the oxide globules. Whilst phosphorus-deoxidised copper may be valuable for processes where welding is involved, it is definitely not suitable for electrical purposes because of the big reduction in electrical conductivity (up to 30% reduction) introduced by the presence of dissolved phosphorus.

An example of the use of copper is in the production of printed-circuitboards. A board, made of a polymer or polymer-composite material, is used as the substrate. The surface is chemically treated so that it has enough conductivity to allow electroplating in which copper is plated over the entire surface to a depth of about 100 μm. A layer of photoresist film is then rolled, in a vacuum to avoid air bubbles, onto the copper. A mask with the required pattern of conductors is then placed over the resist and ultraviolet light shone onto it. Where the mask transmits the UV, the photoresist polymer becomes insoluble in the 'developer', whilst the unexposed areas can be dissolved. The dissolved areas thus reveal the copper. Spraying with nitric acid then

Table 16.1 *Properties of coppers*

Reference	Copper	Condition	Tensile strength (MPa)	Percentage elongation	Hardness (HB)	Electrical conductivity IACS (%)
C101	Electrolytic tough-pitch hc	Annealed	220	50	45	101.5–100
		Hard	400	4	115	
C103	Oxygen-free hc	Annealed	220	60	45	101.5–100
		Hard	400	6	115	
C104	Fire refined tough-pitch hc	Annealed	220	50	45	95–89
		Hard	400	4	115	
C106	Phosphorus deoxidized	Annealed	220	60	45	95–89
		Hard	400	4	115	
C110	Oxygen-free hc	Annealed	220	60	45	101.5–100
		Hard	400	6	115	

Note that the electrical conductivities have been expressed on the IAC scale, the value of 100% on this scale corresponding to the electrical conductivity of annealed copper at 20°C with conductivity 5.800 × 107 S/m.

dissolves the exposed copper and so leaves bare board, i.e. a pattern of insulators, with the pattern of conductors being the unexposed, unattacked copper.

16.4.1 Alloys of copper

The tensile strength of hard-rolled copper reaches about 375 MPa, so for most engineering purposes where greater strength is required, copper must be suitably alloyed. Alloys of copper are less widely used than they were. The growing price of copper relative to its increasing scarcity, coupled with the fact that the quality of cheaper alternatives has improved in recent years, has led to the replacement of copper alloys for many purposes. Moreover, improved shaping techniques have allowed less ductile materials to be employed; thus, deep-drawing quality mild steel is now often used where ductile brass was once considered to be essential. Nevertheless, copper alloys are still of considerable importance and the following are discussed later in this chapter:

1 *Brasses*: copper–zinc alloys
2 *Bronzes*: copper–tin alloys
   (a) *Phosphor bronzes*: copper–tin–phosphorus alloys
   (b) *Gunmetals*: copper–tin–zinc alloys
3 *Aluminium bronzes*: copper–aluminium alloys
4 *Cupro-nickels*: copper–nickel alloys
   (a) *Nickel silvers*: copper–zinc–nickel alloys
5 *Beryllium bronzes*: copper–beryllium alloys

16.5 The brasses

These are copper-base alloys containing up to 45% zinc and, sometimes, small amounts of other metals, the chief of which are tin, lead, aluminium, manganese and iron. The equilibrium diagram (Figure 16.1) shows that plain copper–zinc alloys with up to approximately 37% zinc have a structure consisting of a single phase – that labelled α. Phases occupying such a position on an equilibrium diagram are solid solutions. Such solutions are invariably tough and ductile and this particular one is no exception, being the basis of one of the most malleable and ductile metallurgical materials in common use. Brasses containing between 10% and 35% zinc are widely used for deep-drawing and general presswork – the maximum ductility being attained in the case of 70–30 brass, commonly known as *cartridge metal,* since it is used in the deep-drawing of cartridge- and shell-cases of all calibres. Many of these, however, are now produced in modern low-nitrogen, deep-drawing quality mild steel (see Section 11.5). Figure 16.2A and B shows the microstructures of 70–30 brass as cast and cold-worked.

Brasses with more than 37% zinc contain the phase β'. This is a hard, somewhat brittle substance, so a 60–40 brass lacks ductility. If such a brass is heated to 454°C (Figure 16.1), the phase β' changes to β, which is soft and malleable. As the temperature is increased further, the α phase present dissolves in β, until at X the structure is entirely malleable β. Therefore,

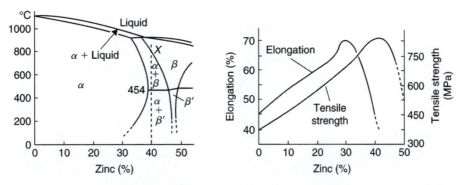

Figure 16.1 *The section of the copper–zinc equilibrium diagram which covers brasses of engineering importance.*

60–40 brasses are best hot-worked at about 700°C. This treatment also breaks up the coarse cast structure and replaces it with a fine granular structure. Figure 16.2C and D shows the microstructures of 60–40 brass, as cast and hot-worked. Thus, brasses with less than 37% zinc are usually cold-working alloys, whilst those with more than 37% zinc are hot-working alloys. A copper–zinc alloy containing more than 50% zinc would be useless for engineering purposes, since it would contain the very brittle phase γ.

Up to 1% tin is sometimes added to brasses, to improve their resistance to corrosion, particularly under marine conditions. Lead is insoluble in both molten and solid brass, and exists in the structure as tiny globules. About 2% lead will improve machinability (see Section 7.5) of brass. Small amounts of arsenic are said to improve corrosion-resistance and inhibit dezincification.

Manganese, iron and aluminium all increase the tensile strength of a brass, and are therefore used in high-tensile brasses. These alloys are sometimes, rather misleadingly, known as *manganese bronzes*. Additions of up to 2% aluminium are also used to improve the corrosion resistance of some brasses.

Table 16.2 shows the composition, typical properties and uses of the more important brasses.

The IACS electrical conductivity of brasses is typically: gilding metal 44%, cartridge brass 27%, Muntz metal 28% and Naval brass 26%.

16.5.1 'Shape memory' alloys

'Shape memory' is a phenomenon associated with a limited number of alloys. The important characteristic of these alloys is the ability to exist in two distinct shapes or crystal structures, one above and one below a critical transformation temperature. As the temperature falls below this critical temperature, a martensitic type of structure forms. As the temperature is raised again, the martensite reverts to the original structure. This *reversible change* in structure is linked to a change in dimensions and the alloy thus exhibits a 'memory' of the high and low temperature shapes.

The best-known commercial alloys are the SME brasses ('Shape memory effect' brasses) containing 55–80% copper, 2–8% aluminium and the balance zinc. By choosing a suitable composition, transition temperatures between

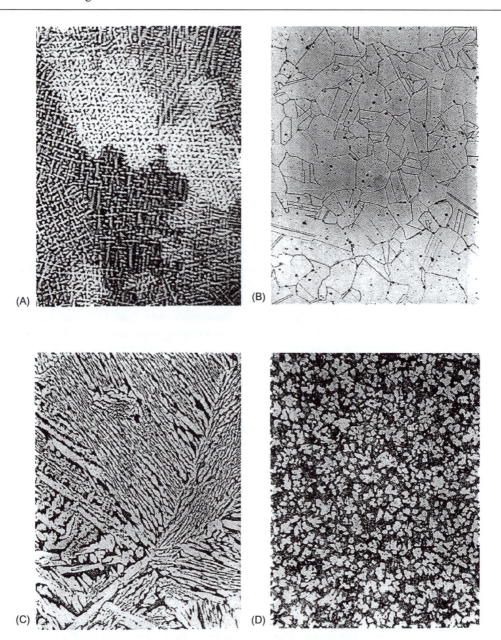

Figure 16.2 *Microstructures of some brasses. (A) 70–30 brass as cast with cored crystals of a solid solution (see also Figure 8.3). (B) 70–30 brass, cold-worked and then annealed at 600°C. The coring of the original cast structure has been removed by this treatment and recrystallisation has produced small crystals (twinned) of the solid solution α. (C) (D) 60–40 brass as cast. This shows a typical Widmanstätten structure and, on cooling, small α crystals (light) have precipitated from the β phase (dark).*

Table 16.2 Brasses

BS specification	Composition (%)			Condition	Typical mechanical properties		Uses
	Cu	Zn	Other elements		Tensile strength (MPa)	Percentage elongation	
CZ 101	90	10		Soft Hard	280 510	55 4	*Gilding metal* – used for imitation jewellery, because of its gold-like colour, good ductility and its ability to be brazed and enamelled.
CZ 106	70	30		Soft Hard	320 700	70 5	*Cartridge brass* – deep-drawing brass of maximum ductility. Used particularly for the manufacture of cartridge- and shell-cases.
CZ 107	65	35		Soft Hard	320 700	65 4	*Standard brass* – a good general-purpose cold-working alloy when the high ductility of 70–30 brass is not required. Widely used for press-work and limited deep-drawing.
CZ 108	63	37		Soft Hard	340 725	55 4	*Common brass* – a general-purpose alloy, suitable for limited cold-working operations.
CZ 123	60	40		Hot-rolled	370	40	*Yellow or Muntz metal* – hot-rolled plate. Can be cold-worked only to a limited extent. Also extruded as rods and tubes.
CZ 120 2Pb	59	39	Pb 2	Extruded-rod	450	30	*Free-cutting brass* – very suitable for high-speed machining, but can be deformed only slightly by cold-work.
CZ 112	62	37	Sn 1	Extruded	420	35	*'Naval brass'* – structural uses, also forgings. Tin raises corrosion-resistance, especially in sea water.
	58	Remaining	Mn 1.5, Al 1.5, Fe 1, Pb 1, Sn 0.6	Extruded	500	15	*High-tensile brass* ('manganese bronze') – pump-rods, stampings and pressings. Also marine castings such as propellers, water-turbine runners, rudders, etc.
	61.5	Remaining	Pb 2.5, As 0.1	Hot-stamping	380	25	*Dezincification-resistant-brass* – water fittings where water supply dezincifies plain α/β' brasses. After stamping, the brass is usually annealed at 525°C and water-quenched to improve the dezincification resistance.

−70°C and 130°C can be achieved. The force associated with the change in shape can be used to operate temperature-sensitive devices, the snap on/off positions often being obtained by using a compensating bias spring. Such devices are used in automatic greenhouse ventilators, thermostatic radiator valves, de-icing switches, electric kettle switches and valves in solar heating systems.

16.6 Tin bronzes

Tin bronze was almost certainly the first metallurgical *alloy* to be used by Man and it is a sobering thought that for roughly half of the 4000 years of history during which he has been using metallurgical alloys, bronze was his sole material. Though tin bronzes have relatively limited uses these days – due in part to the high prices of both copper and tin – they still find application for special purposes. Tin bronzes, contain up to 18% tin, sometimes with smaller amounts of phosphorus, zinc or lead.

The complete copper–tin equilibrium diagram is a rather complex one and its interpretation is rendered more difficult by the fact that tin bronzes need to be cooled extremely slowly indeed – *far more slowly than is ever likely to prevail in a normal industrial casting process* – if microstructural equilibrium is to be attained. The reader may be forgiven for assuming, from a simple interpretation of the diagram (Figure 16.3), that a cast 10% tin bronze which begins to solidify at X (about 1000°C) will be completely solid at X_1 (about 850°C) as a polycrystalline mass of a solid solution, which then cools without further change until at Z particles of the intermetallic compound ε begin to form within the α crystals. Sadly, the transformations which take place are not nearly so simple.

First, the wide range between the liquidus line AL and the solidus AS is an indication that extensive coring takes place in tin bronzes as they solidify. Thus, the molten bronze containing 10% tin begins to solidify at X by forming dendrites of composition Y (about 2% tin). Since these dendrites contain only about 2% tin this means that the remaining liquid has been left much richer in tin, i.e. its composition moves to the right along AL. Cooling is too rapid for the microstructure to attain equilibrium and solidification is not complete until the temperature has fallen to 798°C when the remaining liquid will probably contain about 20% tin. This will then solidify as a mixture of α and β, as indicated by the diagram.

As the now solid alloy cools to 586°C, the $\alpha + \beta$ mixture transforms to $\alpha + \gamma$ which at 520°C transforms further to $\alpha + \delta$. As the temperature falls lower, transformations become increasingly sluggish and so the change to a structure of $\alpha + \varepsilon$ at 350°C does not occur at the speed at which the temperature is falling. So under industrial conditions of cooling, we are left with a final microstructure of heavily cored α solid solution with a network and particles of the brittle intermetallic compound δ ($Cu_{31}Sn_8$) at the α dendrite boundaries.

Sorry if all this sounds a bit of a fudge! It isn't really – it merely illustrates that in interpreting these equilibrium diagrams, a number of complicating influences have often to be considered.

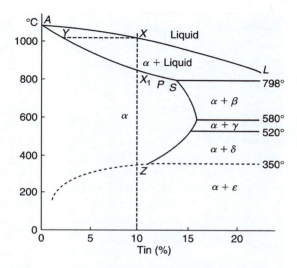

Figure 16.3 *Part of the copper–tin equilibrium diagram. The phase δ is an intermetallic compound* $Cu_{31}Sn_8$ *whilst ε is the intermetallic compound* Cu_3Sn, *but this latter will never be present in a bronze casting which solidifies and cools under normal industrial conditions.*

The networks of the δ phase at the crystal boundaries of the solid solution α makes the 10% tin bronze rather brittle under shock. Commercial tin bronzes can be divided into two groups:

1 Wrought tin bronzes containing up to approximately 7% tin. These alloys are generally supplied as rolled sheet and strip, or as drawn rod, wire and turbine-blading.
2 Cast tin bronzes containing 10–18% tin, used mainly for high-duty bearings.

Figure 16.4 shows the microstructures of 95–5 and 90–10 bronzes under various conditions.

The IACS electrical conductivity of tin bronzes is phosphor bronzes PB101 15–25% and PB102 13–18%, gun metals G1 10–11%, LG2 10–15%. See Table 16.3 for data on the properties of tin bronzes.

During the Second World War, the company where I was employed as chief metallurgist accepted an order for slabs of hard-rolled 12% bronze.

Such material had never been produced commercially at that time, but since – as it transpired later – the material was required for secret development of the jet engine, it was presumably hoped that I could 'wave my magic wand'.

After some weeks of laboratory trials, I discovered that, as the copper–tin equilibrium diagram – though then rather incomplete – predicted, annealing the cast ingots of 12% tin bronze at 740°C for 12 hours was effective in absorbing all traces of the brittle δ phase and also eliminating almost

Figure 16.4 *Microstructures of some tin bronzes. (A) 95–5 tin bronze, as cast, showing cored dendrites of α solid solution. The black areas are voids, i.e. intercrystalline cavities probably caused by shrinkage. (B) 90–10 tin bronze, as cast, and showing a cored α solution (dark) surrounded by δ which has coagulated so that the original α + δ eutectoid structure has been lost. The large black areas are again shrinkage cavities. (C) 95–5 tin bronze, cold-worked and annealed, showing 'twinned' crystals of α solid solution. Coring has been dispersed by the mechanical work and recrystallisation during annealing. The structure is soft. The dark spots are probably voids resulting from the original shrinkage cavities. (D) 95–5 tin bronze, cold-worked, annealed and cold-worked again. Now the α solid solution crystals show extensive strain bands (indicating the positions of the slip planes in the crystals – locked-up strain energy there causes the region to etch more quickly). The structure is much harder because of the cold-work it has received.*

Table 16.3 *Tin bronzes and phosphor bronzes*

BS specification	Composition (%)			Condition	Typical mechanical properties		Uses
	Cu	Sn	Other elements		Tensile strength (MPa)	Percentage elongation (%)	
	95.5	3	Zn 1.5	Soft Hard	320 725	65 5	Coinage bronze – British 'copper' coinage now contains rather less tin (0.5%) and more zinc (2.5%).
PB 101	96	3.75	P 0.1	Soft Hard	340 740	65 15	Low-tin bronze – springs and instrument parts. Good elastic properties and corrosion-resistance.
PB 102	94	5.5	P 0.2	Soft Hard	350 700	65 15	Drawn phosphor bronze – generally used in the work-hardened condition; steam-turbine blading. Other components subjected to friction or corrosive conditions.
PB1–B	89	10	P 0.5	Sand cast	280	15	Cast phosphor bronze – supplied as cast sticks for turning small bearings, etc.
	81	18	P 0.5	Sand cast	170	2	High-tin bronze – bearings subjected to heavy loads – bridge and turntable bearings.
G1–C	88	10	Zn 2, Ni 2 (max.)	Sand cast	290	16	Admiralty gunmetal – pumps, valves and miscellaneous castings (mainly for marine work, because of its high corrosion-resistance); also for statuary, because of good casting properties.
LG2–A	85	5	Zn 5, Pb 5, Ni 2 (max.)	Sand cast	220	13	Leaded gunmetal (or 'red brass') – a substitute for admiralty gunmetal; also where pressure tightness is required.
LB5–B	75	5	Pb 20, Ni 2 (max.)	Sand cast	160	6	Leaded bronze – a bearing alloy; can be bonded to steel shafts for added strength.

completely all traces of coring. At 740°C of course, the structure of these $\alpha + \delta$ areas will have changed to $\alpha + \beta$ and since 740°C is only just below the melting range, the movement of copper and tin atoms within the lattice structure will have accelerated – just as salt dissolves more quickly in hot water than in cold. Therefore, uniform α (as indicated by point P in Figure 16.3) was produced and equilibrium was finally reached.

My efforts bore fruit only temporarily. Not only was 12% tin bronze expensive due to the high costs of tin and copper, but production costs were high because of the long annealing process in a protective atmosphere and the fact that the alloy work-hardened very quickly during rolling and so required frequent inter-stage annealings. Very soon, a cheaper material was sought. Often an R-and-D project serves only to prove that a suggested programme is not feasible!

16.6.1 Phosphor bronze

Most of the tin bronzes mentioned earlier contain up to 0.05% phosphorus, left over from the deoxidation process which is carried out before casting. Sometimes, however, phosphorus is added – in amounts up to 1.0% – as a deliberate alloying element, and only then should the material be termed *phosphor bronze*. The effect of phosphorus is to increase the tensile strength and corrosion-resistance, whilst, in the case of cast bearing alloys, reducing the coefficient of friction. See Table 16.3 for data on phosphor bronzes.

16.6.2 Gunmetal

This contains 10% tin and 2% zinc, the latter acting as a deoxidiser, and also improving fluidity during casting. Since zinc is considerably cheaper than tin, the total cost of the alloy is reduced. Gunmetal is no longer used for naval armaments, but it is used as a bearing alloy, and also where a strong, corrosion-resistant casting is required. See Table 16.3 for data on gunmetals.

16.6.3 Coinage bronze

This is a wrought alloy containing 3% tin and, in the interests of economy, 1.5% zinc. In Britain, even more of the tin has been replaced by zinc (see Table 16.3).

16.6.4 Leaded bronzes

Up to 2.0% lead is sometimes added to bronzes, in order to improve machinability. Some special bearing bronzes contain up to 24% lead, and will carry greater loads than will 'white metal' bearings. Since the thermal conductivity of these bronzes is also high, they can work at higher speeds, as heat is dissipated more quickly. See Table 16.3 for data on a leaded bronze.

16.7 Aluminium bronzes

Like brasses, the aluminium bronzes can be divided into two groups: the cold- and the hot-working alloys.

Cold-working alloys contain approximately 5% aluminium and are ductile and malleable, since they are a completely solid solution in structure. They have a good capacity for cold-work. As they also have a good resistance to corrosion, and a colour similar to that of 22 carat gold, they were widely used for cheap jewellery and imitation wedding rings in those less permissive days when such things were deemed necessary; but for decorative purposes they have been replaced by coloured anodised aluminium and other cheaper materials.

The hot-working alloys contain in the region of 10% aluminium and, if allowed to cool slowly, the structure is brittle, due to the precipitation of a hard compound within it. When this structure is heated to approximately 800°C, it changes to one which is a completely solid solution, and hence malleable; so alloys of this composition can be hot-worked successfully. Similar alloys are also used for casting to shape by both sand- and die-casting methods. To prevent precipitation of the brittle compound mentioned earlier, castings are usually ejected from the mould as quickly as possible, so that they cool rapidly.

An interesting feature of the 10% alloy is that it can be heat-treated in a manner similar to that for steel. A hard martensitic type of structure is produced on quenching from 900°C, and its properties can be modified by tempering. Despite these apparently attractive possibilities, heat-treatment of aluminium bronze is not widely employed, and such of these alloys as are used find application mainly because of their good corrosion-resistance, retention of strength at high temperatures and good wearing properties.

Aluminium bronze is a difficult alloy to cast successfully, because, at its casting temperature (above 1000°C) aluminium oxidises readily. This leads to aluminium oxide dross becoming entrapped in the casting, unless special casting techniques are employed, and an increase in the cost of the process inevitably results.

Compositions and uses of some aluminium bronzes are given in Table 16.4. The IACS electrical conductivity is for a 5% aluminium bronze (CA101) 15–18%, 9% aluminium bronze (CA103) 7%, 10% aluminium bronze (CA104) 8%, aluminium bronze (AB1) 8–12%.

16.8 Copper–nickel alloys

The metals copper and nickel 'mix' in all proportions in the solid state; that is, a copper–nickel alloy of any composition consists of only one phase – a uniform solid solution. For this reason, *all* copper–nickel alloys are relatively ductile and malleable, since there can never be any brittle phase present in the structure. In the cast state, a copper–nickel alloy may be cored (see Section 9.3.1), but this coring can never lead to the precipitation of a brittle phase. In other words, the metallurgy of these alloys is very simple – and not particularly interesting as a result.

Cupro-nickels may be either hot-worked or cold-worked, and are shaped by rolling, forging, pressing, drawing and spinning. Their corrosion-resistance is high, and only the high cost of both metals limits the wider use

Table 16.4 Aluminium bronzes

BS specification	Composition (%)			Condition	Typical mechanical properties		Uses
	Cu	Al	Other elements		Tensile strength (MPa)	Percentage elongation (%)	
	Remaining	7.5	Fe, Mn and Ni up to 2.5 total	Hot-worked	430	45	Chemical engineering, particularly at fairly high temperatures.
CA 104	80	10	Fe 5, Ni 5	Forged	725	20	Forged propeller-shafts, spindles, etc. for marine work. Can be heat-treated by quenching and tempering.
AB1–B	Remaining	9.5	Fe 2.5, Ni and Mn up to 1.0 each (optional)	Cast	520	30	The most widely used aluminium bronze for both die- and sand-casting. Used in chemical plant and marine conditions – pump-castings, valve-parts, gears, propellers, etc.

Table 16.5 *Cupro–nickels and nickel–silvers*

BS specification	Composition (%)			Condition	Typical mechanical properties		Uses
	Cu	*Ni*	*Other elements*		*Tensile strength (MPa)*	*Percentage elongation (%)*	
CN 104	80	20	Mn 0.25	Soft Hard	340 540	45 5	Used for bullet-envelopes, because of high ductility and corrosion-resistance.
CN 105	75	25	Mn 0.25	Soft Hard	350 600	45 5	Mainly for coinage – the current British 'silver' coinage.
NA 13	29	68	Fe 1.25, Mn 1.25	Soft Hard	560 720	45 20	*Monel metal* – good mechanical properties, excellent corrosion-resistance. Chemical engineering plant, etc.
NA 18	29	66	Al 2.75, Fe 1.0, Mn 0.4, Ti 0.6	Soft Hard Heat-treated	680 760 1060	40 25 22	*'K' Monel* – a heat-treatable alloy. Used for motor-boat propeller-shafts.
NS 106	60	18	Zn balance, Mn 0.4				*Nickel–silver* – spoons, forks, etc.
NS 111	60	10	Zn balance, Pb 1.5, Mn 0.25				*Leaded nickel–silver* – Yale-type keys, etc.

of these alloys. Nickel–copper alloys containing about two-thirds nickel and one-third copper are called *Monels*.

Table 16.5 gives the composition and uses of some cupro-nickels. The IACS electrical conductivities for cupro-nickels are CN104 6% and CN105 5%.

16.8.1 Nickel–silvers

Nickel–silvers contain from 10% to 30% nickel and 55–65% copper, the balance being zinc. Like the cupro-nickels, they are uniform solid solutions. Consequently, they are ductile, like the high-copper brasses, but have a 'near-white' colour, making them very suitable for the manufacture of forks, spoons and other tableware, though in recent years these alloys have been generally replaced by the less attractive stainless steels. When used for such purposes, these nickel–silvers are usually silver-plated – the stamp 'EPNS' means 'electroplated nickel–silver'.

The machinability of these – as of all other copper alloys – can be improved by the addition of 2% lead. Such alloys are easy to engrave, and are also useful for the manufacture of Yale-type keys, where the presence of lead makes it much easier to cut the blank to shape. Table 16.5 gives the composition and uses of some nickel–silvers. The IACS electrical conductivities are NS106 6.3% and NS111 7.2%.

16.9 Other copper alloys

The following are some other copper-base alloys, these containing just small amounts of alloying elements which have generally been added to increase tensile strengths.

16.9.1 Beryllium bronze

Beryllium bronze (or copper–beryllium) contains approximately 1.75% beryllium and 0.2% cobalt. It is a heat-treatable alloy which can be precipitation-hardened in a manner similar to that of some of the aluminium alloys. A glance at the equilibrium diagram (Figure 16.5) will show why this is so. At room temperature, the slowly cooled structure will consist of the solid solution α (in this case, almost pure copper), along with particles of the compound CuBe. If this is heated – that is, solution-treated – at about 800°C, the structure becomes completely α, as the CuBe is slowly dissolved by the solid solution. The alloy is then quenched and in this condition is ductile, so that it can be cold-worked. If the alloy is now precipitation-hardened at 275°C for an hour, a tensile strength of up to 1400 MPa is obtained. The addition of 0.2% cobalt restricts grain growth during the heat-treatment at 800°C. Table 16.6 shows typical properties.

Since beryllium bronze is very hard in the cold-worked/heat-treated state, it is useful for the manufacture of non-sparking tools – chisels and hack-saw blades – for use in gas-works, 'dangerous' mines, explosives factories or where inflammable vapours are encountered, as in paint and varnish works. Unfortunately, beryllium is a very scarce and expensive metal, and this limits its use in engineering materials.

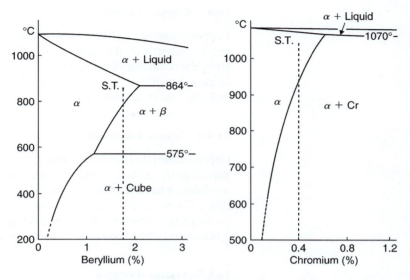

Figure 16.5 *Copper-base alloys which can be precipitation-hardened. In either case, the alloy is solution-treated by heating it to the point ST, at which the structure becomes uniformly α. It is then quenched, to retain this structure, and then precipitation-hardened.*

16.9.2 Copper–chromium

Copper–chromium contains 0.5% chromium and is also a heat-treatable alloy, as is indicated by the equilibrium diagram in Figure 16.5. It is used in some electrical industries and in spot welding electrodes, since it combines a high conductivity of about 80% that of pure copper with a reasonable strength of 550 MPa.

Table 16.6 *Properties of a beryllium bronze*

Reference	Composition (%)	Condition	Tensile strength (MPa)	0.2% proof stress (MPa)	Percentage elongation	Electr. conductivity IACS (%)
CB101	98 Cu, 1.7 Be, 0.2–0.6 Cu + Ni	Solution treated	480–500	185–190	45–50	16–78
		Solution treated + cold worked	730–750	580–620	5	
		Solution treated + precipitation treated	1150–1160	930–940	5	22–32
		Cold worked + precipitation treated	1300–1340	1100–1340	2	

16.9.3 Copper–cadmium

Copper–cadmium (C108) contains about 1% cadmium and this raises the tensile strength of the hard drawn alloy to 700 MPa. Since the fall in electrical conductivity due to the cadmium is very small (to nearly 90% that of pure copper, IACS 75–87%), this material is useful for telephone-wires and other overhead electrical wires.

16.9.4 Copper–tellurium

Copper–tellurium (C109) contains 0.5% tellurium which, being insoluble in pure copper, exists as small globules in the structure. Since tellurium is insoluble, it has little effect on electrical conductivity (only drops to 99.5% that of pure copper, IACS 92–98%) but gives a big improvement in machinability (resembling lead in this respect). It is used for machined high-conductivity items such as terminals and clamps.

16.9.5 Arsenical copper

Arsenical copper contains 0.4% arsenic. This increases from 200°C to 550°C the temperature at which cold-worked copper begins to soften when it is heated. This type of copper was widely used in steam-locomotive fireboxes and boiler-tubes, and still finds use in high-temperature steam plant. It is useless for electrical purposes, however, because of the great reduction in electrical conductivity caused by the presence of the arsenic in the solid solution.

17 Aluminium and its alloys

17.1 Introduction

Minerals containing aluminium are difficult to decompose so that aluminium can be extracted. This is because the metal has strongly electropositive ions and thus has a strong affinity for all of the non-metals with their electronegative ions. For this reason, samples of the metal were not produced until 1825 when the Danish scientist H. C. Oersted used metallic potassium to chemically reduce aluminium from one of its compounds. Consequently, in those days aluminium was very expensive and cost about £250/kg to produce, and was far more expensive than gold. It is reported that the more illustrious foreign visitors to the court of Napoleon III were privileged to use forks and spoons made from aluminium, whilst the French nobility had to be content with tableware in gold and silver. One still meets such cases of 'one-upmanship' even in the metallurgical world – quite recently I was told of a presentation beer tankard produced in the metal zirconium.

This chapter is about the properties of aluminium and its alloys, now no longer the 'luxury' metal.

17.2 Extraction of aluminium

The modern electrolytic process for extracting aluminium was introduced simultaneously and independently in 1886 by Hall (in the USA) and Héroult (in France). Nevertheless, the metal remained little more than an expensive curiosity until the beginning of the twentieth century. Since then, the demands by both air- and land-transport vehicles for a light, strong material have led to the development of aluminium technology and an increase in the production of the metal, until now it is second only to iron in terms of annual world production.

The only important ore of aluminium is bauxite, which contains aluminium oxide (Al_2O_3). Unfortunately, this cannot be reduced to the metal by heating it with coke (as in the case of iron ore), because aluminium atoms are, so to speak, too firmly combined with oxygen atoms to be detached by carbon. For this reason, an expensive electrolytic process must be used to decompose the bauxite and release aluminium. Since each kilogram of aluminium requires about 91 mJ of electrical energy, smelting plant must be located near to sources of cheap hydroelectric power, often at great distances from the ore supply, and from the subsequent markets. Consequently, most aluminium is produced in the USA, and in Canada, and Norway. Hydroelectric power in the Western Highlands enables some aluminium to be smelted in Scotland, but the bulk of aluminium used in the UK is imported.

Crude pig iron can be purified (turned into steel) by blowing oxygen over it, to burn out the impurities (see Section 11.2), but this would not be possible in the case of aluminium, since the metal would burn away first, and leave us with the impurities. Instead, the crude bauxite ore is first purified by means of a chemical process, and the pure aluminium oxide is then decomposed by electrolysis. Since aluminium oxide has a very high melting-point, it is mixed with another aluminium mineral, cryolite, to form an electrolyte which will melt at a lower temperature.

The furnace used consists of a 'tank' some 2.5 m long to contain the molten electrolyte. It is lined with carbon and this constitutes the cathode of the electrolytic cell. Carbon rods dipping into the electrolyte form the anode. In the molten electrolyte, the aluminium oxide (Al_2O_3) dissociates into ions:

$$Al_2O_3 \rightarrow 2Al^{+++} + 3O^{--}$$

The Al^{++} ions are attracted to the cathode, where they receive electrons to become atoms. This aluminium collects at the base of the 'tank' and is tapped off at intervals. The O^{--} ions are attracted to the anode where they discharge electrons to become atoms which immediately combine with the carbon electrodes. This causes the latter to burn away so that they need frequent replacement.

17.3 Properties of aluminium

Although aluminium has a high affinity for oxygen, and might therefore be expected to oxidise – or corrode – very easily, in practice it has an *excellent resistance to corrosion*. This is due largely to the thin, but very dense, film of oxide which forms on the surface of the metal and, since it is very tenacious, effectively protects it from further atmospheric attack. The reader will be familiar with the comparatively dull appearance of the surface of polished aluminium. This is due to the oxide film which immediately forms on fresh metal. The protective oxide skin can be artificially thickened by a process known as *anodising* in which the article is made the anode in an electrolyte of dilute sulphuric acid (the function of the acid is merely to produce ionisation of the water so that it is conductive). Oxygen atoms released at the anode combine with aluminium atoms at the surface to thicken the oxide film already present. Since aluminium oxide is extremely hard, anodising makes the surface more wear-resistant. The anodised film is sufficiently porous to allow it to be dyed with either organic or inorganic dyes.

The fact that aluminium has a *high thermal conductivity* and good corrosion-resistance, and that any corrosion products which are formed are non-poisonous, makes aluminium very suitable for the manufacture of domestic cooking-utensils such as kettles, saucepans and frying pans. In the form of disposable collapsible tubes, it was used to contain a wide range of foodstuffs and toilet preparations ranging from caviare to toothpaste, but for many such purposes it has been replaced by plastics materials. The high malleability of aluminium makes it possible to produce very thin foil which is excellent for the packaging of food and sweets. It is also widely used as 'cooking foil'.

Aluminium, for a metal, has a relatively low density of 2.7 Mg/m³, compared with 7.9 Mg/m³ for iron and 8.9 Mg/m³ for copper. Thus, aluminium which has a *very good electrical conductivity,* though only about half that of copper, when considered weight for weight can make aluminium a better proposition as an electrical conductor in some circumstances than copper. Thus, if conductors of copper and aluminium of equal length and equal weight are taken, that of the aluminium is a better overall conductor, because it has a greater cross-sectional area than that of the copper one (Figure 17.1). So, whilst aluminium might be unsuitable as the windings of small components in the electronic industry, because of the greater volume of wire needed to produce a similar conductive capacity to that of copper, aluminium conductors are used in the grid system, generally strengthened by a steel core.

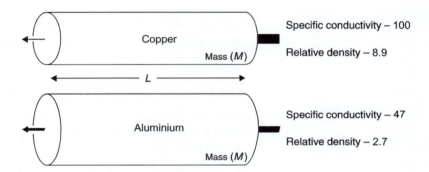

Figure 17.1 *Conductors of copper and aluminium of equal length L and equal mass M. Although copper has the better conductivity, the aluminium conductor passes a greater current under similar conditions, because of its greater cross-section and hence greater conductance (lower resistance).*

For use as a constructional material, pure aluminium lacks strength. In the 'soft' condition, its tensile strength is only 90 MPa, whilst even in the work-hardened state it is no more than 135 MPa. Hence, for most engineering purposes, aluminium is alloyed, in order to give a higher tensile strength and hence a better strength/weight ratio. Some of the high-strength alloys have a tensile strength in excess of 600 MPa when suitably heat-treated.

17.4 Aluminium alloys

Many aluminium alloys are used in the wrought form, i.e. they are rolled to sheet, strip or plate, drawn to wire or extruded as rods or tubes. Other alloys are cast to shape, by either a sand-casting or a die-casting process. In either case, some of the alloys may receive subsequent heat-treatment in order to further improve their mechanical properties by inducing the phenomenon originally known as *age-hardening,* but now more properly termed *precipitation-hardening.*

Thus, the engineering alloys can be conveniently classified into four groups:

1   Alloys which *are not* heat-treated
     (a) Wrought alloys
     (b) Cast alloys
2   Alloys which *are* heat-treated
     (a) Wrought alloys
     (b) Cast alloys

In the pre-war days of the 1930s, a bewildering assortment of aluminium alloys confronted the engineer. Worse still, they were covered by a rather untidy system of specification numbers, and each manufacturer used his own particular brand-name to describe an alloy. Since the Second World War, however, the number of useful alloys has been somewhat streamlined, and a systematic method of specification designation has been established internationally. In Britain, this is operated by the British Standards Institution. In this system, each alloy is identified by a four-digit number, whilst suffix letters are used to show what treatment (if any) it has received.

17.4.1 Designation system for aluminium alloys

Each alloy is identified by a four-digit number which itself carries some basic information about the alloy. The first of the four digits indicates the *major* alloying element(s):

1   Commercial aluminium (more than 99.0% Al)
2   Major alloying element copper
3   Major alloying element manganese
4   Major alloying element silicon
5   Major alloying element magnesium
6   Major alloying elements magnesium and silicon
7   Major alloying element zinc
8   Major alloying element – other elements
9   Unused series

The second digit in the alloy designation denotes the number of modifications which have been made to the alloy specification. If the second digit is zero, it indicates the original alloy – integers 1–9 indicate alloy modifications. The last two digits in the alloy designation for the 1XXX alloys show the aluminium content above 99.0% in hundredths and for the other alloys have no significance other than to identify different alloys in the group. National variations of alloys registered by another country are identified by a serial letter immediately *after* the numerical designations, e.g. A.

17.4.2 Designation system for wrought alloy tempers

The condition in which the alloy is supplied is indicated by suffix letters, similar codes being used in the USA but with F being used for as manufactured and a number, rather than a letter, attached to the T to indicate the heat-treatment process.

M   Material 'as manufactured', e.g. rolled, drawn, extruded, etc.
O   Annealed to its lowest strength.
H   Strain hardened – material subjected to cold work after annealing or to a combination of cold work and partial annealing. With the British system the degree of work hardening is indicated by a number after the H, i.e. H1, H2, H3, H4, H5, H6, H7, H8. H2 is quarter work hardened, H4 is half work hardened, H6 is three-quarters work hardened and H8 full work hardened. The USA use a similar system but with two digits. The first digit is 1 for strain hardened only, 2 for strain hardened and thermally treated, 3 for strain hardened and stabilised. The second digit indicates the degree of strain hardening 2 being quarter hardened, 4 half hardened and 8 fully hardened. For example, H18 is strain hardened and fully hard, H22 is strain hardened and partially annealed to give the quarter hardened condition, H24 is strain hardened and partially annealed to achieve the half-hard condition.
T   Tempers for heat-treated alloys.
   TB   Material solution-treated and aged naturally (USA T4).
   TD   Material solution-treated, cold-worked and aged naturally (USA T3).
   TE   Material cooled from an elevated-temperature shaping process and precipitation treated, i.e. artificially aged (USA T10).
   TF   Material solution treated and precipitation treated, i.e. artificially aged (USA T6).
   TH   Material solution treated, cold-worked and then precipitation treated, i.e. artificially aged (USA T8).

Wrought alloys are covered by British Standard speculations for plate, sheet and strip; drawn tube; forgings; rivet, bolt and screw stock; bars, extruded tube and sections; and wire. For example, BS 1471/6082-TF represents the original specification of an aluminium alloy in the tube form containing magnesium and silicon, which has been solution treated and precipitation-hardened. As an example of the USA designation, 2014-T6 represents an aluminium–copper alloy in the solution-treated and artificially aged condition.

17.4.3 Designation system for cast alloys

Cast alloys are covered by BS EN 604, and specification numbers give information as to:

(a)   Form of material:
      Ingots – the alloy has *no* suffix letter
      Castings – the suffix denoted the condition of the material
(b)   Condition:
      M   'As cast' with no further treatment
      TS   Stress-relieved only

TE   Precipitation treated
TB   Solution treated
TB7 Solution treated and stabilised
TF   Solution treated and precipitation treated
TF7 Full treatment, plus stabilization

All of the alloy numbers representing cast alloys are prefixed by the letters LM. Thus, LM 10 denotes alloy no. 10 in the ingot form, whilst LM 10-TF represents a casting of the same alloy in the fully heat-treated condition.

17.5 Wrought alloys which are not heat-treated

These are all materials in which the alloying elements form a solid solution in the aluminium and this is a factor which contributes to their good corrosion-resistance, high ductility and increased strength, the latter arising from the resistance to slip produced by the solute atoms present. Since these alloys are not heat-treatable, the necessary strength and rigidity can be

Table 17.1 *Typical properties and uses of wrought aluminium alloys not heat-treated*

Alloy	Condition (%)	Condition	Tensile strength (MPa)	Percentage elongation	Uses
1200	99.5 Al	O	90	35	Panelling and moulding, hollow-ware, equipment for chemical-, food- and brewing-plant, packaging.
		H12	125	9	
		H18	155	5	
3004	1.2 Mn, 1.0 Mg	O	110	20	Beverage can bodies
		H18	200	5	
3103	1.2 Mn	O	105	34	Metal boxes, bottle-caps, food-containers, cooking utensils, panelling of land-transport vehicles.
		H18	200	4	
5251	2.25 Mg	O	185	20	Marine superstructures, panelling exposed to marine atmospheres, chemical plant, panelling for road and rail vehicles, fencing-wire.
		H28	255	2	
5154 A	3.5 Mg	O	220	18	Shipbuilding, deep-pressing for car bodies.
		H24	275	6	
5083	5.0 Mg	O	275	16	Shipbuilding and applications requiring high strength and corrosion-resistance, rivets.
		H24	350	6	
5182	0.35 Mn, 4.5 Mg	O	120	22	Tops for beverage cans.
		H34	240	10	

obtained only by controlling the amount of cold-work in the final shaping process. These are available in various stages of hardness between 'soft' (O) and 'full hard' (H).

The commercial grades of aluminium (1200) are sufficiently strong and rigid for some purposes, and the addition of up to 1.5% manganese (3103) will produce a stronger alloy. The aluminium–magnesium alloy with 1.2% manganese and 1% magnesium (3004) is used for beverage cans. It combines strength with ductility so that cans can be deep drawn. The aluminium–magnesium alloys have a very good resistance to corrosion, and this corrosion-resistance increases with the magnesium content, making them particularly suitable for use in marine conditions (Table 17.1). The aluminium – magnesium alloy with 4.5% magnesium and 0.35% manganese (5182) is used for the lids for beverage cans since it is harder than 3004 and so more able to withstand the forces involved with ring pulls.

17.6 Cast alloys which are not heat-treated

This group consists of alloys which are widely used for both sand-casting and die-casting. Rigidity, good corrosion-resistance and fluidity during casting are their most useful properties.

The most important alloys in this group are those containing between 10% and 13% silicon. Alloys of this composition are approximately eutectic in structure (Figure 17.2). This makes them particularly useful for die-casting, since their freezing-range will be short, so that the casting will solidify quickly in the mould and make rapid ejection, and hence a short production cycle, possible.

In the ordinary cast condition (Figure 17.2), the eutectic structure is rather coarse, a factor which causes the alloy to be rather weak and brittle. However, the structure – and hence the mechanical properties – can be improved by a process known as *modification*. This involves adding a small amount of metallic sodium (only 0.01% by weight of the total charge) to the molten alloy, just before casting. The resultant casting has a very fine grain and the mechanical properties are improved as a result (see Table 17.2 and Figure 17.3). The eutectic point is also displaced due to modification, so a 13% silicon alloy, which would contain a little brittle primary silicon in the unmodified condition, now contains tough primary α solid solution instead, since the composition is now to the *left* of the *new* eutectic point. An unmodified 13% silicon sand cast alloy has typically a tensile strength of 125 MPa and a percentage elongation of 2%. After modification the tensile strength is 195 MPa and the percentage elongation 13%. An unmodified 13% silicon die cast alloy has typically a tensile strength of 195 MPa and a percentage elongation of 3.5%. After modification the tensile strength is 220 MPa and the percentage elongation 8%.

High casting-fluidity and low shrinkage make the aluminium–silicon alloys very suitable for die-casting. Alloys of high shrinkage cannot be used, because they would crack on contracting in a rigid metal mould. Since aluminium–silicon alloys are also very corrosion-resistant, they are useful for marine work.

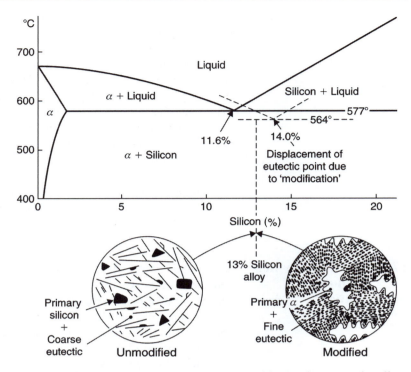

Figure 17.2 *The aluminium–silicon thermal equilibrium diagram. The effects of 'modification' on both the position of the eutectic point and the structure are also shown. See also Figure 17.3.*

The most important property of the aluminium–magnesium–manganese alloys (Table 17.2) is their good corrosion-resistance, which enables them to receive a high polish. These alloys are noted for high corrosion-resistance, rigidity and toughness, making them very suitable for use as moderately stressed parts working under marine conditions. Other casting alloys containing up to 10% of either copper or zinc are now little used, because of their poor resistance to corrosion.

17.7 Wrought alloys which are heat-treated

In 1906, a German research metallurgist, Dr. Alfred Wilm, was investigating the effects of quenching on the mechanical properties of some aluminium alloys containing small amounts of magnesium, silicon and copper. To his surprise, he found that, if quenched test-pieces were allowed to remain at room temperature for a few days, without further heat-treatment, a considerable increase in strength and hardness of the material occurred. This phenomenon, subsequently called *age-hardening,* was unexplained at the time, since no apparent change in the microstructure was detected, and it is only in recent years that a reasonably satisfactory explanation of the causes of 'age-hardening' has been evolved. However, there are many instances in which metallurgical practice has been established long before it could be

Table 17.2 *Cast aluminium alloys – not heat-treated*

Alloy	Composition (%)				Condition	Typical properties		Uses
	Si	Cu	Mg	Mn		Tensile strength (MPa)	Percentage elongation	
LM 2M	10	1.6			Chill cast	250	3	General purposes, particularly pressure die-castings. Moderate strength alloy.
LM 4M	5	3		0.5	Sand cast Chill cast	155 190	4	Sand-, gravity- and pressure die-castings. Good foundry characteristics. Inexpensive general-purpose alloy, where mechanical properties are of secondary importance.
LM 5M			4.5	0.5	Sand cast Chill cast	170 200	6 10	Sand- and gravity die-castings. Suitable for moderately stressed parts. Good resistance to marine corrosion. Takes a good polish.
LM 6M	12				Unmodified Modified	125 200	5 15	Sand-, gravity- and pressure die-castings. Excellent foundry characteristics. Large castings for general and marine work. Radiators, sumps, gear-boxes, etc. One of the most widely used aluminium alloys.

explained in terms of underlying theory – the heat-treatment of steel is an example – and Wilm's discovery was soon developed in the form of the alloy 'duralumin', which found application in the structural frames of the airships of Count von Zeppelin, which bombed England during the First World War.

17.7.1 Heat-treatment

As with carbon steels, the heat-treatment of an aluminium alloy is related to the equilibrium diagram and, in this instance, the important part of the diagram (Figure 17.4) is the sloping boundary line *ABCD,* the significance of which was dealt with in Section 9.4.2. Here, the slope of *ABCD* indicates that as the temperature rises, the amount of copper which dissolves in *solid* aluminium also increases. This is a fairly common phenomenon in the case of ordinary liquid solutions; for example, the amount of salt which will dissolve in water increases as the temperature of the solution increases.

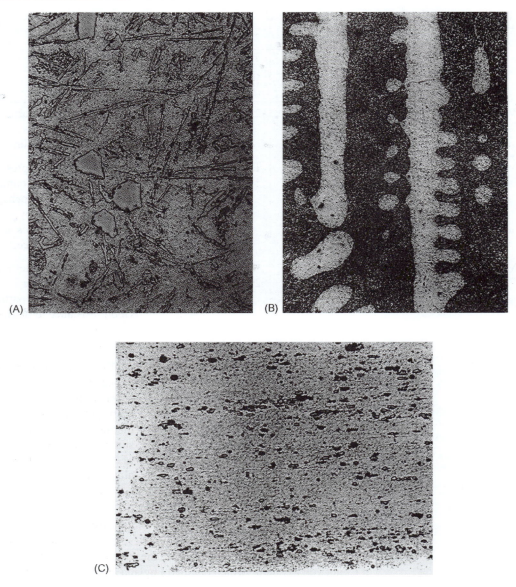

Figure 17.3 *Microstructures of some aluminium alloys. (A) 12% silicon in aluminium – unmodified, as cast. Since this alloy contains more than the eutectic amount (11.6%) of silicon (see Figure 17.2), primary silicon (angular crystals) are present. The eutectic is coarse and brittle and consists of 'needles' of silicon in a matrix of α solid solution because the layers of α in the eutectic have fused together to form a continuous mass (the amount of silicon being only 11.6% of the eutectic so that the layers of α would be roughly 10 times the thickness of those in silicon). (B) The same alloy as (A) but modified by the addition of 0.01% sodium. This has the effect of displacing the eutectic point to 14% silicon so that the structure now consists of primary crystals of α (light) in a background of extremely fine-grained eutectic (dark). The alloy is now stronger and tougher. (C) A duralumin-type alloy in the 'as extruded' condition (unetched). The particles consist mainly of $CuAl_2$ (see Figure 17.4) elongated in the direction of extrusion. Most of this $CuAl_2$ would be absorbed during subsequent solution treatment.*

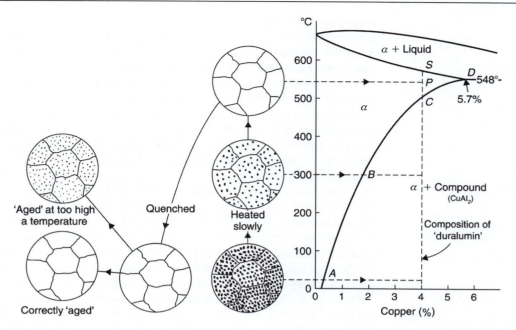

Figure 17.4 *Structural changes which take place during the heat-treatment of a duralumon-type of alloy.*

Similarly, if the hot, 'saturated' salt solution is allowed to cool, crystals of solid salt begin to separate – or 'precipitate', as we say – from the solution, so that 'equilibrium' is maintained. In the case of *solid* solutions, these processes of solution and precipitation take place more slowly, because atoms find greater difficulty in moving in solid solutions than they do in liquid solutions, where all particles can move more freely. Consequently, once a solid solution has been formed at some high temperature by heat-treatment, its structure can usually be trapped by quenching, that is, cooling it rapidly so that precipitation of any equilibrium constituent has no opportunity to take place.

We will assume that we have an aluminium–copper alloy containing 4% copper, since this is the basic composition of some of the 'duralumin' alloys in use. Suppose the alloy has been extruded in the form of rod, and has been allowed to cool slowly to room temperature. This slow cooling will have allowed plenty of opportunity for the copper to precipitate from solid solution – not as particles of pure copper, but as small crystals of an aluminium– copper compound $CuAl_2$ (see Figure 17.3C). Since this inter-metallic compound is hard and brittle, it will render the alloy brittle. Moreover, since only about 0.2% of copper (point A in Figure 17.4) remained behind in the solid solution α, this solid solution will lack strength. Hence, the mechanical properties of the alloy as a whole will be unsatisfactory.

Now suppose the alloy is slowly heated. As the temperature rises, the particles of compound $CuAl_2$ will gradually be absorbed into the surrounding α solid solution (just as salt would be dissolved by water). Thus at, say, 300°C, much of the $CuAl_2$ will have been dissolved, and an amount of copper equivalent to B will now be in solution in the α. The amount of copper which dissolves increases with the temperature, and solution will be complete at C, as the last tiny particles of $CuAl_2$ are absorbed. In industrial practice, a temperature corresponding to P would be used, in order to make sure that all of the compound had been dissolved, though care would be taken not to exceed S, as at this point the alloy would begin to melt.

The heating process is known as *solution-treatment,* because its object is to cause the particles of $CuAl_2$ to be dissolved by the aluminium solid solution (α). Having held the alloy at the solution-treatment temperature long enough for the $CuAl_2$ to be absorbed by the α solid solution, it is then quenched in water, in order to 'trap' this structure and so preserve it at room temperature. In the quenched condition, the alloy is stronger, because now the whole of the 4% copper is dissolved in the α; but it is also more ductile, because the brittle crystals of $CuAl_2$ are now absent. If the quenched alloy is allowed to remain at room temperature for a few days, its strength and hardness gradually increase (with a corresponding fall in ductility) (curve A in Figure 17.5). This was known as *age-hardening,* and is due to the fact that the quenched alloy is no longer in 'equilibrium' and tries its best to revert to its original $\alpha + CuAl_2$ structure by attempting to precipitate particles of $CuAl_2$. In fact, the copper atoms do not succeed in moving very far within the aluminium lattice structure. Nevertheless, the change is sufficient to cause a bigger resistance to movement on slip planes within the alloy, i.e. the strength has been increased. Although the copper atoms have moved closer together, they still occupy positions which are part of the aluminium lattice structure. At this stage, they are said to exist as a *coherent precipitate*.

This internal change within the alloy can be accelerated and made to proceed further by a 'tempering' process, formerly referred to as *artificial age-hardening.* A typical treatment may consist of heating the quenched alloy for several hours at, say, 160°C, when both strength and hardness will be found to have increased considerably (curve B in Figure 17.5). The term *precipitation-hardening* is now used to describe the increase in hardness produced, both by this type of treatment and by that which occurs at ordinary

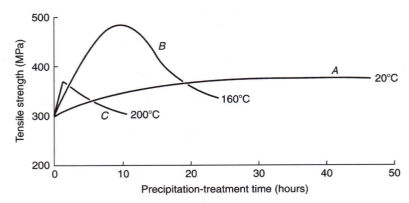

Figure 17.5 *The effects of time and temperature of precipitation-treatment on the strength of duralumin.*

Table 17.3 Wrought aluminium alloys – heat-treatment

Alloy	Composition (%)					Heat-treatment	Tensile strength (MPa)	Percentage elongation	Uses
	Cu	Si	Mg	Mn	Other				
		0.5	0.7			Solution-treated at 520°C; quenched and precipitation-hardened 170°C for 10 hours.	250	10	Glazing bars, window sections; car windscreen and sliding-roof sections. Good corrosion-resistance and surface finish.
6082-TF		1.0	1.0	0.7		Solution-treated at 510°C: quenched and precipitation-hardened 175°C for 10 hours.	310	8	Structural members for road-, rail- and sea-transport vehicles; ladders, scaffold tubes; overhead electrical lines; architectural work.
2L84*	1.5	1.0	0.8			Solution-treated at 525°C; quenched and precipitation-hardened 170°C for 10 hours.	390	10	Structural members for aircraft and road vehicles; tubular furniture.
2L77*	4 1	0.5	0.8	0.7	Ti+ Cr 0.3 optional	Solution-treated at 480°C; quenched and aged at room temperature for 4 days.	400	10	Stressed parts in aircraft and other structures; general purposes. The original 'duralumin'.
2014A-TF	4.3	0.7	0.5	0.7	Ti+ Cr 0.3 optional	Solution-treated at 510°C; quenched and precipitation-hardened 170°C for 10 hours.	510	10	Highly stressed components in aircraft stressed-skin construction; engine parts, e.g. connecting rods.
2L88*	0.3	0.6	1.0	0.5 0.25	Mn or Cr	Solution-treated at 520°C; quenched and precipitation-hardened 170°C for 10 hours.	310	13	Plates, bars and sections for shipbuilding; body panels for car and rail vehicles; containers.
2L95	1.6	0.5	2.5	0.3	Zn 7	Solution-treated at 465°C; quenched and precipitation-hardened 120°C for 24 hours.	650	11	Highly stressed aircraft structures, e.g. booms; items requiring high strength/weight ratio.

* BS Aerospace Series, Section L (aluminium and the light alloys).

temperatures, as mentioned earlier. Care must be taken to avoid using too high a temperature during this precipitation treatment, or visible particles of $CuAl_2$ may form in the structure (see Figure 17.4) – i.e. they now appear as a normal *non*-coherent precipitate. When this happens, the mechanical properties will already have begun to fall again (curve C in Figure 17.5). This is usually referred to as *over-ageing,* or sometimes as *reversion.* Because of the need for accurate temperature control during the heat-treatment of these alloys, solution treatment is often carried out in salt baths, whilst precipitation-treatment generally takes place in air-circulating furnaces.

The aluminium–copper alloys are by no means the only ones which can be precipitation-hardened. Aluminium alloys containing small amounts of magnesium and silicon (forming Mg_2Si) can also be so treated. Recently, aluminium alloys containing up to 2.5% lithium have been developed. Here, precipitation-hardening depends on the presence of the compound Al_3Li. Since lithium has a low relative density (0.51), reductions of up to 10% in the densities of these alloys are achieved, making them important in the aircraft industries. Since lithium is a very reactive element, there are of course difficulties encountered in melting and casting these alloys. Compositions, treatments and properties of some heat-treatable wrought aluminium alloys are given in Table 17.3.

Numerous magnesium-base alloys (see Section 18.4), some copper-base alloys (see Section 16.9), as well as maraging steels (see Section 13.2), can also be precipitation-hardened.

17.8 Cast alloys which are heat-treated

Some of these alloys are of the 4% copper type, as described earlier, but possibly the best known of them contains an additional 2% nickel and 1.5% magnesium. This is Y-alloy, 'Y' being the series letter used to identify it during its experimental development at the National Physical Laboratory, during the First World War. Whilst Germany was concentrating on the production of wrought duralumin for structural members of its Zeppelins, in Britain, research was aimed at the production of a good heat-treatable *casting* alloy for use in the engines of our fighter planes. The airframes were constructed largely of wood and 'doped' canvas, but a light alloy was needed from which high-duty pistons and cylinder-heads could be constructed for use at high temperatures. Y-alloy, only recently obsolescent, was the result. Heat-treatable casting alloys, including Y, are described in Table 17.4.

Before I wax too lyrical about these 'wonderful flying machines' of the early twentieth century, perhaps I should remember that Leonardo da Vinci designed a craft which might well have flown if a suitable power unit had been available to him.

Recently, a series of beryllium–aluminium alloys has been introduced, containing in the region of 60% beryllium. Their main attribute is a low density of 2.16 kg/m^3 combined with a tensile strength of 310 MPa (giving a strength/weight ratio of 150 $kg/m^2/s^2$) with a percentage elongation of 3.5. These alloys are used as investment castings for such components as secondary structural members in US Army helicopters and guidance components for advanced missiles and satellite. Naturally, due to the high cost of beryllium and its processing, these are expensive alloys.

Table 17.4 Cast aluminium alloys – heat-treated

Alloy	Composition (%)					Heat-treatment	Tensile strength (MPa)	Percentage elongation	Uses
	Cu	Si	Mg	Mn	Other				
LM4-TF	3.5	8				Solution-treated 520°C 6 hours; quenched hot water or oil; precipitation-hardened 170°C 12 hours.	320	1	General purposes (sand-, gravity-, and pressure die-casting).
LM4-TF		11.5	0.4	0.5		Solution-treated 530°C 2–4 hours; quenched warm water; precipitation-hardened 160°C 16 hours.	320	18	Suitable for intricate castings, due to fluidity imparted by silicon. Good corrosion-resistance.
4L35*	4		1.5		Ni 2.0	Solution-treated 510°C; precipitation-hardened in boiling water for 2 hours, or aged at room temperature for 5 days.	280		Pistons and cylinder-heads for liquid- and air-cooled engines; general purposes. 'Y' alloy. Now obsolescent.
3L51*	1.4	2.5	0.2		Fe 1.0 Ni 1.3 Ti 0.15	No solution-treated required; precipitation-hardened 165°C for 8–16 hours.	200	3	A good general-purpose alloy for sand- and gravity die-casting. High rigidity and moderate resistance to shock.
LM29-TE	1	23	1		Ni 1.0	Chill cast and precipitation treated.	190 Hardness 140 Brinell, good wear resistance	1	A gravity or pressure die-casting alloy – cylinder blocks and pistons in the automotive industries.
	A hyper-eutectic alloy in which the primary silicon is refined by small additions of red phosphorus.**								

*BS Aerospace Series, Section L (aluminium and the light alloys).
**Specification lays down microscopical examination to control size and distribution of primary silicon.

18 Other non-ferrous metals and alloys

18.1 Introduction

When Columbus discovered the New World, not more than half a dozen metals were known to Man. Now he has isolated all 70 of them and also 'constructed' a number of new ones, such as plutonium and neptunium, which previously did not exist in nature. Some of these metals may never be of use to the engineer, either because they have no desirable physical properties or because they 'corrode' far too readily – some instantaneously on exposure to the atmosphere. Nevertheless, who would have thought, even as late as the 1950s, say, that by the end of the twentieth century we would have used samarium and neodymium (in magnets), hafnium (in nuclear plant), yttrium and lanthanum (in high-temperature alloys) and erbium (in cancer therapy generators)?

This chapter is a brief consideration of nickel and its alloys, titanium and its alloys, magnesium-base alloys, zinc-base alloys and bearing metals.

18.2 Nickel and its alloys

Although the Ancient Chinese may have used alloys similar in composition to our nickel silvers (see Section 16.8), nickel itself was not discovered in Europe until about 1750. At that time, the copper smelters of Saxony were having trouble with some of the copper ores they were using, for, although these ores appeared to be of the normal type, they produced a metal most unlike copper. This metal was given the name of 'kupfer-nickel', which somewhat literally translated from the Old Saxon, meant 'copper possessed of the Devil'. Later the material was found to be an alloy containing a new metal which was allowed to retain the title of 'nickel' – or 'Old Nick's metal'.

18.2.1 Nickel properties and uses

Nickel is a 'white' metal, with a faint greyish tint. Most of the uses of commercially pure nickel depend upon the fact that it has a good resistance to corrosion, not only by the atmosphere, but also by many reagents. Consequently, much nickel is used in the electroplating industries, not only as a finishing coat, but also as a 'foundation layer' for good-quality chromium-plating.

Since nickel can be hot- and cold-worked successfully, and joined by most orthodox methods, it is used in the manufacture of chemical and food-processing plant (see Table 18.1), whilst nickel-clad steel ('Niclad') is used in the chemical and petroleum industries. A greater quantity of nickel, however, is used in the manufacture of alloy steels (see Sections 13.2 and

Table 18.1 *Properties and uses of commercially pure nickel*

Alloy	Specifi-cation	Composition (%)	Condition	Tensile strength (MPa)	0.2% proof stress (MPa)	Percentage elongation	Uses
Nickel 200	NA11	Minimum 99.0 Ni, up to 0.15 C	Cold rolled and annealed	380	105	30–40	Food processing equipment, electrical and electronic parts, equipment for handling caustic soda.
			Hot rolled and annealed	380	105	40	
Nickel 201	NA12	Minimum 99.0 Ni, less than 0.02 C	Cold rolled and annealed	350	85	30–40	Similar to NA11 but preferred for use above 315°C
			Hot rolled and annealed	350	85	30	

13.5), whilst significant amounts are used in other alloys, in particular cupro-nickels (see Section 16.8), nickel silvers (see Section 16.8) and cast irons (see Section 15.8).

18.2.2 Electrical resistance alloys for use at high temperatures

These are generally nickel–chromium or nickel–chromium–iron alloys, the main features of which are:

*   Their ability to resist oxidation at high temperatures.
*   High melting-ranges.
*   High electrical resistivity.

These properties make nickel–chromium alloys admirably suitable for the manufacture of resistance-wires and heater-elements of many kinds, working at temperatures up to bright red heat. Representative alloys (the Henry Wiggin 'Brightray' series) are included in Table 18.2.

18.2.3 Corrosion-resistant alloys

These alloys, for use at ordinary temperatures, all contain nickel, along with varying amounts of molybdenum and iron, and sometimes chromium and copper. Naturally, such alloys are relatively expensive, but their resistance to corrosion is extremely high. Consequently, they are used mainly in the chemical industries, to resist attack by strong mineral acids and acid chloride solutions – conditions which, in terms of corrosion, are about the most severe

Table 18.2 *High-temperature resistance alloys*

Composition (%)			Resistivity (10^{-8} Ωm)	Max working temp (°C)	Uses
Ni	Cr	Fe			
80	20		103	1150	Heaters for electric furnaces, cookers, kettles, immersion-heaters, hairdryers, toasters.
65	15	20	106	950	Similar to above, but for goods of lower quality; also for soldering-irons, tubular heaters, towel-rails, laundry-irons and where operating temperatures are lower.
34	4	62	91	700	Cheaper-quality heaters working at low temperatures, but mainly as a resistance-wire for motor starter-resistances, etc.

Table 18.3 *Corrosion-resistant nickel-base alloys*

Trade names	Composition (%)				Uses
	Ni	Mo	Fe	Others	
Corronel B	66	28	6		Resists attack by mineral acids and acid chloride solutions. Produced as tubes and other wrought sections for use in the chemical and petroleum industries, for constructing reaction-vessels, pumps, filter parts, valves, etc.
Ni-O-Nel	40	3	35	Cr 20 Cu 2	A 'Wiggin' alloy with characteristics similar to those of austenitic stainless steel, but more resistant to general attack, particularly in chloride solutions.
Hastelloy 'A'	58	20	22		Transporting and storing hydrochloric acid and phosphoric acid, and other non-oxidising acids.
Hastelloy 'D'	85			Si 10 Cu 3 Al 1	A casting alloy – strong, tough and hard, but difficult to machine (finished by grinding). Resists corrosion by hot concentrated sulphuric acid.

which are likely to be encountered industrially. A number of these alloys are described in Table 18.3.

18.2.4 High-temperature corrosion-resistant alloys

A well-known high-temperature alloy introduced many years ago is 'Iconel'. It contains 80% nickel, 14% chromium and 6% iron, and is used for many purposes, including food-processing plant, hot-gas exhaust manifolds and heating elements for cookers. It is quite tough at high temperatures, because of the very low-grain growth imparted by nickel (see Section 13.2), and it does not oxidise appreciably, because of the protective film of chromium oxide which forms on the surface (see Section 13.5).

The properties of materials like Iconel were extended in the 'Nimonic' series of alloys introduced by Messrs Henry Wiggin – alloys which played a leading part in the development of the jet engine. These 'Nimonics' are basically nickel–chromium alloys which have been strengthened – or 'stiffened' – for use at high temperatures by adding small amounts of titanium, aluminium, cobalt and molybdenum in suitable combinations. These elements form phases within the structure, and have the effect of raising the limiting creep stress at high temperatures. A few of the better-known Nimonic alloys are given in Table 18.4.

18.2.5 Low-expansion alloys

Most materials expand when they are heated and contract again as they cool, the indicator of the amount of length change being the coefficient of thermal expansion (the change in length per °C per metre length of material). Some iron–nickel alloys, however, have extremely low coefficients of thermal expansion, making them useful in numerous types of precision equipment operating under conditions of varying temperature.

The coefficient of expansion is at a minimum for an iron–nickel alloy containing 36% nickel. This alloy, originally known as *Invar,* was often used for the pendulums of clocks requiring great accuracy, as in astronomical

Table 18.4 *Some 'Nimonic' high-temperature alloys*

Nimonic alloy	BS specifi- cation*	Approximate composition (%)**						Tensile strength (MPa)			Uses
		C	Cr	Ti	Al	Co	Mo	600°C	800°C	1000°C	
75	HR5	0.1	19.5	0.4				590	250	90	Gas-turbine flame tubes and furnace parts.
80A	2HR1	0.07	19.5	2.2	1.4			1080	540	75	Gas-turbine stator-blades, after-burners and other stressed parts working at high temperatures.
90	2HR2		19.5	2.5	1.5	18.0		1080	850	170	Rotor blades in gas turbines.
115	HR4	0.16	15.0	5.0	5.0	14.2	4.0	1080	1010	430	Excellent creep-resistant properties at high temperatures.

*Aerospace series.
**The balance is nickel.

observatories. As the amount of nickel increases above 36%, so the coefficient of expansion increases and it is possible to produce alloys having a useful range of coefficients of expansion. Thus, alloys containing between 40% and 50% nickel have coefficients of expansion similar to those of many types of glass, so that efficient metal–glass seals can be produced in electric lamps, TV tubes and the like.

Many domestic and industrial thermostats depend for their operation on the uneven expansion of two different layers of metal in a bimetallic strip. One layer is usually of brass, which has a considerable coefficient of expansion, whilst the other is one of these iron–nickel alloys, with a very low coefficient of expansion. An increase in temperature will cause the brass layer to expand, whilst the nickel–iron layer will barely change in length. This causes the bimetallic strip to curve, with the greater expansion brass layer on the outside of the curve, and so operate an electrical make/break contact. Many readers will be familiar with such thermostats in domestic central heating systems and in home tropical aquaria. Some low-expansion alloys are detailed in Table 18.5.

Table 18.5 *Iron–nickel low-expansion alloys*

Trade name	Composition (%)			Coefficient of expansion at $20°C \times 10^{-6}$	Uses
	Ni	Fe	Co		
Nilo * 36 (Invar, Nivar)	36	64		0.9	Pendulum-rods, standard lengths, measuring-tapes, delicate precision sliding-mechanisms, thermostats for low-temperature operation.
Nilo 40	40	60		6.0	Thermostats for electric and gas cookers, heater-elements.
Nilo 42	42	58		6.2	Thermostats, also the core of copper-clad wire for glass seals in electric lamps, TV tubes.
Nilo 50	50	50		9.7	Thermostats, also for sealing with soft glasses used in electronic equipment.
Nilo K	29	54	17	5.7	Glass/metal seals in medium-hard glasses used in X-ray tubes and various electronic equipment.

* *The trade name for these alloys, coined by Messrs Henry Wiggin.*

18.2.6 Shape-memory alloys

Shape-memory alloys (SMAs) have the ability to return to some previously defined shape when heated (see Section 16.5). When the temperature falls below its transformation temperature a martensitic structure is formed, the structural change then being reversed to the original structure when the alloy is heated above the transformation temperature. Nickel–titanium alloys are one of the principal commercial forms of such alloys. The transformation temperature can be adjusted by varying the alloy composition.

Nickel–titanium alloys are extremely corrosion-resistant and biocompatible, so have an increasing number of uses in biomedical applications, e.g. a blood-clot filter – a wire that is shaped to anchor itself in a vein and catch passing blood clots.

18.3 Titanium and its alloys

Titanium is probably the most important and most widely used of those metals whose technology was developed in the latter half of the twentieth century. As well as being strong and corrosion-resistant, it is also a very light metal and since it has a relative density of 4.5, only just half that of steel, this gives it an excellent strength-to-weight ratio (termed *specific strength*). It is as corrosion-resistant as 18/8 stainless steel but will also withstand the extreme corrosiveness of salt water. It has a high melting-point of 1668°C. It is this combination of high strength, low density and excellent corrosion resistance which has led to the expansion of the use of titanium in the aerospace, chemical and engineering industries since the late 1940s. It is now no longer a 'new' metal but quite a commonly used metal.

Although discovered in Cornwall by an English priest, W. Gregory, as long ago as 1791, the industrial production of titanium did not begin until the late 1940s and that was to fulfil the needs of aerospace engineers. This was not due to any scarcity of its ores – it is in fact the tenth element in order of abundance in the Earth's crust and 50 times more abundant than copper – but principally because molten titanium reacts chemically with most other substances, making it difficult, and hence extremely expensive, to extract, melt, cast and shape commercially. Nevertheless, although chemically very reactive, once it has been successfully shaped, titanium has an excellent corrosion resistance because, like aluminium, its surface becomes coated with a dense impervious film of oxide which effectively seals it from further atmospheric attack.

18.3.1 Properties of titanium

Titanium is a white metal with a fracture surface like that of steel. In the pure state, it has a maximum tensile strength of no more than 400 MPa, but, when alloyed with small amounts of other metals such as aluminium, tin and molybdenum, strengths of 1400 MPa or more can be obtained, and, most important, maintained at much higher temperatures than is possible with aluminium alloys. In view of the properties mentioned earlier – low relative density, high strength-to-weight ratio, good corrosion resistance and good strength at high temperatures – it is obvious that titanium will be used in increasing amounts in both aircraft and spacecraft. Titanium alloys have been used for some time for parts in jet engines, and in some modern aircraft about a quarter of the weight of the engine is taken up by titanium alloys. In structural members of aircraft, too, these alloys are finding increasing use because of their high strength-to-weight ratio, e.g. the four engines of Concorde used some 16 tonnes of titanium alloys.

18.3.2 Titanium structure and its alloys

Titanium is a polymorphic element, being normally a hexagonal close-packed (CPH) structure, the alpha phase, which transforms at 882°C to a body-centred cubic (BCC) structure, the beta phase. As in the case of iron, the addition of alloying elements to titanium influences the transformation temperature and can, in many alloys, result in beta phase being retained at room temperature. Thus, at room temperature, it is possible to have alpha alloys, alpha plus beta alloys and purely beta alloys. The relative amounts of alpha and beta phases in any particular alloy has a significant effect on its tensile strength, ductility, creep properties, weldability and ease of formability. It is common practice to refer to a titanium alloy by its structure:

1 *Alpha titanium alloys,* e.g. alloys with aluminium and tin which are alpha stabilisers, are composed entirely of the alpha phase and are strong, maintain their strength at high temperatures but are difficult to work. Generally, alpha alloys cannot be heat-treated to increase their strength.

2 *Near alpha-titanium alloys* are almost entirely alpha phase with a small amount of the beta phase dispersed through the alpha. Such alloys are obtained by adding small amounts of molybdenum and vanadium to what otherwise would have been an alpha alloy. They are normally used in the annealed state. For maximum creep resistance above 450°C, near alpha alloys are used.

3 *Alpha–beta titanium alloys* have sufficient quantities of beta-stabilising agents for there to be significant amounts of beta phase at room temperature. These alloys can be solution treated, quenched and aged for increased strength. They tend to be high-strength materials with tensile strengths of 900–1250 MPa and have good creep resistance up to about 350–400°C.

4 *Beta-titanium alloys* are entirely beta phase at room temperature as a result of the addition of sufficient beta-stabilising elements. They can be cold-worked in the solution treated and quenched condition, subsequently being aged to give higher strength.

Some titanium alloys, with their composition, heat-treatment, typical properties and uses, are listed in Table 18.6.

18.3.3 Uses of titanium

Titanium alloys are used in both airframe and engine components of modern supersonic aircraft, because of the combination of high strength-to-weight ratio and corrosion resistance they offer. Thus, titanium alloys are used in gas-turbine engines for compressor blades, discs, casings, engine cowlings and exhaust shrouds. On a smaller scale, expensive precision cameras employ titanium alloy components requiring low inertia, such as shutter blades and

Table 18.6 *Some titanium-base alloys*

BS specification	Composition (%), balance Ti	Heat-treatment	Typical mechanical properties			Typical uses
			0.2% proof stress (MPa)	Tensile strength (MPa)	Percentage elongation	
2Ta6	Commercially pure Ti, 0.2 maximum Fe (alpha alloy)	Annealed 650–750°C	460	650	15	Chemical plant where resistance to acids and chloride solutions is required. Fire, water systems, sea-water lift pipes, in off-shore oil and gas installations.
	Al 8, Mo 1, V 1 (near alpha)	Duplex annealed – 790°C for 8 hours, furnace cool, reheat to 790°C for 1/4 hour, air cool	950	1000	15	Air frame and jet engine parts where high strength to 450°C required, tough, good weldability, good creep properties.
2Ta10 sheet and strip	Al 6 V 4 (alpha–beta)	Annealed 700–900°C	900	1120	8	*Air frame*: engine nacelles, fuselage skinning, wings. *Jet engines*: blades and discs. *Off-shore oil and gas installations*: stress joints, drilling risers in pipelines. High-pressure heat exchangers.
Ta48	Al 4, Mo 4, Sn 2, Si 0.5 (alpha–beta)	Solution-treated at 900°C, air cool. Heat at 500°C for 24 hours.	920	1125	9	*Air frame*: flaps and slat tracks, wing brackets, engine pylons. *Jet engines*: blades and discs, fan casings.
	V 13, Cr 11, Al 3 (beta alloy)	Solution-treated at 775–800°C, air or water cool. Age 4–10 hours at 425–480°C	1172	1220	8	High-strength fasteners, aero-space components, honeycomb panels.

blinds. The high corrosion resistance of the metal has led to its use as surgical implants as well as in the chemical industries, e.g. for heat-exchanger coils in metal-finishing baths.

In Japan, titanium has become established as a roofing material, particularly for 'prestige 'buildings. Since weight-for-weight titanium is some times more expensive than competitors like copper and stainless steel, this may seem surprising. However, since titanium is strong, light-gauge material can be used, thus reducing the cost per unit area covered. More important still, this light-gauge material coupled with the low relative density of the metal means that much less massive support members are necessary in roof construction. Titanium has a high resistance to corrosion and is able to cope with the very corrosive marine atmospheres prevalent in most of Japan. This reduces maintenance costs relative to other metals, so that when all of these points are considered, it is claimed that titanium can compete, particularly when the architecturally pleasing result is taken into account.

18.4 Magnesium-base alloys

Magnesium is a fairly common metal in the Earth's crust and is a principal constituent of dolomite limestone. Of the metallurgically useful metals, only iron and aluminium occur more abundantly. However, magnesium is relatively expensive to produce, because it is difficult to extract from its mineral ores. Electrolysis is used in a process similar to that employed for extracting aluminium. Because of its great affinity for oxygen, magnesium burns with an intense hot flame and was used as the main constituent of incendiary bombs during the Second World War. This great affinity for oxygen makes it also more difficult to deal with in the foundry than other light alloys. To prevent it taking fire, it is melted under a layer of molten flux and, during the casting process, 'flowers of sulphur' are shaken on to the stream of molten metal as it leaves the crucible. The sulphur burns in preference to magnesium.

18.4.1 Magnesium alloys

Apart from lithium, magnesium has the lowest relative density, at 1.7, of the metals used in general engineering. Although pure magnesium is a relatively weak metal (tensile strength approximately 180 MPa in the annealed state), alloys containing suitable amounts of aluminium, zinc and thorium can be considerably strengthened by precipitation-hardening treatments. Consequently, in view of these properties, magnesium-base alloys are used where weight is a limiting factor. Castings and forgings in the aircraft industry account for much of the magnesium alloys produced, e.g. landing-wheels, petrol-tanks, oil-tanks, crank-cases and air-screws, as well as other engine parts in both piston and jet engines. The automobile industry of course makes use of magnesium-base alloys. The one-time popular VW 'Beetle' was a pioneer in this respect.

Table 18.7 *Some magnesium-base casting alloys*

BS specification	Composition (%), balance Mg	Condition	Typical mechanical properties		
			0.2% proof stress (MPa)	Tensile strength (MPa)	Percentage elongation
MAG.1	Al 8, Zn 0.7, Mn 0.3	As die-cast	85	185	4
		Solution-treated	80	230	10
MAG.5	Zn 4.2, Rare earths 1.2, Zr 0.7	Die-cast and precipitation treated	135	215	4
MAG.8	Th 3.2, Zn 2.1, Zr 0.6	Die-cast and precipitation treated	85	185	5
DTD 5035A	Rare earths 2.5, Ag 2.5, Zr 0.6	Die-cast, solution treated and precipitation-hardened	185	240	2

Table 18.8 *Some wrought magnesium-base alloys*

BS specification	Composition (%), balance Mg	Condition	Typical mechanical properties		
			0.2% proof stress (MPa)	Tensile strength (MPa)	Percentage elongation
MAG.101	Mn 1.5	As manufactured	70	200	5
MAG.121	Al 6, Zn 1, Mn 0.3	As manufactured	180	270	8
MAG.151	Zn 3.2, Zr 0.6	As manufactured	180	265	8
MAG.161	Zn 5.5, Zr 0.6	Extruded and precipitation-treated	230	315	8

In addition to the alloying elements mentioned earlier, small amounts of manganese, zirconium, and 'rare earth' metals are added to some magnesium alloys. Manganese helps to improve corrosion resistance, whilst zirconium acts as a grain-refiner. The 'rare earths' and thorium give a further increase in strength, particularly at high temperatures. These alloys include both cast and wrought materials, examples of which are given in Tables 18.7 and 18.8.

18.5 Zinc-base alloys

The growth of the die-casting industry was helped to a great extent by the development of modern zinc-base alloys. These are rigid and reasonably strong materials which have the advantage of a low melting-point, making it possible to cast them into relatively inexpensive dies. Alloys of high melting-point will generally require dies in rather expensive heat-resisting alloy steels (see Table 13.7).

A very wide range of components, both for the engineering industries and for domestic appliances, is produced in a number of different zinc-base alloys sold under the trade name of 'Mazak'. Automobile fittings such as door handles and windscreen-wiper bodies account for possibly the largest consumption, but large quantities of these alloys are also used in electrical equipment, washing machines, radios, alarm-clocks and other domestic equipment.

During the development of these alloys, difficulty was experienced owing to the swelling of the casting during subsequent use, accompanied by a gradual increase in brittleness. These faults were found to be due to inter-crystalline corrosion, caused by the presence of small quantities of impurities, such as cadmium, tin and lead. Consequently, very high-grade zinc of 'four-nines' quality, i.e. 99.99% pure, is used for the production of these alloys.

Good-quality zinc-base alloy castings undergo a slight shrinkage, which is normally complete in about 5 weeks (Table 18.9). Where close tolerances are necessary, a 'stabilising' anneal at 150°C for about 3 hours should be given before machining. This speeds up any volume change which is likely to occur.

18.5.1 High-strength zinc-base alloys

In recent decades, these relatively low-strength zinc-base alloys (Table 18.9) have been replaced for many purposes by plastics mouldings. This, accompanied by a move away from expensive chromium-plated finishes, led

Table 18.9 *Zinc-base die-casting alloys*

BS specification	Composition (%), balance Zn		Shrinkage after 5 weeks normal 'ageing' (mm/mm)	Shrinkage after 5 weeks following stabilising (mm/mm)
	Al	Cu		
Alloy A	4		0.00032	0.00020
Alloy B	4	1	0.00069	0.00022

Table 18.10 *High-strength zinc-base die-casting alloys*

Alloy designation	Composition (%), balance Zn			Condition	Typical mechanical properties		
	Al	Cu	Mg		Tensile strength (MPa)	Percentage elongation	Hardness Brinell
ZA8	8.4	1.0	0.02	Sand-cast	262	1.5	85
				Pressure die-cast	375	8	103
ZA12	11.0	0.85	0.02	Sand-cast	396	1.5	100
				Pressure die-cast	403	5.5	100
ZA27	26.5	2.3	0.015	Sand-cast	420	4.5	115
				Pressure die-cast	424	1	119

to a decline in the use of these 4% aluminium alloys. However, a new series of zinc-base alloys has been developed, containing rather higher quantities of aluminium. Like those alloys covered by BS 1004, the new series are die-casting alloys but have increased strength and hardness.

As might be expected, these alloys are sensitive to the presence of impurities so that these are kept to the same low limits as those alloys covered by BS 1004. The new alloys (Table 18.10) have found use in the automobile industry, e.g. ZA27 replacing cast iron for engine mountings. The use of die-casting allows the component to be produced to closer tolerances than does sand-casting. Consequently, the amount of subsequent machining is reduced.

18.6 Bearing metals

The most important properties of a bearing metal are that it should be hard and wear-resistant, and have a low coefficient of friction. At the same time, however, it must be tough, shock-resistant and sufficiently ductile to allow for 'running in' processes made necessary by slight misalignments. Such a contrasting set of properties is almost impossible to obtain in single metallic phase. Thus, whilst pure metals and solid solutions are soft, tough and ductile, they invariably have a high coefficient of friction, and consequently a poor resistance to wear. Conversely, intermetallic compounds are hard and wear-resistant, but are also brittle, so that they have a negligible resistance to mechanical shock. For these reasons, bearing metals are generally compounded so as to give a suitable blend of phases, and generally contain small particles of a hard compound embedded in the tough, ductile background of a solid solution. During service, the latter tend to wear away slightly, thus providing channels through which lubricants can flow, whilst the particles of intermetallic compound are left standing 'proud', so that the load is carried with a minimum of frictional loss.

18.6.1 White bearing metals

White bearing metals are either tin-base or lead-base. The former, which represent the better-quality high-duty white metals, are known as *Babbitt* metals, after Isaac Babbitt their originator.

All white bearing metals contain between 3.5% and 15% antimony, and much of this combines chemically with some of the tin, giving rise to an intermetallic compound, SbSn. This forms cubic crystals ('cuboids'), which are easily identified in the microstructure (Figure 18.1). These cuboids are hard, and have low-friction properties; consequently, they constitute the necessary bearing surface in white metals.

The background (or 'matrix') of the alloy is a tough ductile solid solution, consisting of tin with a little antimony dissolved in it. In the interest of economy, some of the tin is generally replaced by lead. This forms a eutectic structure with the tin–antimony solid solution. The lead-rich white metals are intended for lower duty, since they can withstand only limited pressures. Some white bearing metals are detailed in Table 18.11.

Figure 18.1 *A 'white' bearing metal (40% tin, 10% antimony, 4% copper, the balance being lead). A copper–tin compound (Cu_6Sn_5) crystallises out first as needle-shaped crystals (light) which form an interlocking network. This network prevents the SbSn cuboids (light), which have a lower relative density, from floating to the surface when they, in turn, begin to crystallise out. Finally, the matrix solidifies as a eutectic of tin–antimony and tin–lead solid solution.*

18.6.2 Aluminium–tin alloys

Aluminium–tin alloys containing 20% tin are now used as main and big-end bearings in automobile design. Aluminium and tin form a eutectic containing only 0.5% aluminium so that the final structure of these bearings consists of an aluminium network containing small areas of soft eutectic tin which wear and so assist lubricant flow. Because of the long freezing range of this alloy, segregation is a danger so that the cast material is usually cold-rolled and annealed to break up the eutectic leaving small islands of tin in an aluminium matrix. Bearing shells of this type are usually carried on a steel backing strip.

Table 18.11 *White bearing metals*

Type	BS specification	Composition (%)					Characteristics
		Sb	Sn	Cu	As	Pb	
Tin-base Babbitt metals	3332/1	7	90	3			These are generally heavy-duty materials.
	3332/3	10	81	5		4	
	3332/6	10	60	3		27	
Lead-base bearing alloys	3332/7	13	12	0.75	0.2	Bal.	These alloys are generally lower-duty materials.
	3332/8	15	5	0.5	0.3	Bal.	
		10			0.15	Bal.	

18.6.3 Copper-base bearing metals

Copper-base bearing metals include plain tin bronzes (10–15% tin) and phosphor bronzes (10–13% tin, 0.3–10% phosphorus) (see Section 16.6). Both of these alloys follow the above mentioned structural pattern of bearing metals. Some of the tin and copper combine to form particles of a very hard intermetallic compound ($Cu_{31}Sn_7$), whilst the remainder of the tin dissolves in the copper to form a tough solid-solution matrix. These alloys are very widely used for bearings when heavy loads are to be carried.

For many small bearings in standard sizes, sintered bronzes are often used. These are usually of the self-lubricating type, and are made by mixing copper powder and tin powder in the proportions of a 90–10 bronze. Sometimes some graphite is added. The mixture is then 'compacted' at high-pressure in a suitably shaped die, and is then sintered at a temperature which causes the tin to melt and so alloy with the copper, forming a continuous structure – but without wholesale melting of the copper taking place. The sintered bronze retains its porosity, and this is made use of in storing lubricant. The bearing is immersed in lubricating oil, which is then 'depressurised' by vacuum treatment, so that oil will be forced into the pores. In many cases, sufficient oil is absorbed to last for the lifetime of the machine. Self-lubricating sintered bearings are used widely in the automobile industries and in other applications where long service with a minimum of maintenance is required. Consequently, many are used in domestic equipment such as vacuum cleaners, washing-machines, extractor fans and audio equipment.

Leaded-bronzes (see Section 16.6) are used in the manufacture of main bearings in aero-engines and for automobile and diesel crankshaft bearings. They have a very high wear-resistance and a good thermal conductivity, which helps in cooling them during operation. Brasses are sometimes used as low-cost bearing materials. They are generally of a low-quality 60–40 type, containing up to 1.0% each of aluminium, iron and manganese.

Table 18.12 shows the composition and properties of copper-base alloys used as bearing metals.

18.6.4 Plastics bearing materials

Plastics bearing materials are also used, particularly where oil lubrication is impossible or undesirable. The best-known substances are nylon (see Section

Table 18.12 *Copper-base bearing alloys and their properties*

Alloy composition (%)	Manufacturing process	0.1% proof stress (MPa)	Strength (MPa)	Hardness (HV)
90 Cu, 10 Sn, 0.5 P	Cast on steel	233	420	120
75 Cu, 20 Pb, 10 Sn	Cast on steel	124	233	70
80 Cu, 10 Pb, 10 Sn	Sintered on steel	249	303	120
73.5 Cu, 22 Pb, 4.5 Sn	Sintered on steel	81	121	46

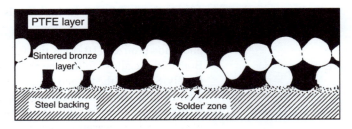

Figure 18.2 *The structure of a 'filled' PTFE bearing (magnification ×50). This sintered bronze (see Section 7.4) is applied to the steel backing support as a mixture of copper and tin powders. During the sintering process, some of the tin 'solders' the bronze to the backing. Since it is porous, it acts as a mechanical key to the PTFE-rich layer. The latter contains some particles of lead which help to form a suitable bearing surface.*

Table 18.13 *Properties of polymeric bearing materials sliding against steel*

Material	Maximum load pressure (MPa)	Maximum temperature (°C)	Maximum speed (m/s)
Phenolics	30	150	0.5
Nylon	10	100	0.1
Acetal	10	100	0.1
PTFE	6	260	0.5
Nylon with graphite filler	7	100	0.1
Acetal with PTFE filler	10	100	0.1
PTFE with silicon filler	10	260	0.5
Phenolic with PTFE filler	30	150	0.5

19.3 and Table 19.2) and polytetrafluoroethylene (see Section 19.3 and Table 19.2), both of which have low coefficients of friction. Polytetrafluoroethylene (PTFE, or 'Teflon') is very good in this respect, and in fact feels greasy to the touch. It is also used to impregnate some sintered-bronze bearings (Figure 18.2). Table 18.13 shows the properties of polymeric bearing materials when sliding against steel.

18.7 Other metals

I began this chapter by referring to a few of the more uncommon metals which have lately found use in engineering practice and it may be appropriate to close with some details of them in the form of Table 18.14 where these metals are listed in order of atomic number.

Of these metals, gold is by no means 'new'. It is in fact the earliest metal to be used by Man since it was there for the taking, 'native' – or uncombined

Table 18.14 *Some of the less well-known metals*

Metal	Relative density	Melting point (°C)	Characteristics	Uses
Lithium (Li)	0.53	180.5	Soft silvery metal. Lowest relative density of the metals.	Because of low density used in some aluminium alloys for aircraft (17.7). Also in high-energy batteries.
Beryllium (Be)	1.85	1284	Steely grey colour. Scarce and difficult to process – hence expensive. Low relative density, good strength-to-weight ratio.	Limited use in high-speed aircraft and rockets. As a moderator in some nuclear reactors. Mainly used as an alloying element (16.9).
Gallium (Ga)	5.91	29.8	Bluish-white metal.	High-temperature thermometers. Low melting-point alloys. Semiconductors.
Germanium (Ge)	5.36	937	A 'metalloid' – silvery lustre and very brittle. Chemically similar to C and Si.	Used in semiconductors but has largely been replaced by silicon.
Yttrium (Y)	5.51	1500	Soft, malleable silvery metal.	As the oxide, alloyed with zircona in some ceramics.
Zirconium (Zr)	6.5	1860	Soft, malleable ductile metal. Chemically very reactive but good corrosion resistance because of tenacious oxide skin.	Lighter flints. In chemical industries as agitator pump and valve parts. Also in steels, magnesium (18.4) and zinc (18.5) alloys.
Niobium (Nb)	8.57	2468	Originally 'Columbium' in the USA. White, ductile metal. Scarce. Very corrosion-resistant.	Fuel canning in nuclear industry. Used mainly in steels (13.5). Use limited by high cost.
Rhodium (Rh)	12.44	1960	Rare, white metal of the 'platinum group'.	Alloy element to harden platinum. Thermocouples.
Indium (In)	7.31	114.8	Soft, malleable grey-white metal. Corrosion-resistant.	Corrosion-protection coatings. Semiconductor. Low melting-point alloys.
Barium (Ba)	3.5	710	Soft, white metal. Very reactive – takes fire on exposure to moist air.	High absorption of X-rays, hence compounds used as screen for X-rays and barium sulphate as 'barium meal' for diagnosis by X-rays.
Lanthanum (La)	6.16	920	Malleable, silvery white metal. Most abundant of the 'rare earths'.	Used in some optical glasses.
Cerium (Ce)	6.77	804	Steel-grey ductile metal. High affinity for oxygen – hence, powerful reducing agent.	A 'getter' (removal of last traces of oxygen from a vacuum). Lighter flints.

continued

Table 18.14 *continued*

Metal	Relative density	Melting point (°C)	Characteristics	Uses
Neodymium (Nd)	7.01	1024	Toxic, silvery white 'rare earth' metal.	Modern permanent magnets. Also as magnesium alloys (18.4).
Samarium (Sm)	7.54	1052	A silvery white 'rare earth' metal.	Modern permanent magnets. Also as neutron absorber in nuclear industry and in laser crystals.
Dysprosium (Dy)	8.56	1500	A 'rare earth' metal.	Neutron absorber in nuclear control rods. Also in laser crystals.
Thulium (Tm)	9.32	1525	Malleable, silver-grey metal.	Radioisotope Tm_{170} is used as electron source in portable X-ray equipment.
Hafnium (Hf)	11.4	2220	Bright metal with good corrosion resistance. Absorbs neutrons effectively.	Neutron absorption control rods in nuclear power. Cost limits use in non-nuclear fields, e.g. lamp filaments, cathodes, etc.
Tantalum (Ta)	16.6	3010	Steel-blue colour. Corrosion-resistant because of dense tenacious oxide film. Very malleable and ductile.	Chemical plant (very high corrosion resistance). Plugging material for repair of glass-lined tanks. Gauze for surgical implants. Cutting-tool carbide.
Osmium (Os)	22.5	3045	Very hard platinum-type metal. The densest of all metals.	Pen-nib tips. Instrument bearings (alloyed with iridium as 'Osmiridium').
Indium (Ir)	22.4	2443	A platinum-group metal. Very hard.	See osmium. Its radioactive isotope[192] It is used as a gamma-ray source.
Platinum (Pt)	21.45	1769	Bright silvery white metal. High resistance to corrosion. High electrical resistivity.	Mainly as a catalyst in processing petroleum and other materials. Gem setting in jewellery trade. Windings for high-temperature furnaces. Laboratory equipment.
Gold (Au)	19.32	1063	Earliest metal known to Man. Very malleable – can be beaten to film 1.4×10^{-7} mm thick.	Mainly as jewellery and as a system of exchange. Specialised uses in spacecraft (18.7).
Uranium (U)	18.7	1130	Lustrous silver grey metal. Tarnishes in air. Naturally radioactive element.	Although discovered in 1789, its first serious use was in the 'atom bomb' and subsequently nuclear power.

– in the beds of mountain streams and rivers. Gold is too soft to be of much use in engineering and is used principally in jewellery and as a system of exchange (for which reason it fills the vaults of Fort Knox). Being the most malleable of workable metals, it can be beaten down to a film only 1.4×10^{-7} mm thick and it was presumably with such 'gold leaf' that Solomon covered his temple some 3000 years ago. Extremely thin films of gold leaf are transparent and will transmit green light. In 1911, Rutherford invented the atomic model with a nucleus containing most of the mass of an atom in order to explain the scattering of alpha particles when they were used to bombard gold leaf.

There have been a few engineering uses of gold. Some 37 kg (1312 oz) of the metal were used in the construction of the *Columbia* space shuttle in the form of brazing solders, fuel cell components, electrical contacts and reflective insulation. Nevertheless, on a more sober note, we are now making use of more than two-thirds of the 70 metallic elements in contrast to some half-a-dozen available to metal workers in the time of Columbus.

19 Plastics materials and rubbers

19.1 Introduction

Nylon, one of the most important man-made fibres, was developed in the USA in the years just before the Second World War but made its glamorous debut to the public in the form of ladies' stockings in the spring of 1940. The production of polyethylene (polythene) started in England in 1941 and who today has not come across the 'polythene bag'. Both nylon and polythene are typical of the materials generally referred to as *plastics*, or more properly as *plastics materials* – but *never* as plastic materials because of course many substances, including metals, undergo *plastic* deformation when the applied stress is great enough.

Though both polythene and nylon are relatively new materials, plastics and rubber development began much earlier. Prior to Columbus, the American Indians had discovered and made use of rubber, the white milky sap which oozed out of some trees when the bark was cut. The technology of rubber probably began in 1820 when a British inventor, Thomas Hancock, developed a method for shaping raw rubber; whilst some 20 years later, in the USA, Charles Goodyear established the 'vulcanisation' process by which raw rubber was made to 'set' and produce a tough, durable material, later to be so important in the growth of the motor-car industry.

Shortly afterwards, the plastics cellulose nitrate and celluloid were developed from ordinary cellulose fibre, and when, in the early years of the twentieth century, Dr. Lee Bakeland, a Belgian chemist, introduced the material which was ultimately named after him – 'bakelite' – the plastics industry could be said to have 'arrived'. It is interesting to note that since this book was first published, the *volume* of plastics materials produced annually has exceeded the volume of steel.

At present, the use of plastics continues to increase rapidly as new materials are developed. For example, modern cars include many items made from plastics and could we now exist without the plastic carrier bags and plastics bottles. Currently, polythene and polypropylene between them account for almost half of the plastics produced in the European Union. They are followed by PVC, polystyrene and polyethylene terephthalate and then the more expensive ABS, polyurethanes and nylons.

19.2 Types of plastics

Bakelite differs from either polythene or nylon in one important respect. Whereas the latter substance will soften repeatedly whenever they are heated to a high enough temperature, bakelite does not. Once moulded to shape, it remains hard and rigid and reheating has no effect – unless the temperature used is so high as to cause it to decompose and char.

Plastics materials can be classified into three groups (Figure 19.1):

1   *Thermoplastic materials* – substances which lose their rigidity whenever they are heated, so that they can be moulded under pressure with the new shape being retained on cooling.
2   *Thermosetting materials* – substances which undergo a definite chemical change during the moulding process, causing them to become permanently rigid and incapable of being softened again. The process is not reversible. Materials used in conjunction with glass-fibre for the repair of the bodywork of decrepit motor cars fall into this class.
3   *Elastomers* – substances which are characterised by very high elasticity but a very *low modulus of elasticity,* i.e. great extensibility when stressed, and the capacity for very large rapidly recoverable deformation. Think of a rubber band.

The term *polymer* is used for the very large organic molecules which are involved in plastics.

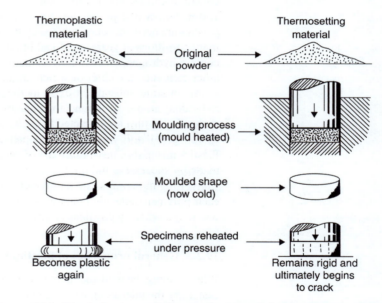

Figure 19.1 *The behaviour of thermoplastic and thermosetting materials when reheated under pressure.*

19.2.1 Raw materials

Raw materials used in the manufacture of plastics traditionally come from two main sources:

1   Animal and vegetable by-products such as casein (from cow's milk), cellulose (mainly from cotton fibres too short for spinning) and wood pulp, common products being cellulosics.

2   Petroleum by-products obtained during the refining and 'cracking' of crude oil, common products being polythene, PVC and polystyrene. This method is responsible for the bulk of plastics manufacture.

Commercial plastics generally involve the addition to basic polymers of additives to modify properties, e.g. plasticisers to enhance flexibility (see later in this chapter) and antioxidants to inhibit degradation.

19.2.2 Composition of plastics

All plastics are polymers. Polymers are very large organic molecules, based on the element carbon, and made by the successive linking together of small molecules. The term *organic* in chemical nomenclature has a similar connotation to its current use in farming practice in that it implies that only living materials and substances derived therefrom are involved. Organic chemical compounds are associated mainly with those derived from either living or once-living matter and can be regarded as the material which constitutes the skeletons of plants, from blades of grass to conifers some 100 m tall, whilst petroleum, from the by-products of which most of our plastics are derived, was produced by the decay and fossilisation of vegetable matter, millions of years ago. In addition to the element carbon, most plastics contain hydrogen, whilst many contain oxygen. A smaller number contain other elements, the chief of which are nitrogen, chlorine and fluorine.

All organic substances exist in the form of molecules, within which individual atoms are bound together by very strong covalent bonds. The molecules in turn are attracted to each other by much weaker forces known as *van der Waals forces*. See Chapter 2 for a discussion of such bonds. Plastics materials consist of a mass of very large molecules, termed *polymers*, made by successive linking together of small molecules, called *monomers*. Converting monomers into a polymer is termed *polymerisation*. As a result, each polymer molecule contains several thousands of atoms tightly bound one to the other by covalent bonds.

19.2.3 General properties of plastics materials

When visiting the wild and more remote parts of Europe's coastline one finds, cast up by the tide, along with the seaweed and driftwood, a motley collection of plastics bottles. Because of its low relative density and comparative indestructibility, this plastics junk will presumably congregate in increasing quantity, as yet another example of Man's careless pollution of his environment.

This fact illustrates some of the more important properties of plastics generally:

1   They are resistant both to atmospheric corrosion and to corrosion by many chemical reagents.
2   They have a fairly low relative density – a few will just float in water, but the majority are somewhat more dense.

3   Many are reasonably tough and strong, but the strength is less than that of metals. However, since the relative density of plastics is low, this means that many have an excellent strength-to-weight ratio – or *specific strength* as it is often termed.

4   Most of the thermoplastic materials begin to soften at quite low temperatures, and few are useful for service at temperatures much above 100°C. Strength falls rapidly as the temperature rises.

5   Most plastics have a pleasing appearance, and can be coloured if necessary. Some are transparent and completely colourless.

19.3 Thermoplastics

In thermoplastics materials, the carbon atoms are attached to each other by covalent bonds in the form of a long chain – this in polythene is a chain of about 1200 carbon atoms in length. Strong binding forces operate between the molecules in the chains with just van der Waals forces between chains. These van der Waals forces will be stronger the closer the molecules are together. Furthermore, it is easy to imagine that a considerable amount of entanglement will exist among these long chain-like molecules (Figure 19.2). When such a solid is heated, vibration within the molecules becomes greater so that the distances between molecules becomes greater and so the forces of attraction between them will decrease (in a similar way, the force acting between two magnets becomes smaller the further they are moved apart). Then such a material will be weaker and less rigid, so that it can be moulded more easily – hence the term *thermoplastic* for such materials.

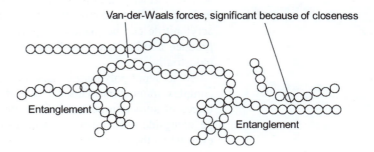

Figure 19.2 *Thermoplastics molecular chains.*

The household plastics material we call *polythene* is more correctly known as *polyethylene*. It is made from the gas 'ethylene', a by-product of the petroleum industry. In the ethylene molecule, there is a weak double covalent bond (indicated in figures as = in Figure 19.3A) between two carbon atoms as a result of the sharing of four electrons between them. Such a double bond is not a sign of extra strength but of greater instability and suitable chemical treatment causes this bond to break. When this occurs simultaneously amongst many molecules of the gas, the resultant units (Figure 19.3B) are able to link up, forming long chain molecules of

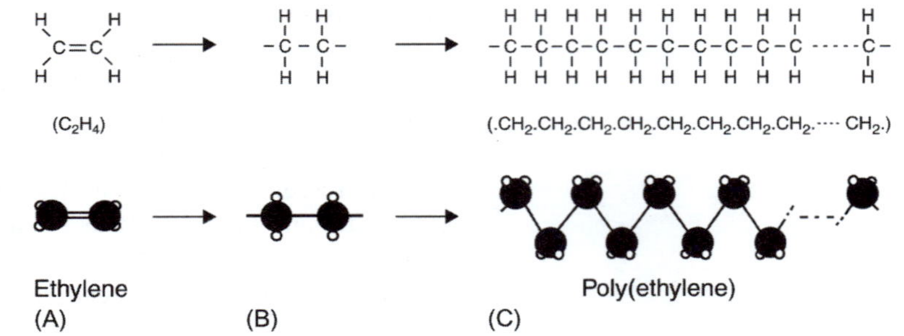

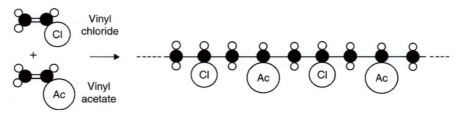

Figure 19.3 *The polymerisation of ethylene to polyethylene. The lower part of the illustration shows the arrangement of the carbon and hydrogen atoms.*

polyethylene (Figure 19.3C). This type of chemical process is known as *polymerisation,* and the product is called a *polymer.* The simple substance from which the polymer is derived is called a *monomer,* so in this case the gas ethylene is the monomer of polyethylene.

Sometimes the term *mer* is used in this context. A mer is a single unit which, though it may not exist by itself, occurs as the simplest repetitive unit in the chain molecule. Figure 19.4 thus shows the mer in polythene.

Frequently, a polymer molecule is built up from different monomers, arranged alternately in the chain. Such a substance is called a *copolymer.* In this way, molecules of vinyl chloride and vinyl acetate can be made to polymerise (Figure 19.5). The result is the plastics material polyvinyl chloride acetate, formerly used in the manufacture of LP gramophone records. It is in fact possible to use more than two monomers in making a copolymer. Thus acrylonitrile can be made to copolymerise with butadiene and styrene to produce acrylonitrile–butadiene–styrene or *ABS* as it is usually known industrially.

Some super-polymers consist of chain molecules built up from very complex monomers. This often leads to the operation of considerable forces of attraction between the resultant molecules at points where they lie alongside each other; consequently, such a material lacks plasticity, even when the temperature is increased. Cellulose is a natural polymer of this type.

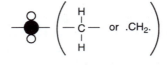

Figure 19.4 *The polythene mer.*

Figure 19.5 *The formation of a copolymer (polyvinyl chloride acetate).*

19.3.1 Plasticisers

In 1846, Dr. Frederick Schönbein, of the University of Basle, produced cellulose nitrate by treating ordinary cellulose with nitric acid. When heated, it proved to be slightly more plastic than ordinary cellulose. This is because the 'nitrate' side branches (Figure 19.6B), which became attached to the chain molecules as a result of the treatment with nitric acid, acted as 'spacers' which separated the chain molecules and so reducing the forces of attraction between them.

In 1854, Alexander Parkes, the son of a Birmingham industrialist, began to experiment with cellulose nitrate, with the object of producing a mouldable plastic from it. He found that, by adding some camphor to cellulose nitrate, a mixture was produced which passed through a mouldable stage when hot. Not long afterwards, this substance was produced commercially under the name of 'celluloid'. The bulky camphor molecules separate the chain molecules of cellulose nitrate by even greater distances, so that the forces of attraction between them are reduced further. Thus, the camphor is termed a *plasticiser* (Figure 19.6C). Celluloid is a very inflammable material, and this fact has always restricted its use. Before other plastics materials were freely available, much celluloid was used in the manufacture of dolls and other children's toys, often with tragic results when these articles were brought too near a fire.

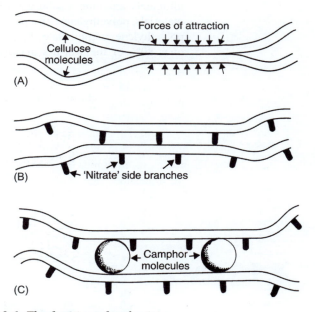

Figure 19.6 *The function of a plasticiser.*

19.4 Thermoplastic materials

In general terms, the properties of thermoplastic polymers depend on:

- The lengths of individual molecular chains.
- Whether the chains have side branches.
- Whether the chains include 'lumpy' molecules.
- The strengths of the bonds within chains.
- The strengths of the bonds between chains.

Linear polymer chains give quite a flexible polymer but the presence of side branches or 'lumpy' molecules have a stiffening effect. Side branches, 'lumpy' molecules and strong forces between chains all increase the melting temperature.

The following are commonly encountered thermoplastic materials.

19.4.1 Vinyl plastics

Presumably, the term *vinyl* is derived from 'vin' (as in 'vineyard') since the aromatic nature of some vinyl compounds is reminiscent of that of wine. Some of the vinyl compounds undergo polymerisation of their own accord, 'monomers' linking together spontaneously at ambient temperatures to form 'polymers'. Often they begin as 'watery' liquids, but, if allowed to stand for some time, become increasingly viscous as polymerisation proceeds, ultimately attaining a solid glass-like state (Figure 19.7). Of these plastics materials, polyethylene (PE), polyvinyl chloride (PVC), polypropylene (PP), polyethylene terephthalate (PET) and polystyrene (PS) in total account for over 80% of the world's plastics market. In this section, the structural

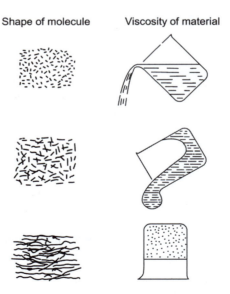

Figure 19.7 *Stages in the spontaneous polymerisation of some vinyl compounds.*

formulae of the units derived from the monomers are given to show the relationship of these polymers to ethylene.

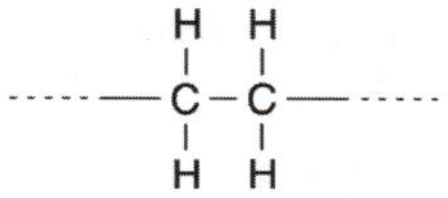

H H
| |
——C – C——
| |
H H

Figure 19.8 *Polyethylene.*

1 *Polyethylene (polythene) PE (Figure 19.8)*
This is probably the best known of the thermoplastics materials. It has an excellent resistance to corrosion by most common chemicals and is unaffected by foodstuffs. It is tough and flexible, has a high electrical resistivity, has a low density and is easily moulded and machined. Since it is also comparatively cheap to produce, it is not surprising that polythene finds such a wide range of applications.

Polythene is available in several different modifications. Low-density polythene (LDPE) consists of molecules containing many side branches (Figure 19.9A) and is usually about 50% crystalline (see Section 20.2), i.e. about half of it can be considered to be packed in an orderly manner in the solid – like atoms in a crystal. High-density polythene (HDPE) is approximately 90% crystalline, because it consists principally of linear, unbranched molecules which can more easily be packed together to give the orderly arrangement necessary for a crystalline structure. Being harder, it is used principally for the manufacture of bottles and similar utensils by blow-moulding (see Section 20.5.3). Its particular feature is that it has a superior resistance to environmental stress cracking.

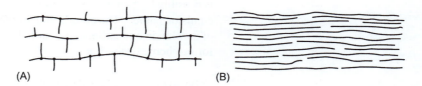

(A)   (B)

Figure 19.9 *Different molecular structures in polythene. (A) LDPE – here side-branching of the molecules increases the distance between adjacent chains. This reduces van der Waals forces so that LDPE is softer and weaker than HDPE (B) where there is no side-branching. In (B) the unbranched molecules lie closer together in a more orderly manner, so that crystallinity can be greater.*

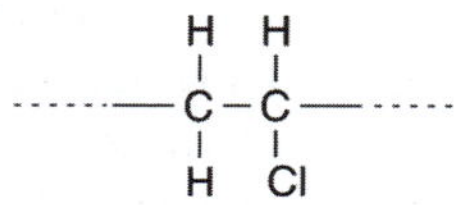

H H
| |
——C – C——
| |
H Cl

Figure 19.10 *Polyvinyl chloride.*

2 *Polyvinyl chloride (PVC) (Figure 19.10)*
The gas vinyl chloride was discovered more than a century ago. It was found that, on heating, it changed to a hard, white solid, later identified as polyvinyl chloride. Having rather a high softening temperature, PVC was difficult to mould, and it was not until the late 1920s that it was discovered that it could be plasticised. By adjusting the proportion of liquid plasticiser used, a thermoplastic material can be produced which varies in properties from a hard rigid substance to a soft rubbery one. Consequently, during the Second World War, and since then, PVC has been used in many instances to replace rubber. Protective gloves, raincoats and garden hose are examples of such use. PVC can be shaped

Table 19.1 *Mechanical properties (average) of PVC related to the degree of plasticisation*

Degree of plasticisation	Tensile strength (MPa)	Percentage elongation
None	55	20
Low	35	200
High	17	400

by injection moulding, extrusion and the normal thermoplastic processes. It can also be compression-moulded without a plasticiser, to give a tough, rigid material such as is necessary in miners' helmets. The effects on strength and ductility of PVC relative to the degree of plasticisation are indicated in Table 19.1.

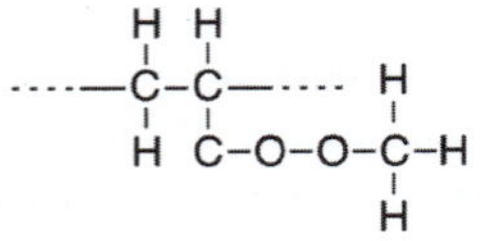

Figure 19.11 *Polyvinyl acetate.*

3   *Polyvinyl acetate (PVA) (Figure 19.11)*
This polymer softens at too low a temperature for it to be much use as a mouldable plastic, but it is useful as an adhesive, since it will stick effectively to almost any surface. It is supplied as an emulsion with water; when used in the manner of glue, the water dries off, leaving a strong adhesive polyvinyl acetate film for bonding the parts together. It is possibly best known as an emulsion with water, as the basis for 'vinyl emulsion paints'. The introduction of these was certainly a boon to those of us who, in our earlier days, battled with 'glue-size' based 'distemper' for our home decoration.

4   *Polyvinyl acetate/chloride copolymers*
These have already been mentioned (Section 19.3). In these, the very high softening temperature of PVC is reduced by combining it with vinyl acetate to form a copolymer of lower softening temperature. The LP gramophone records made from this have been superseded by poly-carbonate compact discs.

5   *Polyethylene–vinyl acetate (EVA)*
This is a copolymer of ethylene and vinyl acetate in which the proportions of vinyl acetate are varied between 3% and 15% to produce materials with different ranges of properties. The principal features of EVA are high flexibility, toughness, clarity and a high resistance to stress cracking, together with resistance to damage from ultraviolet radiation and ozone.

The rubbery nature and glossy finish of some forms of EVA make it useful for meat packaging and for cling-wrap purposes. Other forms are used for the moulding of automotive parts, ice-cube trays, road-marker cones, medical equipment, turntable mats and garden hose, whilst a harder cross-linked form is used for shoe soles. EVA is also the basis for a number of hot-melt adhesives.

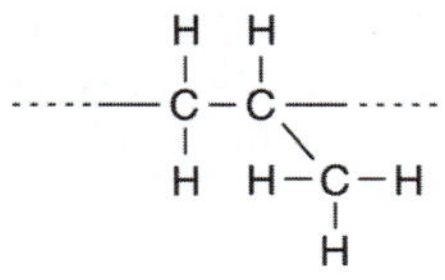

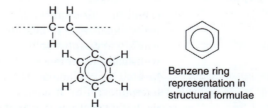

Figure 19.12 *Polypropylene.*

6   *Polypropylene (PP) (Figure 19.12)*

This is similar in structure and properties to polyethylene but it has a higher temperature tolerance. Whilst PE is produced from the gas ethylene, PP is polymerised from the gas propylene. PP is stronger than PE and is used for a wide variety of mouldings where greater strength and rigidity, than can be obtained with PE, are required. The higher melting point of PP as compared with HDPE makes it more suitable for fibre manufacture, whilst large amounts are also produced as clear film for wrapping cigarettes and crisps.

7   *Polypropylene–ethylene copolymers*

A disadvantage of PP is that it becomes brittle at about 0°C as the amorphous regions of the polymer become glassy (see Section 20.2). To reduce this tendency PP is copolymerised with PE. If 5–15% PE is used, toughness is increased without loss of rigidity. Such copolymers are used for pipes and pipe fittings, cable insulation and ropes. Increasing the amount of PE to 60% produces a rubbery copolymer, such elastomers being resistant to atmospheric attack.

8   *Polystyrene (PS) (Figure 19.13)*

PS was developed in Germany before the Second World War and is made by the polymerisation of styrene, a chemical substance known since 1830. PS has become one of the most important of modern thermoplastic materials. It is a glassy, transparent material similar in appearance to polymethyl methacrylate (PMMA) and has a high electrical resistivity. In the homo-polymeric state (just the single PS polymer) it is brittle and for this reason toughened versions of PS are used for most engineering and many other applications. *High-impact polystyrene* (HIPS) is produced by the addition of some polybutadiene – a rubber – which is present in the structure as tiny spheres. However, PS is probably best known in its 'expanded' or foamed form (see Section 20.4), such a material being widely used for packaging.

PS hollow-ware is fairly easy to identify as such since it emits a resonant 'metallic' note when tapped smartly. PS can be cast, moulded, extruded and shaped easily by the ordinary methods of sawing, filing

Benzene ring
representation in
structural formulae

Figure 19.13 *Polystyrene: the 'ring' of carbon atoms is the nucleus of a benzene molecule and the significance of the circle in its centre is to indicate that all the valency electrons joining its six carbon atoms are shared equally by the six.*

and drilling. Mouldings in HIPS have a lower tensile strength than those in ordinary PS but a much higher impact value, particularly at low temperatures, for which reason they are extremely useful as refrigerator parts.

9   *Acrylonitrile–butadiene–styrene (ABS)*

The 'rubber toughening' of styrene has been described earlier and the same principle is carried further in ABS plastics materials which are copolymers of acrylonitrile, butadiene and styrene (Figure 19.14). Some of the butadiene polymerises as globules in HIPS but the remainder co-polymerises with acrylonitrile and styrene to produce a material of outstanding resistance to fracture by impact, combined with a high tensile strength and abrasion resistance. So much so that luggage made from ABS resists all but the most serious attacks by airport baggage-handling systems! ABS is resistant to acids and alkalis as well as to some organic solvents and is available in the form of moulding powders for use in injection-moulding and extrusion; sheet is also available for vacuum-forming processes.

Large amounts of ABS are used in the automobile industry where glass-fibre reinforced plastics are increasingly used for panel work. An added bonus for ABS and PP was that they could be chromium plated when used in radiator grilles and similar components. In recent years, the fashion has turned away from 'chrome' in favour of more sombre hues so that now everything from car bumpers to hi-fi- and TV cabinets and cameras are black. Of course, carbon black is much cheaper than increasingly expensive chromium. However, I suppose I must suppress this unwarranted cynicism!

Table 19.2 summarises the properties and uses of vinyls.

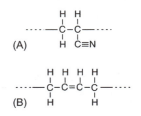

Figure 19.14 *ABS: (A) shows acrylonitrile and (B) butadiene. These are copolymerised to give ABS.*

19.4.2 Fluorocarbons

The most important of these materials is *polytetrafluoroethylene* (PTFE). It can be regarded as polythene in which all of the hydrogen atoms have been replaced by atoms of the extremely reactive gas fluorine (Figure 19.15). Since fluorine is so chemically reactive, its compounds, once formed, are difficult to decompose. Thus PTFE will resist attack by all solvents and corrosive chemicals; for example, hot concentrated sulphuric acid does not affect it, whilst *aqua regia* (a mixture of concentrated nitric and hydrochloric acids which will dissolve gold and platinum) leaves PTFE unharmed. It will also withstand a wider range of temperatures than most plastics materials and can be used for continuous working between –260°C and +250°C (+300°C for intermittent use). It is an excellent electrical insulator but above all it has the lowest coefficient of friction of any solid material, a feature which also makes it difficult to coat onto other surfaces.

The high cost of fluorine – and hence of PTFE – at present restricts the use of this material except where its particular properties of low coefficient of friction (bearings and non-stick coatings for frying pans) and resistance

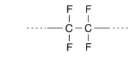

Figure 19.15 *PTFE.*

Table 19.2 *Properties and uses of vinyl thermoplastics materials*

Compound		Relative density	Tensile strength (MPa)	Chemical resistance	Safe working temperature	Relative cost	Typical uses
Polythene (PE)	HD	0.95	30	Excellent	120	Low	Acid-resisting linings. Babies' baths, kitchen and other household ware. Piping, toys, fabric filaments.
	LD	0.92	13	Excellent	80	Low	Sheets, wrapping material, polythene bags, squeeze bottles, electrical insulation, bottle caps, tubing for ball-point pens, ink cartridges.
Polyvinyl chloride (PVC)	Un-plasti-cised	1.40	55	Good	70	Low	Domestic/industrial piping (rainwater, waste, etc.), light fittings, curtain rail (with metal insert) radio components, safety helmets, ducting, plating vats.
	Plasti-cised	1.30	35	Good	100	Low	Artificial leather cloth, gloves, belts, raincoats, curtains, packaging, cable covering, protective clothing.
Polyvinyl chloride/acetate		1.3	25	Fairly good	70	Moderate	Gramophone records, screens, containers, chemical equipment, protective clothing.
Ethylene vinyl acetate (EVA)		0.94	25	Good	70	Moderate	Meat packaging, cling wraps, turntable mats, automobile parts, garden hose, car-door protectors, shoe-soles, road cones, surgical ware. Adhesives.
Polypropylene (PP)		0.90	33	Excellent	100	Low	Packaging, pipes and fittings, cable insulation, battery boxes, refrigerator parts, sterilisable bottles and other uses where boiling water is involved, cabinets for TV and radio sets, fan blades, crates and containers, stackable chairs.
Polystyrene (PS)	General purpose	1.05	45	Fairly good	80	Low	Moulded containers (food and cosmetic), boxes, toilet articles. *Foams:* ceiling tiles, heat insulation, packaging for fragile equipment.
	HIPS	1.02	20	Fairly good	80	Low	Radio and TV cabinets, vacuum cleaners, kitchen equipment, refrigerator parts, vending machine cups, cases for cheap cameras.
Acrylonitrile—butadiene—styrene (ABS)		1.01	35	Very good	80	Moderate	Pipes, radio cabinets, tool handles, protective helmets, textile bobbins, pumps, battery cases, luggage, typewriter and camera cases, telephone handsets, hairdryers, large amounts in automobile bodywork.

Table 19.3 *Properties and uses of fluorocarbon thermoplastics*

Compound	Relative density	Tensile strength (MPa)	Chemical resistance	Safe working temperature (°C)	Relative cost	Typical uses
Polytetrafluoroethylene (PTFE) ('Teflon')	1.01	25	Excellent	250	Very high	Gaskets, valve packaging, inert laboratory equipment, chemical plant, piston rings, bearings, non-stick coatings (frying pans), filters, electrical insulation.

to chemical attack (chemical plant) are utilised. Because of its very low coefficient of friction, bearing surfaces coated with PTFE can be used without lubrication at temperatures approaching 250°C. Similarly, chemical plant carrying corrosive liquids can be used at that temperature.

Table 19.3 summarises the properties and uses of PTFE.

19.4.3 Cellulose-base plastics (cellulose esters)

Cellulose-base plastics are derived from natural cellulose, one of the world's most plentiful raw materials. It occurs in many forms of plant life but much of the raw cellulose used in the plastics industry is as cotton linters (cotton fibres too short for spinning to yarn). These 'cellulosics' were amongst the first thermoplastics materials to be developed. In the early days of the cinema, the film base was of cellulose nitrate. This was so inflammable that the unfortunate projectionist was often incarcerated, along with his film and a powerful carbon-arc 'lamp', in a fire-proof cabin in order to safeguard the audience.

This very dangerous 'nitrate' film has since been replaced by *cellulose acetate* (CA) for this, and other purposes, since CA is virtually nonflammable. Although only moderately strong and tough, CA is fairly cheap and is used for moulding a wide range of articles from pens and pencils to toys and toothbrush handles.

Cellulose acetate–butyrate (CAB) is tougher and more resistant to moisture than is ordinary CA and at the same time retains the other useful properties of CA. Because of its greater resistance to moisture, it is useful for the manufacture of handles for brushes and cutlery. Cellulose acetate–propionate (CAP) is slightly tougher and more ductile than the other cellulosics but is used for similar purposes. Table 19.4 summarises the properties and uses of cellulosics.

Table 19.4 *Properties and uses of cellulosic thermoplastics*

Compound	Relative density	Tensile strength (MPa)	Chemical resistance	Safe working temperature (°C)	Relative cost	Typical uses
Cellulose acetate (CA)	1.3	35	Fair	70	Fairly high	Artificial leather, brush backs, combs, spectacle frames, photographic film base, mixing bowls, lamp-shades, toys, laminated luggage, knobs, wire- and cable-covering.
Cellulose acetate–butyrate (CAB)	1.18	35	Fair	70	Fairly high	Illuminated road signs, extruded pipe, containers.
Cellulose acetate–propionate (CAP)	1.21	55	Fair	70	Fairly high	Steering wheels, packaging, toothbrushes, door knobs.

19.4.4 Polyamides (PA)

Polyamides (PA) include a number of compounds better known by the collective name of *nylon*. The more common members of this series of polymers are designated nylon 6 (Figure 19.16), nylon 6.6, nylon 6.10, nylon 11 and nylon 12, of which the most popular forms are 6 and 6.6. In each case, the first digit indicates the number of carbon atoms in the repeating unit or mer in the polymer chain. Where there are two numbers, the full stops separating them are sometimes omitted. Where there are two numbers, it means there are two different blocks which constitute the full repeating unit.

Nylon 6.6 has a higher melting point than nylon 6 and is also stronger and stiffer. Nylon 11 has a lower melting point and is more flexible. In general, nylons are strong, tough materials with relatively high melting points, but they do absorb moisture.

Although nylon 6.6 will always be associated with ladies' hosiery and underwear, it is in fact one of the most important engineering thermoplastics.

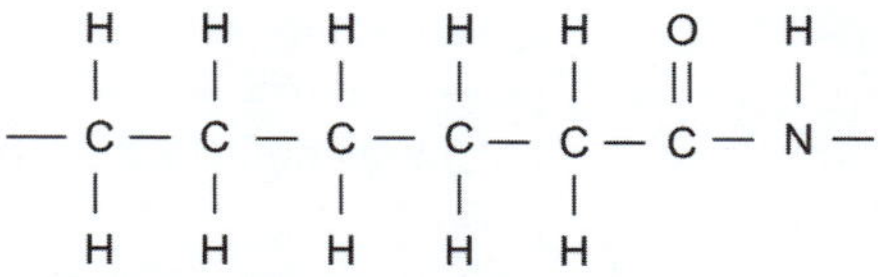

Figure 19.16 *The repeating unit for nylon 6.*

It is a strong, tough, hard-wearing material with a low coefficient of friction. Its softening temperature is relatively high so that it can be safely heated in boiling water. However, nylons do absorb considerable amounts of moisture which reduces strength whilst increasing toughness. Nylons have very good resistance to most organic solvents, oils and fuels and are inert to inorganic reagents with the exception of mineral acids and chloride solutions containing Cl^- ions.

The high strength of nylon fibre claimed my personal interest and attention about half way through my rock-climbing career when nylon ropes were introduced. I soon realised that a nylon 'line' with an equivalent 'potential life-saving capacity' to that of the old hemp rope was but a fraction of the weight of the latter – a factor to be considered when a long trek in high mountain country was involved. The conservative 'hemp rope' school warned me that friction against rock could cause nylon to melt! However, it was never my intention to find myself dangling at the end of a long rope swinging pendulum-wise across a cliff face whilst wondering what the heck to do next. Hemp and manila ropes are still used on ships and in dockyards, but modern climbing ropes now are invariably synthetic fibres like nylon and polypropylene.

In addition to the use of nylon 6.6 in the fibre and textile field, nylons have many applications in engineering industries, particularly as wearing and moving parts where high impact strength is necessary. Small gears operating at low noise levels in music-reproduction equipment and electric clocks are examples. These polymers are injection moulded or extruded as rod, sheet or tube from which components can be machined. Some forms of nylon 6 are cast. The high-pressure lubricant molybdenum disulphide MoS_2 is sometimes added to nylon 6 destined for use as a bearing material. Nylon 6.6 often has glass fibres included in it (see Chapter 24) to give a composite material which has good properties under impact loading, good resistance to oils and solvents and can be used at relatively high temperatures. Such a material is injection moulded to form power-tool housings. Table 19.5 summarises the properties and uses of polyamides.

Table 19.5 *Properties and uses of polyamide thermoplastics*

Compound	Relative density	Tensile strength (MPa)	Chemical resistance	Safe working temperature (°C)	Relative cost	Typical uses
Nylon 66 (PA66)	1.12	Moulded 60 Filaments 350	Good	140	High	Raincoats, yarn (clothing), containers, cable-covering, gears, bearings, cams, spectacle frames, combs, bristles for brushes, climbing ropes, fishing lines, shock absorbers.

19.4.5 Polyesters

These are sometimes of the 'setting' form (see Section 19.5) but one important member of the group is thermoplastic. This is polyethylene terephthalate (PET) – better known as *Terylene*. Like nylon, this was originally produced as a 'man-made' fibre developed in Britain during the Second World War and adopted by the textile industry. Since then, uses of PET have developed further, first in the injection moulding of electrical components and then to the manufacture of bottles for a wide range of beers and soft drinks. When these drinks contain dissolved gases, such as CO_2 to give the 'fizz', under pressure, diffusion of the gas through the walls of the PET bottle can take place. Such diffusion is more rapid through an amorphous (glassy) structure than through a crystalline structure, hence the forming process is controlled to produce a highly crystalline structure. Thus, the process of 'stretch blow moulding' was developed. In this process, extruded PET tube is heated and then simultaneously stretched and blown into the mould. The temperature and speed are controlled so that just below the glass transition temperature (see Section 20.2) very small crystals are formed from the amorphous structure. Because the crystals are very small, the bottle retains its transparent

Table 19.6 *Properties and uses of polyester thermoplastics*

Compound	Relative density	Tensile strength (MPa)	Chemical resistance	Safe working temperature (°C)	Relative cost	Typical uses
Polythene terephthalate ('Terylene', 'Dacron') (PET)	1.38	Moulded 60 Fibres 175	Moderate	85	Moderate	*Fibres:* a wide range of clothing. *Tape and film:* music and recording tapes, insulating tape, gaskets. *Moulded*: electrical plugs and sockets. *Blow moulding:* bottles for beer and soft drinks.
Polycarbonate (PC)	1.2	66	Good	140	High	Very tough – protective shields (police vehicles), hairdryer bodies, telephone parts, automobile tail-light lenses, tool handles, machine housings, babies' bottles, vandal-proof street light covers, safety helmets, CDs (music and data storage).

appearance and is also strong. The stretching of the PET in two directions during the blow moulding promotes the crystallisation process. Nevertheless, to make such bottles completely gas tight, they can be given an outer coating of polyvinylidene chloride (PVDC). This is important to prevent the ingress of oxygen which would spoil the contents of beer bottles.

Polycarbonates (PC) are structurally line-chain thermoplastic polyesters. They have excellent mechanical properties – particularly high strength and high impact toughness. Since they are also transparent, they are useful for the manufacture of vandal-proof light globes and for babies' feeding bottles. Their temperature tolerance is good so that they are used in high-temperature lenses, coffee pots and the like. A combination of suitable optical properties, along with mechanical toughness, a high heat-distortion temperature, ease of processing and good solvent resistance has led to the use of PC for the manufacture of CDs and similar data-storage products.

Table 19.6 summarises the properties and uses of thermoplastic polyesters.

19.4.6 Polyacetals

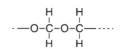

Figure 19.17
Polyoxymethylene.

'Acetal resin' in its simplest form (homo-polymer) structure is a highly crystalline form of polymerised formaldehyde or polyoxymethylene (POM) (Figure 19.17). It has a high yield strength, both at ambient and at high temperatures, associated with a high modulus of elasticity. Resistance to creep is greater than with most other plastics. Impact toughness is high and is only slightly less at –40°C than at ambient temperatures but it is notch sensitive so that sharp corners in design are best avoided. It is a hard material with good dimensional stability whilst abrasion resistance is good and its coefficient of friction low. Moisture absorption is very low, chemical resistance high and electrical insulation properties good. Heat distortion temperature is very high – near its melting point. POM can be shaped by injection moulding or extrusion, though wrought forms cannot be shaped by heating and stretching. However, if necessary, it can be reground and used again. Its machining properties are excellent and are similar to those of alpha/beta brasses. But it is expensive.

Table 19.7 *Properties and uses of polyacetal thermoplastics*

Compound	Relative density	Tensile strength (MPa)	Chemical resistance	Safe working temperature (°C)	Typical uses
Homo-polymer	1.4	70	Fairly good	95	Bearings, cams, gears, flexible shafts, office machinery, carburettor parts, pump impellers, car instrument panels, knobs, handles, water pumps, washing machine parts, seat-belt buckles.
Copolymer	1.4	60	Fairly good	95	

Copolymer polyacetals contain extra CH_2 units in the polymer chain. They are generally less strong than the homo-polymer but tend to retain strength at elevated temperatures over longer periods. Polyacetals are particularly useful in applications requiring a high degree of dimensional accuracy, especially under conditions of varying humidity. Their combination of properties enables them to replace metals in many cases, e.g. die-casting.

Table 19.7 gives the properties and uses of the homo-polymer and the copolymer.

19.4.7 Acrylics

Acrylics are a group of vinyl plastics of which the most important is polymethyl methacrylate (PMMA). It is a clear, glass-like, plastics material better known as *Perspex* in Britain or *Plexiglass* in the USA and was developed during the Second World War for use in aircraft. Not only is it much tougher than glass, but it can easily be moulded. It is produced by the polymerisation of methyl methacrylate. Table 19.8 gives the properties and uses of PMMA.

Since it will transmit more than 90% of daylight, and is much lighter and tougher than glass, it can be used in the form of corrugated sheets, interchangeable with those of galvanised iron, for use in industrial buildings. Lenses can be made from PMMA by moulding from powder. This is an inexpensive method of production as compared with the grinding and polishing of a glass blank. Unfortunately, PMMA lenses scratch very easily.

19.4.8 High-temperature thermoplastics

The rapid development of plastics materials following the Second World War led to a demand for thermoplastics with higher temperature tolerance. Some

Table 19.8 *Properties and uses of acrylic thermoplastics*

Compound	Relative density	Tensile strength (MPa)	Chemical resistance	Safe working temperature (°C)	Relative cost	Typical uses
Polymethyl methacrylate (PMMA) ('Perspex', 'Plexiglass')	1.18	55	Fairly good	95	Moderate	Aircraft glazing, building panels, roof lighting, baths, sinks, protective shields, advertising displays, windows and windscreens, automobile tail lights, lenses, toilet articles, dentures, knobs, telephones, aquaria, double glazing, garden cloches, shower cabinets.

of these materials are of complex structure and are produced by complicated processes; these matters will not be discussed here. One important point, however, which needs mention is that the linear chain molecules are built up from bulky ring-type units. Van der Waals forces, acting between these large units in adjacent chain molecules, are therefore relatively high and this results in a high softening temperature, high thermal stability and exceptional heat resistance. The following are three such plastics:

1   *Polyimides*
    These are, structurally, similar to the polyamides (nylons) and were the first of the high-temperature thermoplastics of this group to be developed in the early 1960s. As described earlier, these have bulky, complex ring structure units along the polymer chain and these give rise to increased van der Waals forces between the long-chain molecules so that a higher temperature is necessary to cause their separation – greater heat input causes the molecules to vibrate more vigorously until they separate. Thus, polyimides have a very high heat resistance and are useful for high-temperature service. Even after exposure for 1000 hours in air at 300°C, about 90% of the original tensile strength is retained.

2   *Polysulphones*
    These are, in some respects, similar in structure to the polycarbonates, and have linear molecular chains containing the bulky benzene ring structure (Figure 19.18). These large units contribute to increased van der Waals forces and so higher softening temperatures. The polysulphones are more expensive than polycarbonates and are therefore only used when the properties of the latter are inadequate. The creep resistance of polysulphones is excellent and superior to that of polycarbonates. They have good temperature resistance and rigidity, also transparency. They do not burn readily and then do not present a smoke hazard. As a result, they are widely used in aircraft for parts of passenger-service units, e.g. air ducts and vents, call buttons, lights, loudspeakers. Polysulphones can be injection moulded, extruded and thermoformed.

3   *Polyether ether ketone (PEEK)*
    These have outstanding heat-resistant properties as a consequence of the presence of the benzene ring as a unit in the linear polymer chains (Figure 19.19). In addition to its high softening temperature, it is resistant to oxidation and has low flammability. It is useful as a moulding material

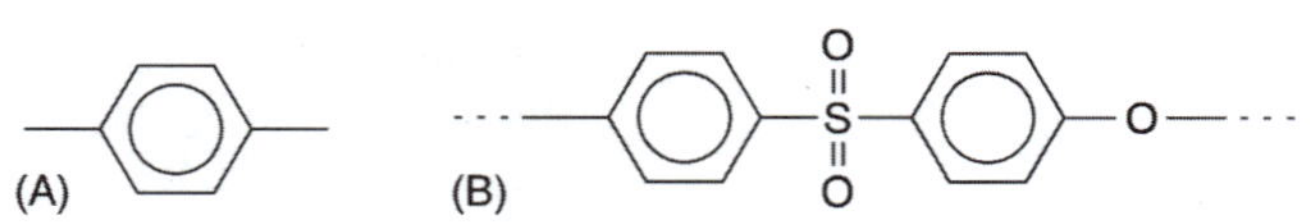

Figure 19.19 *PEEK.*

Figure 19.18 *The linear molecular chains contain the bulky unit shown in (B) to give the polysulphone structure.*

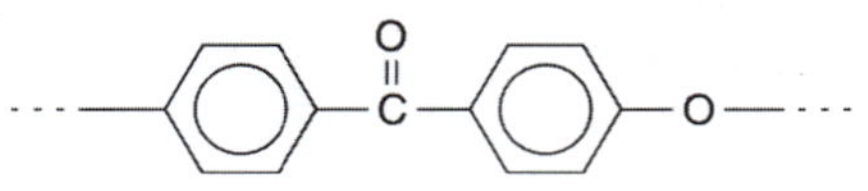

Table 19.9 *Properties and uses of high-temperature thermoplastics*

Compound	Relative density	Tensile strength (MPa)	Chemical resistance	Safe working temperature (°C)	Relative cost	Typical uses
Polyimides	1.42	90	Good	350	High	Seals, gaskets, piston rings, jet engine compressor seals, pressure discs, bearings, friction elements.
Polysulphones	1.24	70	Good	175	High	Hairdryers, oven, iron and fan heaters, microwave oven parts, pumps for chemical plant, transparent pipelines, hot-water dispenser cups for drinks machines, passenger-service unit parts in jet aircraft, integrated circuit boards.
Polyether ether ketone (PEEK)	1.28	92	Good	150	Very high	Aggressive environments (nuclear plant), oil and geothermal wells, high-pressure steam valves, aircraft and automobile engine parts, wire covering, filament for weaving high-temperature filtration cloth.

for aggressive environments. It has superior toughness and as a coating for wire it resists cuts and fracture caused by sharp corners. Fatigue resistance is very high.

Table 19.9 gives the properties and uses of these high-temperature thermoplastics.

19.5 Thermosets

In thermosets, a chemical change is initiated during the moulding process, and 'cross-links', in the form of strong covalent bonds, are formed between adjoining molecules. The powerful forces associated with these covalent bonds are greater than the van der Waals forces operating between the chain molecules in a thermoplastic substance. Consequently, the molecules are unable to slide over each other and a strong, rigid three-dimensional network is formed (Figure 19.20). Thus, *thermosetting* has taken place.

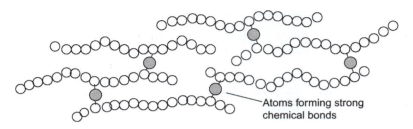

Figure 19.20 *Covalent bonds between neighbouring chains in a thermoset, resulting in a three-dimensional network.*

19.6 Thermoset materials

The following are examples of thermoset materials.

19.6.1 Phenolics

These are probably the best known of the thermosetting group of materials:

1 *Phenol formaldehyde (PF)*

The Crimean War (1853–1856) is remembered chiefly for the military blunders which led to the decimation of the 'Gallant Six Hundred' in the Charge of the Light Brigade. Nevertheless, in contemplating this melancholy prospect – as Sir Winston Churchill might have described it – we should reflect on the heroic works of one, Florence Nightingale, whose efforts in the reorganisation and discipline of field hospitals reduced the death rate of wounded soldiers from 42% to a little over 2%. In those days, the badly wounded had little hope of survival as the inevitable *septicaemia* took control. The introduction of a new disinfectant known as *carbolic acid* was, however, an important factor in this clinical revolution.

Carbolic acid – or phenol as we now know it – is a colourless, crystalline solid and a weak acid. It was used in 1907 by Dr. Leo Bakeland to develop the first synthetic plastics material 'bakelite'. This name became synonymous with plastics materials generally as the plethora of 'brown-bakelite-and-chrome' artefacts of the 1930s filled our homes, but we can't blame Dr. Bakeland for that!

The phenol molecule is based on that of benzene in that one hydrogen atom has been replaced by the –OH radical (Figure 19.21A). If phenol is mixed with a *limited* amount of formaldehyde (H.CHO), then the molecules of each react to form long-chain molecules (Figure 19.21C). In this condition, the material is brittle but thermoplastic, and can be ground to a powder suitable for moulding. This powder (or 'novolak') contains other materials, which on heating will release more formaldehyde, so that during the final moulding process, cross-links are formed between the chain molecules (Figures 19.21D and 19.22). The resulting 'network' structure represented in Figure 19.21D can only be shown two-dimensionally on flat paper, but it is in fact a three-dimensional

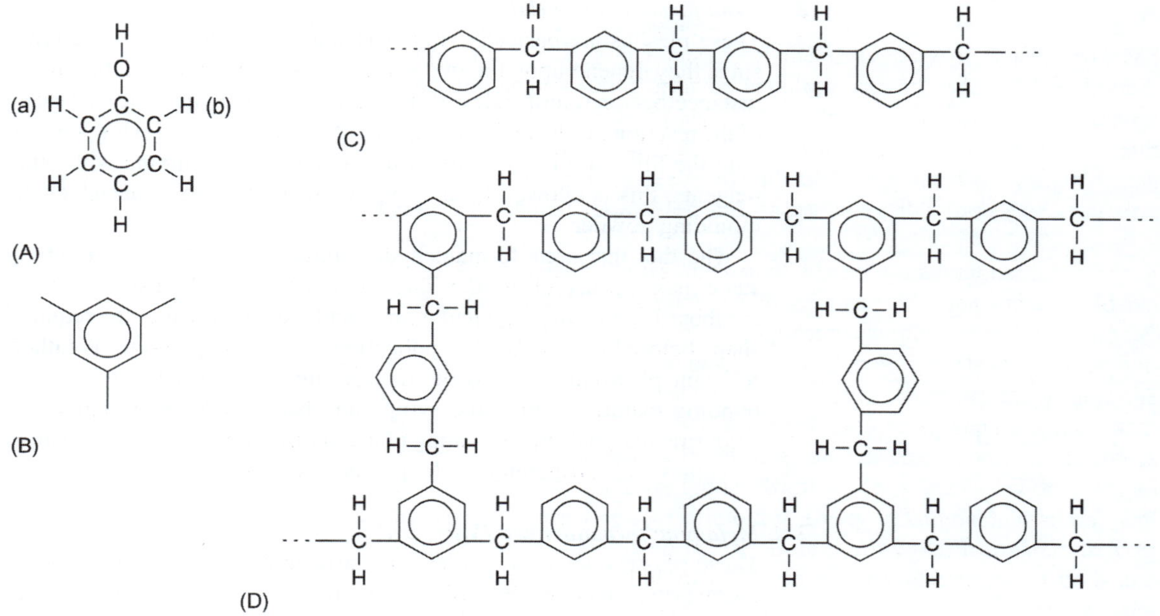

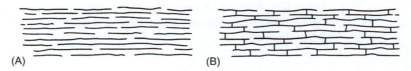

Figure 19.21 *(A) The phenol molecules and (B) its representation in the following molecules, the –OH and H atoms which take no part in the reaction have, for clarity, been omitted. (C) The long chain molecule when phenol reacts with formaldehyde, (D) the network structure produced on heating with more formaldehyde during moulding.*

Figure 19.22 *Thermosetting in a polymer material: (A) illustrates the structure at the 'novolak' stage, here the material is still thermoplastic and the chain molecules are held together only by van der Waals forces, (B) shows the effect of thermosetting, here the linear molecules are now held firmly together by powerful covalent bonds ('cross-links').*

cage-like network in which the units have 'set' *permanently*. Since a non-reversible chemical reaction has taken place, phenol formaldehyde cannot be softened again by heat.

This type of plastic is hard and rigid, but tends to be rather brittle in thin sections. However, it has a good electrical resistivity, so it is not surprising that the 'bakelite' industry grew in conjunction with the electrical industries from the late 1920s onwards. Various mouldings for wireless-cabinets, motor-car parts, bottle-tops, switchgear, electric plugs, door-knobs and a host of other articles were commonly made from bakelite. In most of these articles, a wood-flour filler is used; but, if greater strength is required a fibrous filler (paper, rags, jute, or sisal) is employed.

2   *Urea formaldehyde (UF)*

These plastics are basically similar to the phenol–formaldehyde types since they depend upon the two-stage cross-linking reaction, but in this instance between molecules of urea and formaldehyde. The first stage of the reaction results in the formation of a syrupy material. This is mixed with the filler, to give a moist, crumbly mass. After mixing with other reagents, this is allowed to dry out, producing the urea–formaldehyde moulding powder.

The fact that urea–formaldehyde mixture passes through a syrupy stage makes it useful in other directions. For example, paper, cloth or cardboard can be impregnated with it, and then moulded to the required shape before being set by the application of heat and pressure. Weather-resistant plywood can also be made using urea formaldehyde as the bonding material, whilst the syrup can also be used as a vehicle for colouring materials in the coating of metal furniture, motor cars, washing machines and refrigerators with a layer of enamel.

3   *Melamine formaldehyde (MF)*

These plastics are of a similar type structurally to the two foregoing thermosets, but are much harder and more heat resistant. Moreover, they are most resistant to water and consequently find use in the household as cups and saucers, baths and sundry kitchen utensils, particularly where greater heat-resistance is necessary.

Table 19.10 gives the properties and uses of the above phenolic thermosets. In general, thermosets have high thermal stability, high stiffness, low densities, high electrical insulating properties, good thermal insulation properties and high dimensional stability. Thermosets are stronger and stiffer than thermoplastics.

19.6.2 Polyester resins

Polyester resins of the thermoplastic type have already been mentioned in this chapter. These are of the 'straight-chain' molecule form, which are incapable of linking up with other adjacent chain molecules; they are therefore thermoplastic.

However, by using special monomers, polyesters can be produced with side-chains capable of forming cross-links between adjacent molecules and these therefore are 'setting plastics', of which some are of the cold-setting type. For example, if styrene is added to the liquid polyester, it provides the necessary cross-links, and the liquid gradually sets without the application of heat. Used with glass-fibre to provide reinforcement, these materials are useful in building up such structures as the hulls of small boats, wheelbarrows and car bodies. The glass fibre, impregnated with the liquid mixture, is built up a layer at a time on a suitable former. Pressure is not required, except to keep the material in position until it has set. These glass fibre/polyester resin composites are very strong and durable. See Table 19.11 for properties and uses.

Table 19.10 *Properties and uses of phenolics thermosetting materials*

Compound	Relative density	Tensile strength (MPa)	Chemical resistance	Safe working temperature (°C)	Relative cost	Uses
Phenol formaldehyde ('bakelite', 'phenolic') (PF)	1.45	50	Very good	120	Low	Electrical equipment, radio-cabinets, vacuum-cleaners, ashtrays, buttons, cheap cameras, automobile ignition systems, ornaments, handles, instrument-panels, advertising displays, novelties and games, dies, gears, bearings (laminates), washing-machine agitators.
Urea formaldehyde ('urea') (UF)	1.48	45	Fair	80	Moderate	Adhesives, plugs and switches, buckles, buttons, bottle-tops, cups, saucers, plates, radio-cabinets, knobs, clock-cases, kitchen equipment, electric light-fittings, surface coatings, bond for foundry sand.
Melamine formaldehyde ('melamine') (MF)	1.49	50	Good	130	Moderate	Electrical equipment, handles, knobs, cups, saucers, plates, refrigerator coatings, trays, washing-machine agitators, radio-cabinets, light-fixtures, lamp-pedestals, switches, buttons, building-panels, automotive ignition-blocks, manufacture of laminates.

Alkyd resins constitute a further group of polyesters. They are thermosetting materials originally introduced as constituents of paints, enamels and lacquers. They have a high resistance to heat and a high electrical resistivity and are dissolved by neither acids nor many organic solvents. The resins are now available in powder form, which is generally compression-moulded. Liquid alkyds are used in enamels and lacquers for automobiles, stoves, refrigerators and washing machines (Table 19.11).

Table 19.11 *Properties and uses of polyester (setting types) thermosetting materials*

Compound	Relative density	Tensile strength (MPa)	Chemical resistance	Safe working temperature (°C)	Relative cost	Uses
Polyester	1.3	40	Fairly good	95	Moderate	Adhesives, surface coatings, corrugated and flat translucent lighting-panels, lampshades, radio grilles, refrigerator parts; polyester laminates are used for hulls of boats, car bodies, wheelbarrows, helmets, swimming-pools, fishing-rods and archery-bows.
Alkyd resins	2.2	25	Fair	230	Moderate	Enamels and lacquers for cars, refrigerators and stoves, electrical equipment for cars, light-switches, electric-motor insulation, television-set parts.

19.6.3 Polyurethanes

Polyurethanes constitute a group of very adaptable plastics, comprising both thermoplastic and thermosetting materials. They are generally clear and colourless. One type is used in the manufacture of bristles, filaments and films.

In another group of polyurethanes, carbon dioxide is evolved during the chemical process necessary for establishing cross-links between the chain molecules. This carbon dioxide is trapped by the solidifying polymer, thus producing a foam. The mechanical properties of the foam can be varied by using different materials to constitute the polymer. Thus, some of the foams become hard and rigid, whilst others are soft and flexible. Rigid thermosetting polyurethane foams are used as heat-insulators, and for strengthening hollow structures, since they can be poured into a space where foaming and setting will subsequently occur. Aircraft wings can be strengthened in this manner. Flexible sponges, both for toilet purposes and for seat upholstery, are generally of polyurethane origin. See Table 19.12 for properties and uses.

19.6.4 Epoxy resins

These are produced in a similar way to polyesters, i.e. by being mixed with a cross-linking agent which causes them to set as a rigid network of polymer molecules. They are used for manufacturing laminates, for casting and for the 'potting' of electrical equipment. Excellent adhesives can be derived from

Table 19.12 *Properties and uses of polyurethane thermosetting materials*

Compound	Relative density	Tensile strength (MPa)	Chemical resistance	Safe working temperature (°C)	Relative cost	Uses
Polyurethane (PUR)	1.2	Mainly foams	Good	120	High	Adhesives (glass to metal), paint base, wire-coating, gears, bearings, electronic equipment, handles, knobs. Foams are used for insulation, upholstery, sponges, etc. Rigid foams are used for reinforcement of some aircraft wings.

Table 19.13 *Properties and uses of epoxy resin thermosetting materials*

Compound	Relative density	Tensile strength (MPa)	Chemical resistance	Safe working temperature (°C)	Relative cost	Uses
Heat-resistant type	1.15	70	Good	200	Fairly high	Adhesives (metal gluing), surface coatings, casting and 'potting' of specimens. Laminates are used for boat hulls (with fibreglass), table surfaces and laboratory furniture, drop-hammer dies. Epoxy putty is used in foundries to repair defective castings.
General purpose	1.15	63	Good	80	Fairly high	

the epoxy resins available as syrups, and they are particularly useful for metal gluing. The reader will no doubt be familiar with those two-component adhesives available at DIY stores – and be aware of how effective the cold-setting process really is if he gets the tube caps mixed! Table 19.13 gives the properties and uses of epoxy resin thermosets.

Solid epoxy resins are sometimes mixed with phenolic resins for moulding purposes.

19.6.5 Polyimides

These are high-temperature resistant polymers which can be either thermoplastic or thermosetting according to their formulation. They are available as film or as solid parts which retain mechanical properties at 300°C

Table 19.14 *Properties and uses of polyimide thermosetting materials*

Compound	Relative density	Tensile strength (MPa)	Chemical resistance	Safe working temperature (°C)	Relative cost	Uses
Polyimide	1.43	40	Good	300	Fairly high	Bearings, compressor valves, piston rings, diamond abrasive wheel binders.

over long periods. Even at 500°C, properties are retained for a short time. Table 19.14 gives their properties and uses.

19.6.6 Silicones

The polymers mentioned so far in this chapter are based on the element carbon, and are known as *organic compounds*. Unfortunately, they all soften, decompose or burn at quite low temperatures. However, more than a century ago, it was realised that there was a great similarity in chemical properties between carbon and the element silicon. Since then, chemists have been trying to produce long-chain molecules based on silicon, instead of carbon, hoping that in this way materials would be discovered which lacked many of the shortcomings of carbon polymers. Common silicon materials include quartz, glass and sand – all substances which are relatively inert, and which remain unchanged after exposure to very high temperatures. Consequently, the more fanciful of our science-fiction writers have speculated on the possibility of a system of organic life, based on silicon instead of carbon, existing on the hot planets Venus and Mercury. Whilst their 'bug-eyed monsters' may remain a figment of the imagination, some progress has taken place here on Earth in producing polymers based on silicon, and called *organo-silicon compounds,* or *silicones*.

Some of these compounds – of which sand is the basic raw material – have properties roughly midway between those of common silicon compounds, such as glass, and the orthodox organic plastics. Silicones are based on long-chain molecules in which carbon has been replaced by silicon and oxygen. They are available as viscous oils, greases, plastics and rubbers. All are virtually non-combustible, and their properties remain constant over a very wide temperature range. Thus, silicone lubricating oils retain their fluidity more or less unchanged at temperatures low enough to make ordinary oils congeal. Similarly, silicone rubbers remain flexible from low sub-zero temperatures up to temperatures high enough to decompose ordinary rubbers. Silicones are water repellent, and are widely used for waterproofing clothes, shoes and other articles, whilst silicone jelly is useful as a moisture-proof coating and sealing compound.

By forming still longer chain molecules, silicone plastics can be produced. These are very useful as an insulating varnish for electrical equipment designed to work at high temperatures. As moulding plastics, silicones are used in the manufacture of gaskets and seals for engineering purposes where high temperatures are involved, since they retain their plasticity and sealing efficiency under such conditions.

Silicone-resin paints provide durable finishes which clean easily, and which do not deteriorate appreciably. Such a finish applied to a motor car would normally outlast the car, and with a minimum of attention.

19.7 Elastomers

Christopher Columbus and his sailors may well have been the first Europeans to handle natural rubber since it was reported that some tribes of South American Indians played ball games well before the days of Pele. Natural rubber is in fact derived from the sap of the rubber tree *Hevea brasiliensis.* Following the work of Goodyear in vulcanising rubber, production grew rapidly in Brazil, the home of the rubber tree, and the city of Manaus, complete with an opera house and many other trappings of civilised living, mushroomed in the middle of the Amazon jungle as the capital of a Rubber Empire. In 1876, however, some seeds of the rubber tree were smuggled out of Brazil by British botanists and planted in greenhouses in Kew Gardens. The young plants were sent to the Dutch East Indies (now Indonesia) and Malaya so that Britain was able to establish and control its own rubber industry. This naturally led to a severe contraction of the Brazilian rubber industry and a rather shabby Manaus is all that remains of its former glory.

19.7.1 Long chain molecules in rubber

In natural rubber, the long chain molecules are more complex in structure than in the simpler polymers such as polythene, and consist of chains of 44 000 or more carbon atoms in length. The monomer present in these rubber molecules is the carbon–hydrogen compound *isoprene,* so that natural rubber is really *polyisoprene.* When the isoprene mers are joined to form the large molecular chain (Figure 19.23), the side group makes the molecule rather 'lopsided' and this causes the molecule to bend into a folded or coiled form in order to accommodate these large side groups. Since these rubber molecules are folded and coiled, they possess elasticity in a similar manner to that of a coiled spring and immediately stress is applied there is an elastic response. Nevertheless, natural rubber stretches like dough when stressed, because the chain molecules slide past each other into new positions (Figure 19.24). This plastic flow takes place slowly because weak van der Waals

Isoprene
molecule

Mer

Poly(isoprene)

Figure 19.23 *The 'side groups', consisting of one carbon atom (the solid black circles) and three hydrogen atoms (the smaller open circles) cause the polyisoprene molecule to 'bend' into a coiled form.*

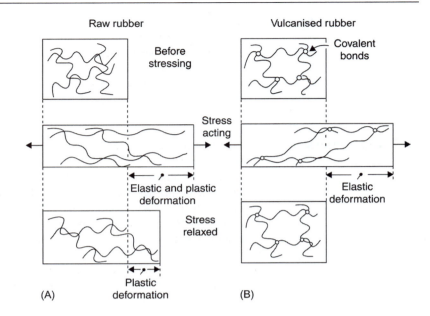

Figure 19.24 *Due to their folded or coiled form, rubber molecules become extended in tension but return to their original shapes when the stress is removed. In raw rubber (A) a steady tensile force will also cause separate molecules to slip slowly past each other into new positions because only weak van der Waals forces attract them to each other; so that when the force is relaxed some plastic deformation remains, though elastic deformation has disappeared. By vulcanising the raw rubber (B) the chain molecules are covalently bonded at certain points so that no permanent plastic deformation can occur and only elastic deformation is possible. This disappears when stress is relaxed.*

forces acting between molecules are overcome progressively as entangled molecules become very slowly disentangled. For this reason, a piece of unvulcanised natural rubber is both *elastic and plastic* at the same time, and, once stretched, will not return to its original shape.

19.7.2 Vulcanisation

When up to 3% sulphur is added to rubber prior to moulding, 'vulcanisation' will occur if the moulding temperature is high enough. Atoms of sulphur form strong covalent bonds between the rubber molecules so that they are permanently anchored to each other. Thus, whilst the vulcanised rubber retains its elasticity due to the folded or coiled nature of its molecules, it retains its shape permanently because of the cross-linking. If the vulcanisation process is carried further – that is, if more sulphur is used – so many cross-links are produced between adjoining molecules that the whole mass becomes rigid and loses most of its elastomer properties. The product is a black plastics material known as *ebonite* or *vulcanite* and was used extensively in the early days of plastics for the manufacture of such materials as fountain-pens, before other more suitable materials were developed.

The Japanese invasion of Britain's Far East rubber plantations in 1942 led to the development of synthetic rubber substitutes. In fact, synthetic poly-isoprene was produced. Since those days, the properties of these synthetic rubbers have considerably improved and rubbers with new properties developed, so that the industrial output of synthetic rubbers now exceeds that of natural rubber. As with natural rubber, sulphur is used to provide the cross-linking atoms in the vulcanisation process for many of these synthetic rubbers, but in some cases oxygen cross-linking atoms are used, the oxygen atoms being provided by magnesium oxide or zinc oxide.

To be of value in engineering, a rubber must have a low 'hysteresis', that is, it must return within very close limits to its original shape following each successive deformation cycle of loading and unloading. It must also have a low heat build-up, i.e. heat generated as a result of friction (braking in a motor vehicle) or as a result of absorbing vibration energy, must be dissipated adequately.

19.7.3 Engineering elastomers

In addition to natural rubber, the more important elastomers in general use include:

1 *Natural rubber (NR) and polyisoprene (IR)*
   The latter is manufactured from petrochemicals and has a molecular structure similar to that of natural rubber. Both are relatively cheap and used for commercial vehicle tyres, anti-vibration and anti-shock mountings and dock fenders. They have good mechanical properties in respect of strength, high elasticity and low hysteresis in the important working range. The low hysteresis makes it very 'bouncy'. Creep properties are also good. Unfortunately, the resistance to attack by oxygen and ozone is poor, though this can be overcome by the use of additives for most applications. Its properties make it useful in the range −50°C to +115°C.

   The tyres of the equipment transporter used by the Apollo 14 mission to the moon were made from a specially purified isoprene rubber.

2 *Styrene–butadiene rubber (SBR)*
   This is possibly the best known and most widely used of elastomers and was originally a replacement for natural rubber following the Japanese occupation of Malaya during the Second World War. It is a copolymer of styrene and butadiene with physical properties slightly inferior to those of natural rubber. Although ageing properties and abrasion resistance of SBR are slightly superior to those of natural rubber, it suffers more from heat build-up. Nevertheless, it is widely used for car tyres where adhesion and wear resistance are of paramount importance. Heat build-up is not a great problem here since the modern car tyres are

heavily reinforced and consequently have thin walls which lose heat more quickly. Hot-water bottles are often made from a blend of styrene–butadiene and polyisoprene rubbers.

3   *Butadiene rubber (BR)*
Butadiene rubber – monomer butadiene – is noted for its high resilience but poor tensile strength and poor tear resistance, for which reasons it is seldom used on its own but generally blended with other elastomers. Since it has a low heat build-up, it is generally used for vehicle tyres – blended with natural rubber for truck tyres and with SBR for motor-car tyres. BR imparts good tread wear to tyres and also has good low-temperature properties. It is also used as a sealant compound in caulking and as the cores for golf balls.

4   *Polychloroprene rubber (CR)*
Polychloroprene rubber – the monomer of which is chloroprene – is also known as *neoprene*. It has the good physical properties of natural rubber coupled with a much better resistance to oxygen, ozone and some oil products. CR retains its properties well up to 90°C but tends to become hard and stiff below –10°C. It finds use in a variety of products, such as diver's wet suits, anti-vibration mountings, conveyor belts, tank linings, hoses and cable sheathing. Inflatable life rafts used with ships are waterproofed with polychloroprene.

5   *Acrylonitrile–butadiene rubbers (NBR)*
These are copolymers of acrylonitrile and butadiene and are usually called *nitrile rubbers*. The normal grades have quite good physical properties though slightly inferior to those of natural rubber and SBR. Nitrile rubber is relatively expensive and is used principally for sealing applications (gaskets) where its excellent resistance to oil products and its good temperature resistance (95°C) are essential. Its oil-resistance also gives it a use as a covering for rollers used in colour printing. It is also used for hoses, conveyor belts and cable sheathing where similar conditions prevail.

6   *Butyl rubber (IIR)*
This is a copolymer of isobutane and small amounts of isoprene. It is cheaper than natural rubber to which its physical properties are, however, inferior. Since it has a very low permeability to air and other gases, it is used mainly for tyre inner tubes and also for the liners of tubeless tyres. It has excellent resistance to ozone and to ageing generally so that it is less apt to perish. *Butyl* – as it is generally called – is also used for air-bags and supporting equipment, as well as for more general purposes such as hoses and tank linings since it is impervious to many chemicals.

7   *Ethylene–propylene rubber (EPM)*
This is a copolymer of the gases ethylene and propylene in which a quantity of either monomer can vary up to 65%. During polymerisation,

any weak double-bonds are 'used up' and the resulting rubber is therefore chemically inert so that, like polythene, it has good resistance to ageing, weathering and ozone attack. Its service-temperature range of $-50°C$ to $+125°C$ is also good but it has poor resistance to some solvents and a rather high creep rate. One use is in the construction industry for bridge expansion joints. Since EPM has good electrical insulation properties, it is often used for cable covering. A further -*diene* material is often added to provide extra cross-linking sites (thus toughening the polymer and increasing its heat tolerance). It is then designated EPDM and has a number of uses in the automobile industry.

8   *Silicone rubbers (SI)*
In these materials, silicon and oxygen replace carbon in the polymer chains. They are expensive but their high temperature tolerance ($200°C$) makes them very useful for many purposes. Their resistance to low temperatures too is good but resistance to some oils and other organic liquids is only moderate. This can be improved by replacing some of the hydrogen atoms by fluorine, but at a great increase in cost. Chemically, they are extremely inert and have the non-stick properties associated with most silicones. They thus find a use as conveyor belting in the food processing industry, for seals in freezers, and in the tubing and valves used in surgical applications.

Few of other synthetic rubbers are of such engineering significance. *Polysulphide rubber* has a good resistance to oils and solvents and is used as an internal coating for oil and paint piping. *Styrene–butadiene–styrene rubber* (SBS) is thermoplastic at temperatures above that of glass transition (see Chapter 20) so that it can be moulded by conventional methods. For this reason, it is used as a carpet-backing material and in some adhesives.

Table 19.15 summarises some of the properties of elastomers. The term *resilience* is used for the capacity of a material to absorb energy in the elastic range – a high-resilience rubber ball will bounce to a much higher height than a low-resilience one.

19.8 Recycling of polymers

What is to be done with waste plastic, the carrier bag when finished with, the packing material round the fruit or vegetables bought at the supermarket, etc.? The way it has been disposed of in the past is to bury it in a landfill site. Increasingly, now alternative ways are being looked at and beginning to be used. The simplest method is to collect all the plastic and burn it to provide energy to generate steam and hence electricity. This is the simplest method as there is no need to distinguish between the different types of plastic. Such a method is used in countries such as Switzerland and Japan. Another method of disposal is to grind up the waste plastic and then reform it into another plastic product. This requires the type of plastic gathered to be identifiable. This can now be done at recycling centres and thus I can put out for collection quite a wide variety of plastics and they are recycled.

Table 19.15 *Properties of elastomers*

Elastomer	Tensile strength (MPa)	Percentage elongation	Service temperatures (%)	Resistance to oil and greases	Resilience	Uses
Natural rubber	28	700	−50 to +100	Poor	Good	Tyres, engine mountings, flexible hoses, foamed padding, balloons, rubber bands.
Styrene–butadiene	25	600	−60 to +100	Poor	Good	Tyres, transmission belts.
Polychloroprene	25	1000	−50 to +100	Good	Good	Car radiator hoses, gaskets and seals, conveyor belts.
Butyl	20	900	−50 to +100	Poor	Fair	Inner tubes, steam hoses and diaphragms.
Ethylene–propylene	20	300	−50 to +100	Poor	Good	Resistance to hot water and superheated steam, steam hoses, conveyor belts.
Silicone	10	700	−90 to +200	Moderate		Good electrical properties. Used for electrical insulation seals, shock mounts, vibration dampers. Used in food processing industry and in surgical applications.
Polysulphide	9	500	−50 to +80	Good	Fair	Printing rolls, cable coverings, coated fabrics, sealants in building work.
Styrene–butadiene–styrene	14	700	−60 to +80	Poor		Properties controlled by the ratio of styrene and butadiene. Used for footwear, carpet backing and adhesives.

20 Properties of plastics

20.1 Introduction

In Chapter 19, we considered a polymer material as consisting of an intertwined mass of giant molecules each containing thousands of atoms. In thermoplastics materials, these molecules are attracted to each other by relatively weak van der Waals forces, whilst in thermosetting plastics they are joined to each other by strong, permanent covalent bonds. This means that whilst thermosetting materials are sometimes stronger than thermoplastics, they are inevitably more rigid and much more brittle. This chapter is about the properties of plastics, their testing, and the methods used to shape them.

20.2 Crystal and glass states

Most solid substances – metals, solids and minerals – exist in a purely crystalline form, i.e. the atoms or ions of which they are composed are arranged in some regular geometric pattern. This is generally possible because the atoms (or ions) are small and very easily manoeuvrable. With polymers, however, we are dealing with very large molecules which entangle with each other and are consequently much less manoeuvrable. As a result, only a limited degree of crystalline arrangement is possible with most polymers as they cool, and only restricted regions occur in which linear molecule chains can arrange themselves in an ordered pattern (Figure 20.1). These ordered regions are called *crystallites*.

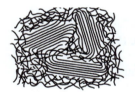

Figure 20.1 *'Crystallites' in a solid plastics material.*

Thus in a solid state, polymers can consist of both crystalline and amorphous regions. Highly crystalline polymers contain up to 90% crystalline regions, whilst others are almost completely amorphous. Polymers with side branches, or bulky groups in the chains, are less likely to be crystalline than linear chain polymers since the branches or bulky bits get in the way of orderly packing of the polymer. For example, polythene in its linear form can show up to 95% crystallinity, whereas in its branched form no more than 60% crystallinity is possible. The term *high density polyethylene* is used for the polyethylene with crystallinity of the order of 95% with low density being used for the polyethylene with crystallinity of about 50%. The high density form has a density of about 950 kg/m^3, the low density form about 920 kg/m^3. The higher density occurs because crystallinity allows the polymer chains to be more closely packed.

20.2.1 Melting points of polymers

A pure, completely crystalline, solid such as a metal or inorganic salt, melts at a single definite temperature called the melting point; this is because *single* atoms (or ions) are easily mobile and can change from the orderly crystalline state to the amorphous state of being a liquid.

Amorphous and partly crystalline materials, however, start off with some degree of being amorphous and so they merely become progressively less rigid when heated and show no clear transition from solid to liquid, both being amorphous in this case. This is what we would expect since liquids become less viscous as the temperature rises – viscous liquids like treacle and tar flow more readily when they are heated. With highly crystalline polymers on the other hand, a sudden change in structure at a particular temperature, is shown by a change such as the rate of expansion. This is the point at which all crystalline regions have become amorphous – as happens of course when all crystalline metals melt to form amorphous liquids. For such polymer materials, this temperature is designated the melting point T_m (Table 20.1). As will be noted in Table 20.1, the greater the degree of crystallinity of polythene, the higher its melting point. Presumably, this is because van der Waals forces are much greater in the crystalline regions, since the molecules are closer together there.

Table 20.1 *Melting points of some crystalline polymers*

Polymer	Melting point (°C)
Polythene (PE) 50% crystalline	120
Polythene (PE) 80% crystalline	135
Polypropylene (PP)	170
Polyvinyl chloride (PVC)	212
Polyethylene terephthalate (PET)	255
Polyamide (Nylon 6.6)	265
Polytetrafluoroethylene (PTFE)	327
Polyether ether ketone (PEEK)	334

Table 20.2 *Glass transition temperatures of some polymers*

Polymer	Glass transition temperature (°C)
Polythene (PE)	−120
Natural rubber (NR)	−73
Polypropylene (PP)	−27
Polymethyl methacrylate (PMMA)	0
Polyamide (Nylon 6.6)	60
Polyethylene terephthalate (PET)	70
Polyvinyl chloride (PVC)	87
Polystyrene (PS)	100
Cellulose acetate (CA)	120
Polytetrafluoroethylene (PTFE)	126
Polyether ether ketone (PEEK)	143
Polycarbonate (PC)	147

20.2.2 Glass transition temperature

At ambient temperatures, polymers may be either soft and flexible or hard, brittle and glassy. When a soft, flexible polymer is cooled to a sufficiently low temperature, it becomes hard and glassy. For example, a soft rubber ball when cooled in liquid air shatters to fragments if any attempt is made to bounce it whilst it is still at that very low temperature. The temperature at which this soft-to-glassy change occurs is known as the *glass transition temperature* T_g (Table 20.2).

For some plastics, T_g is above ambient temperature so that they only become soft and flexible when heated past T_g. Polystyrene is a polymer of this type. The polystyrene kitchen hollow-ware is hard and rigid at ambient temperatures and gives a characteristic 'ring' when lightly struck, but on heating above its glass transition temperature of 100°C it loses that rigidity. Below the glass transition temperature amorphous polymers have molecular chains which are unable to move when the material is heated and so it is stiff and rather brittle. When heated to above T_g, there is sufficient thermal energy for some motion of segments of the polymer chains to occur and so impart flexibility to the polymer.

20.2.3 Vicat softening temperature

This is a value which can be determined more precisely in numerical terms than either T_m or T_g, so that it offers a useful means of comparison between plastics materials as far as their response to temperature change is concerned (Table 20.3). It measures the temperature at which a standard indentor penetrates a specific distance into the surface of the plastics material using a standard force. During the test, the temperature is raised at a specified uniform rate and the penetration is recorded by a dial gauge micrometer. The *vicat temperature* is the temperature when the specified penetration (usually 1 mm) has been attained.

Table 20.3 *Vicat temperatures for some polymers*

Polymer	Vicat softening temperature (°C)
Ethylene vinyl acetate (EVA)	51
Cellulose acetate (CA)	72
Polyvinyl chloride (PVC)	83
Low-density polythene (LDPE)	85
Polymethyl methacrylate (PMMA)	90
Polystyrene (HIPS)	94
Polystyrene (PS)	99
Acrylonitrile–butadiene–styrene (ABS)	104
High density polythene (HDPE)	125
Polypropylene (PP)	150
Polycarbonate (PC)	165
Polyacetal	185
Polyamide (Nylon 6.6)	185

20.3 Mechanical properties

The strength of plastics materials is generally much lower than that of most other constructional materials. Nevertheless, plastics are light materials with a relative density between 0.9 and 2.0 so that when considered in terms of strength/weight ratio they compare favourably with some metals and alloys. Figure 20.2 indicates the types of stress/strain relationship obtained for different groups of polymers.

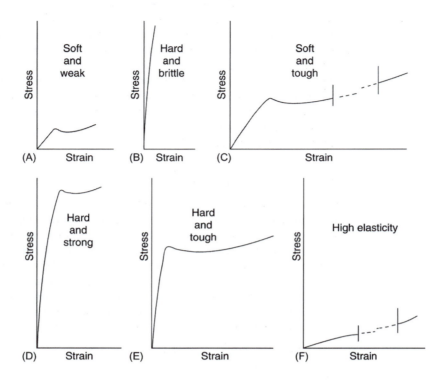

Figure 20.2 *Types of stress–strain diagram for different polymer materials (after Carswell and Nason). (A) Low elastic modulus, low yield stress, e.g. PVA and PTFE. (B) High elastic modulus, low elongation, e.g. PF, PMMA and PS. (C) Low elastic modulus, low yield stress but high elongation and high stress at break, e.g. PE and plasticised PVC. (D) High elastic modulus, high yield stress, high tensile strength and low elongation, e.g. rigid PVC and modified PS. (E) High elastic modulus, high yield stress, high tensile strength and high elongation, e.g. nylons and polycarbonates. (F) Very low elastic modulus, low yield stress and low tensile strength but very high elastic elongation, e.g. natural rubber and other elastomers.*

The mechanical properties of most engineering materials and alloys vary very little within the range of ambient temperatures encountered in service. This is to be expected since no structural changes occur (other than thermal expansion) until the recrystallisation temperature (see Section 6.3.2) is reached and this is usually well above 100°C. However, with many polymer materials – particularly thermoplastic – mechanical properties vary considerably with temperature in the ambient region. The gradual reduction in van der Waals forces with rise in temperature will affect mechanical properties; at both T_g and T_m mechanical properties change. Thus, a thermoplastic polymer may have a tensile strength of, say, 70 MPa at 0°C, falling to 40 MPa at 25°C and to no more than 10 MPa at 80°C. As tensile strength falls with rise in temperature, there is a corresponding increase in percentage elongation (Figure 20.3).

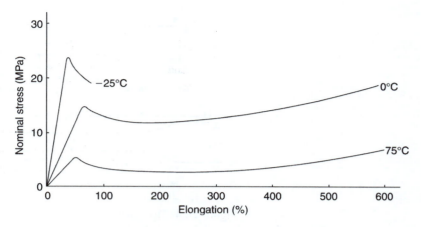

Figure 20.3 *Tensile stress–strain curves for low-density polythene at –25°C, 0°C and 75°C.*

20.3.1 Tensile testing

Clearly, if mechanical tests on plastics are to have any meaning, then a *fixed testing temperature* must be specified. Thus, BS 2782 (Part 3) requires a testing temperature maintained at 23 ± 2°C with an atmospheric humidity of 50 ± 5% for many thermoplastics materials; moreover, the test-pieces must be maintained under these conditions for 88 hours prior to the test. In some cases, e.g. thermosetting plastics, this conditioning is unnecessary since the permanent covalent bonds between the polymer chains render these materials much less temperature sensitive.

Plastics materials are also very time-sensitive as far as mechanical testing is concerned. This is because the total deformation depends upon:

1   Bond bending of the carbon–carbon covalent bonds in the polymer chain – this is manifested as the ordinary elasticity and is an *instantaneous* deformation.
2   Uncoiling of the polymer chains – this gives rise to high elasticity and is *very time dependent*.
3   Slipping of polymer chains past each other – this produces *irreversible* plastic flow and is also *very time dependent*.

The sum total of deformation arising from (1), (2) and (3) is generally termed *viscoelastic,* and the rate of strain has a considerable effect on the recorded mechanical properties. Generally, as the strain rate increases so does the recorded yield stress and again BS 2782 (Part 3) lays down different conditions for different plastics as far as tensile testing is concerned. The speed of separation of the test-piece grips varies between 1 and 500 mm/minute for different materials and forms of material.

The geometry of test-pieces also differs from that used in metals testing. Abrupt changes in shape cause stress concentrations which are likely to

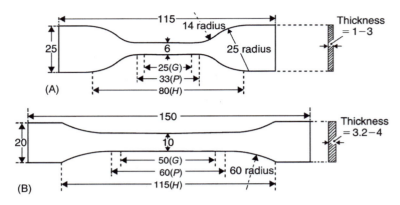

Figure 20.4 *Two forms of tensile test-piece for plastics. Gauge length G, parallel section P, distance between grips H (measurements are in mm).*

precipitate failure so tensile test-pieces are generally of the form shown in Figure 20.4

Many plastics do not obey Hooke's law, i.e. elastic strain produced is not proportional to the stress applied, so that it becomes impossible to derive the tensile modulus of elasticity (Young's modulus), since this value applies only to materials with Hookean characteristics. As an alternative, it is usual to derive the *secant modulus* of the material. This is defined as the ratio of stress (nominal) to corresponding strain at some specific point on the stress/strain curve. In Figure 20.5, the secant modulus associated with a strain of 0.2% is shown and is, in fact, the slope of the line *OS*.

At the commencement of the tensile test, an initial force of w (usually about 10% of the expected force necessary to produce 0.002 strain) is applied

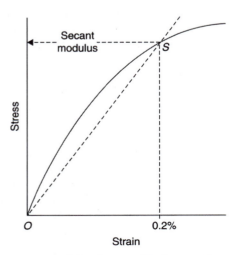

Figure 20.5 *The secant modulus for non-Hookean polymer materials.*

to 'take up slack' and straighten the test-piece. With this force applied, the extensometer is set to zero. The force is then gradually increased (in accordance with the specified straining rate) until the necessary force W is reached to produce 0.2% strain in the gauge length. The stress is then $(W-w)/A$, where A is the initial cross-sectional area of the test-piece at the gauge length used. Thus:

$$\text{Secant modulus} = \frac{\text{stress}}{\text{strain}} = \frac{(W-w)/A}{0.002}$$

Other values of strain between 0.1% and 2.0% may be used, depending upon the type of material. The strain value must, therefore, be stated when quoting the secant modulus.

20.3.2 Hardness tests

Ball-indentation tests similar to those used for metals are applied to some of the harder plastics, e.g. ebonite and hard rubbers, and the hardness index is derived from the same principle, i.e.

$$\text{hardness} = \frac{\text{applied force}}{\text{surface area of impression}}$$

When using the Rockwell test (see Section 3.3.3) for these hard plastics materials, scales M, L or R are appropriate and the hardness index is prefixed with the letter denoting the test, i.e. M, L or R.

For other plastics, a form of indentation hardness test is used in which the depth of penetration is measured under the action of a standard force. In the *durometer* (which measures Shore hardness) two different indentors are available. Type A (Figure 20.6) is used for soft plastics materials and type D for hard plastics.

As is the case when hardness testing metals, the thickness and width of test-pieces must be adequate, as indicated in Figure 20.6. The load is applied for 15 seconds whilst the conditions of temperature and humidity (23°C and

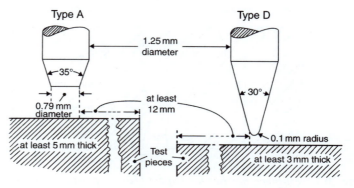

Figure 20.6 *Indentors used in the durometer (Shore hardness).*

50% humidity) specified in the case of the tensile test for plastics also apply here. In the Shore hardness test, the hardness index is related inversely to the depth of penetration. It was suggested earlier in this book (see Section 3.3) that values described as *hardness indices* are in fact generally measurements of the resistance of a material to penetration rather than the resistance of its surface to abrasion. This is particularly true of plastics. Thus, high-density polystyrene will resist penetration in tests of this type better than will polythene, yet the surface of polystyrene can be *scratched* much more easily.

20.3.3 Impact tests

Both Izod and Charpy tests are used to assess the toughness of plastics materials. Plastics fall into three groups as far as toughness is concerned (Table 20.4). In this table, those materials described as *brittle* will break even when unnotched, whilst those described as *notch-brittle* will not break unnotched but will break when sharp-notched. The 'tough' materials will not break completely even when sharp-notched. For this reason, a variety of notches are used in impact testing of plastics (Figure 20.7), in order to suit the material under test and to make possible the production of comparable test results.

Table 20.4 *Relative impact brittleness of some common plastics*

Brittle	Notch-brittle	Tough
Acrylics	Polyvinyl chloride (PVC)	Low-density polythene (LDPE)
Polystyrene	Polypropylene (PP) (PS)	Some acrylonitrile–butadiene–styrene (ABS)
Phenolics	High-density polythene (HDPE); some acrylonitrile–butadiene–styrene (ABS)	Polyethylene–propylene copolymers

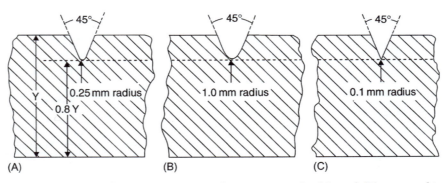

Figure 20.7 *Test-piece notches for impact tests on plastics materials. (A) and (B) are used in the Izod test whilst all three are available for the Charpy test.*

20.3.4 Creep

Creep, the gradual extension under a constant force, must be considered seriously for metals if they are to work at high temperatures (see Section 3.5). Plastics materials, however, are much more prone to this type of deformation since they undergo the viscoelastic type mentioned earlier (Section 20.3.1). Creep rates, like other mechanical properties for polymers, are very temperature dependent. Even for a plastics raincoat, creep must be considered since it must be capable of supporting its own weight so that it can hang in a wardrobe possibly for long periods without slowly flowing towards the floor during a period of warm weather.

The creep curves shown in Figure 20.8 are typical of those for most thermoplastics materials. These show that the initial rate of creep is rapid but then the rate decreases to a lower constant rate, which for small stresses might mean no further change in creep. Naturally, the greater the applied force the greater the creep rate and elongation and quite small rises of temperature will produce much greater rates of creep.

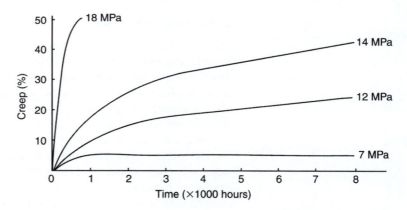

Figure 20.8 *Creep curves for cellulose acetate at 25°C. At a stress of 18 MPa, rapid creep soon leads to catastrophic failure, whilst at a stress of 7 MPa a small initial amount of creep ceases after about 1000 hours and no further measurable creep occurs.*

20.3.5 Other mechanical tests

Other mechanical tests are used to measure compressive properties, tear strength, shear strength and deflection in bend under an applied load. All of these tests, together with those described briefly here, are dealt with in BS 2782: Part 3.

20.4 Additives

Various materials can be introduced into a polymer at the moulding stage by mixing the additive thoroughly with the moulding powder.

20.4.1 Fillers

Fillers are used principally in the interests of economy when up to 80% of the moulding powder might be filler. Generally, fillers are finely powdered, naturally occurring, non-polymeric substances which are readily available at low cost. Phenolic (bakelite)-type resins are nearly always compounded or filled with substances such as sawdust, wood flour, short cellulose fibres from wood pulp or cotton flock; fillers like paper, rag and cotton fibres will increase the strength of bakelite. Some soft thermoplastics materials are blended with up to 80% mineral solids such as crushed quartz or limestone. Here, the plastics materials operate in much the same way as does cement with the aggregate in concrete. Thus, some fillers serve to increase the strength and hardness and to reduce creep and thermal expansion.

20.4.2 Anti-static agents

The act of walking across a polypropylene fibre carpet can easily produce a considerable charge of static electricity, whilst most readers will be aware that the simple act of removing a synthetic-fibre garment, particularly in a dry atmosphere, will generate 'static' – a charge more than sufficient to create a spark capable of igniting a petrol–air mixture. One wonders how many unexplained fires are started in this way.

Since the valence electrons, i.e. the outer shell electrons, are held captive within the covalent bonds throughout a plastics materials, such a material is a non-conductor of electricity. In fact, most polymers are excellent insulators. It follows therefore that a plastics material can become highly charged with 'static' electricity because this static is unable to be conducted away. This may be undesirable or indeed hazardous and in such cases an anti-static substance is added to the mix. Additives to achieve this result are substances which will lower the electrical resistivity of the surface of the polymer, generally by attracting moisture to its surface since moisture is sufficiently conductive to provide a path by which the 'static' charge can 'leak' away.

Many such additions are 'waxes' in which the molecules are strong dipoles (see Section 1.4.8), i.e. they have a negatively charged end and a positively charged end. They therefore attract water molecules which also are dipoles, so that the two are held together by van der Waals forces. Such a strategy is of course not effective in very dry atmospheres (below 15% humidity). Such 'waxes' can be added to PE and PS since their moulding temperatures are low enough to avoid decomposition of such additives. Other organic compounds (amines and amides), also phosphoric acid derivatives and sulphonic acids, are sometimes used for this purpose.

20.4.3 Flame retardants

Many plastics materials constitute a fire risk as witnessed by a number of home disasters in past years. Plastics foam fillings of cushions can be a fire hazard. To reduce such fire risks, flame retardants are used with the object of interfering with those chemical reactions underlying combustion. One such substance is hexabromocyclodecane (HBCD) which is incorporated in materials like polypropylene. If combustion of the plastics material begins, hydrogen bromide is released from HBCD as a vapour which immediately combines with those portions of the plastics molecule, generally –OH groups, which would otherwise burn and lead to a rapidly spreading conflagration. Thus combustion is stifled.

Some organic chlorides have a similar action in releasing hydrogen chloride which similarly neutralises the –OH groups. Their effectiveness can often be improved by the addition of antimony oxide. This reacts with the organic chloride in the flame producing antimony trichloride which increases flame retardation.

20.4.4 Friction modifiers

For two apparently flat surfaces in contact, whether metals or plastics, there will only be contact at a few high spots between them. At such points there is 'intimate' atomic contacts. Thus, friction between two plastics surfaces occurs because of the van der Waals forces at such contact points acting between molecules of two adjacent surfaces. In order to move one surface over another, such adhesion bonds have to be broken. Coefficients of friction for different plastics materials vary widely since the different groupings of atoms at molecule surfaces affect the magnitude of these van der Waals forces, such coefficients being as low as 0.04 for polytetra-fluoroethylene (PTFE) and as high as 3.0 for natural rubber. Consequently, PTFE is widely used as a bearing material, as is nylon with a coefficient of 0.15 and ultra-high molecular weight polythene with coefficient varying from 0.08 to 0.2.

The addition of graphite to a plastics material will reduce the coefficient, as will particles of added PTFE and the high-pressure lubricant molybdenum disulphide. Such materials will reduce wear in bearings and bushings.

20.4.5 Other additives

Other additives include *colourants* which are either insoluble, finely divided pigments or soluble dyes, usually chosen for colour stability, inertness and non-toxicity. In some instances, pigments may act as a filler or serve some other purpose such as that of a stabiliser.

The addition of *asbestos* to a plastics material will increase its heat tolerance, whilst *mica* improves its electrical resistance.

20.4.6 Foamed or 'expanded' plastics materials

Foamed rubber was developed many years ago; now a number of polymers are available as 'foams'. 'Expanded' polystyrene is probably the best known of these. It is used as a very lightweight packing material for fragile equipment. Without it, one wonders how many of the cameras, computers and TV sets from the Far East would reach us intact. Up to 97% of its volume consists of air bubbles so that it is useful as a thermal insulator and for the manufacture of DIY ceiling tiles, as well as being an excellent flotation material which, unlike cork, does not become waterlogged.

Polyurethane foams are available in two different forms, either as flexible foams for the production of sponges and upholstery, or as rigid foams used for the internal reinforcement of some aircraft wings. Both thermoplastic and thermosetting polymers (including some elastomers) can be foamed and four basic methods are available for introducing the gas packets into the polymer:

1   By mechanically mixing air with the polymer, whilst the latter is still molten.
2   By dissolving a gas such as carbon dioxide into the mix under pressure. As hardening proceeds, the pressure is reduced so that the gas is released from solution, rather like the bubbles of carbon dioxide which are released from mineral water, or of course champagne, when the stopper is removed. The bubbles in this case are trapped in the hardening polymer.
3   A volatile component may be added to the hot mix. This vaporises during hardening, forming bubbles.
4   In some processes, a gas is evolved during the actual polymerisation process. This can be utilised to cause foaming. Alternatively, a 'blowing agent' may be added which will release gas when the mix is heated for the moulding process. This is useful for thermoplastics materials. A simple reagent of this type is sodium bicarbonate which releases carbon dioxide when heated. It behaves in much the same way when used as 'baking powder' with bubbles of carbon dioxide released during baking making a cake 'rise'.

20.5 Shaping plastics

Most of the larger plastics manufacturers supply raw materials in the form of powders, granules, pellets and syrups. These are purchased by the fabricators for the production of finished articles. As with metals, plastics can be purchased in the form of extrusions and sheet on which further work can be carried out.

The principal methods of fabrication are calendering, extrusion, moulding, blow moulding and vacuum forming, but raw materials available in liquid form may be cast.

20.5.1 Calendering

This is used to manufacture thermoplastics sheet, mainly PE and PVC. The calendering unit consists of a series of heated rolls into which the raw

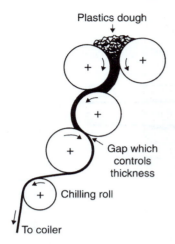

Plastics dough

+ +

+

+ Gap which
controls
thickness

+ Chilling roll

To coiler

Figure 20.9 *The principles of calendering.*

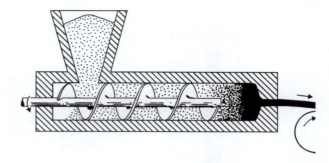

Figure 20.10 *The extrusion of plastics. Often known as screw-pump extrusion.*

material is fed as heated 'dough' (Figure 20.9). The formed sheet is cooled by a chill roller. Fabric, paper and foil are coated with a film of plastics material using a similar process.

20.5.2 Extrusion

The principle of extrusion is shown in Figure 20.10. Here, the plastic material is carried forward by the screw mechanism, and, as it enters the heated zone, it becomes soft enough to be forced through the die. The die aperture is shaped according to the cross-section required in the product.

Man-made fibres are extruded using a multi-hole die, or 'spinneret'. Wire can be coated with plastics by extrusion, assuming that the machine is adapted so that the wire can pass through the die in the manner of a mandrel.

20.5.3 Moulding

Several important hot-moulding processes are commonly used. They are as follows.

1 *Compression-moulding* (Figure 20.11)
   This is probably the most important, and is used for both thermoplastic and thermosetting materials, though it is particularly suitable for the latter. In either case, the mould must be heated, but for thermoplastic substances it has to be cooled before the work-piece can be ejected. A careful measured amount of powder is used and provisions are made to force out the slight excess necessary to ensure filling of the mould cavity. Machine and tooling costs are low compared to most other permanent mould casting processes.

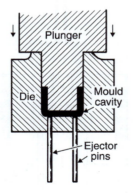

Plunger

Die

Mould
cavity

Ejector
pins

Figure 20.11 *One system of compression-moulding.*

2 *Injection-moulding* (Figure 20.12)
   This is a very rapid response process and is widely used for moulding such materials as polythene and polystyrene. The material is softened by heating it in the injection nozzle. The mould itself is cold, so that the plastic soon hardens and can be ejected. The work-piece generally

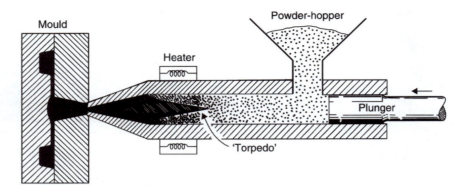

Figure 20.12 *The principle of injection-moulding.*

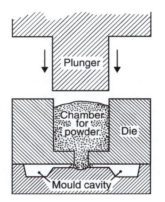

Figure 20.13
Transfer-moulding.

consists of a 'spray' of components, connected by 'runners' which are subsequently broken off. The operating cost is high due to the cost of machines and moulds.

3   *Transfer-moulding* (Figure 20.13)
This is used for thermosetting plastics. The material is heated to soften it, after which it is forced into the heated mould, where it remains until set. Rather more intricate shapes can be produced by this process than by compression-moulding.

4   *Blow moulding* (Figure 20.14)
This is used to produce hollow articles. The plastic is first softened by heating and is then blown by air pressure against the walls of the mould. In the figure, extruded tube (parison) is being used. A variety of

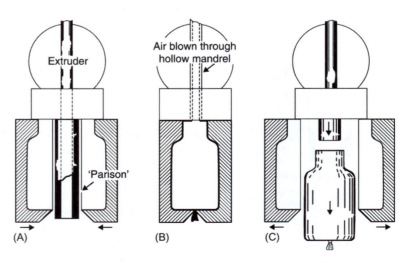

Figure 20.14 *Blow-moulding by interrupted extrusion.*

containers, e.g. bottles and coffee jars, are produced in this way. It is mainly used with the thermoplastics of PET and polyalkenes.

A similar process, known as *stretch-blow moulding,* is used in the manufacture of PET drinks bottles. Here, a preform is stretched and blown within the mould so that crystallisation of the PET occurs within the amorphous structure. This process provides increased yield strength and stability in bottles which can be filled with liquids at around 90°C.

5   *Film-blowing* (Figure 20.15)
Film can be blown continuously from heated thermoplastics delivered straight from the extruder. This material has a nominal thickness of not more than 0.25 mm. The process is used for the manufacture of PE film and 'tube' for parting off as bags or other forms of packaging.

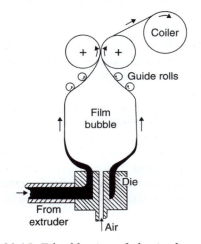

Figure 20.15 *Film-blowing of plastics bags.*

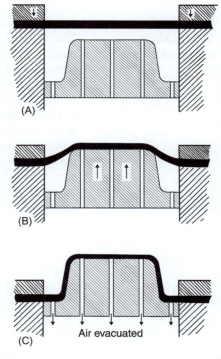

Figure 20.16 *Stages in a vacuum-assisted forming process.*

6   *Vacuum-forming* (Figure 20.16)
This is used in producing simple shapes from thermoplastics materials in sheet form. The heated sheet is clamped at the edges and is then stretched by the mould as it advances into position. The ultimate shape is produced by applying a vacuum, so that the work-piece is forced into shape by the external atmospheric pressure. It is used for shapes with essentially constant section thickness.

20.5.4 Casting

This is limited to those plastics materials whose ingredients are in liquid form, i.e. as monomers which polymerise in the mould. The mixed liquids are poured into the open mould and are allowed to remain at atmospheric pressure until setting has taken place. It is mainly used for simple shapes; operating costs can be very low.

Slush-moulding is essentially a casting process rather than the moulding process its name implies. It is used mainly for PVC, the mix being prepared as a thin paste which is injected into the mould. The mould is then rotated and heated until the paste forms a solid 'gel' on the inner surface of the mould. The excess fluid is then poured out and the *hollow* formed article removed from the mould. The process is used for the manufacture of dolls' heads and other semi-soft toys. The 'slush casting' of aluminium teapot spouts and of lead soldiers in former days, followed a similar method.

20.6 Machining polymers

The processes used to shape polymers generally produce the finished article with no, or little, need for any further process such as machining. However, with injection moulding, compression moulding and blow moulding there is a need to cut off gates and flashing.

Because polymers have low melting points, machining conditions have to be such that high temperatures are not produced if the material is not to soften and deform. Single point and multi-point cutting tools can be used. However, roughness and cracking can occur round the point of contact of the tool if discontinuous chips are produced. This can occur with a polymer such as polystyrene if the shear strain at the machining surface becomes larger than the limiting rupture strain. Thus, the cutting conditions have to be chosen carefully.

21 Ceramics

21.1 Introduction

The term *ceramics* is derived from the Greek *keramos* – 'potter's clay'. Gradually, this term has been extended to include all products made from fired clay, such as bricks, tiles, fireclay refractories, wash-hand basins and other sanitary ware, electrical porcelains and ornaments, as well as pottery tableware. Many substances now classed as ceramics in fact contain no clay though most are relatively hard, brittle materials of mineral origin with high fusion temperatures. Thus, hydraulic cement is usually classed as a ceramic material, whilst a number of metallic oxides such as alumina, magnesia, zirconia and beryllia form the basis of high-temperature ceramic refractories. Some of the latter, in particular alumina and zirconia, along with recently developed materials like silicon nitride and the 'sialons', boron nitride and boron carbide, have more sophisticated uses.

Ceramics can be grouped as:

- Domestic ceramics, e.g. china, earthenware, stoneware and cement.
- Natural ceramics, e.g. stone.
- Engineering ceramics, e.g. alumina, zirconia, boron nitride, etc. which are widely used in engineering as furnace components, tool tips and grinding tools.
- Glasses and glass ceramics. Most of the ceramics materials mentioned earlier are either completely crystalline in structure or are a mixture of crystalline regions cemented together by an infilling of amorphous networks. Glasses, on the other hand, are materials which, at ambient temperatures, are still in an amorphous state, i.e. they are virtually still in a liquid condition. Glass comprises a range of substances, from boiled sweets to window panes and beer tankards. Glass ceramics are fine-grained polycrystalline materials produced by the controlled crystallisation of glasses. Glasses are discussed in more detail in Chapter 22.
- Electronic materials, e.g. semiconductors and ferrites.

Even some metallic materials which can be cooled quickly enough from the molten state form glasses; speeds of about $10^6 °C$/second are necessary to prevent them from crystallising. At the other extreme, plastics materials are either completely 'glassy' or contain both crystalline and glass regions (see Figure 20.1) because they are composed of very large and cumbersome linear molecules which can 'wriggle' into position only with difficulty so that the process is overtaken by a fall in temperature.

21.2 Silicate-based ceramics

Those ceramic materials derived from sand, clay or cement contain the elements silicon and oxygen (the two most abundant elements on planet Earth) in the form of *silicates,* which also contain one or more of the metals sodium, potassium, calcium, magnesium and iron. The most simple silicon–oxygen unit in these compounds is the group SiO_4^{-4} in which a small silicon atom is covalently bonded to four oxygen atoms (Figure 21.1), the –4 indicating, as will be seen in Figure 21.1B, that each oxygen atom in this unit has an 'unused' valency bond. Thus, these units can link up, rather like the 'mers' in a polymer structure, to form a continuous structure. A number of different basic structures can be formed in this way, but two of the more important are described here.

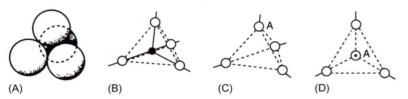

Figure 21.1 (A) The structure of the SiO_4^{-4} unit in silicates. (B) The small central silicon atom surrounded by and covalently bonded to four oxygen atoms. (C) Here, the silicon atom is omitted to simplify the diagram of the unit. (D) A 'plan' view of the SiO_4^{-4} unit, i.e. with the 'apex' oxygen atom A on 'top', again with the silicon atom omitted.

21.2.1 'Chain' type arrangements

Figure 21.2 represents a 'chain' type arrangement in which silicon–oxygen units are found combined in a large number of naturally occurring minerals. However, although the basic tetragonal form of the 'SiO$_4$' unit is retained, we are no longer dealing with a series of 'SiO$_4$ ions' linked together since oxygen atoms along the chain are shared between successive units so that in a single-chain structure (Figure 21.2A) we are dealing with 'backbone chains' of a formula $n[(SiO_3)^{-2}]$, whilst in 'double-chain structures', since sharing of oxygen atoms goes a stage further, the backbone chains will have a general formula $n[(Si_4O_{11})^{-6}]$ (Figure 21.2B).

Those 'unused' oxygen valencies along the 'edges' and at the 'apices' of the units are in fact bonded ionically to either metallic or –OH ions. When the metallic ions concerned are of metals such as calcium or magnesium (both with valencies of two), the spare valency links the chain to a neighbouring chain.

However, the point of all this is that the greatest strength of the structure is *along the covalently bonded backbone* of tetragonal silicon–oxygen units, long strands of which are held together by weaker ionic bonds. This shows some resemblance to the structure of thermoplastic polymers, except that the van der Waals forces operating in polymers are replaced by ionic bonds here. This type of structure is present in the mineral *asbestos* and is convincingly

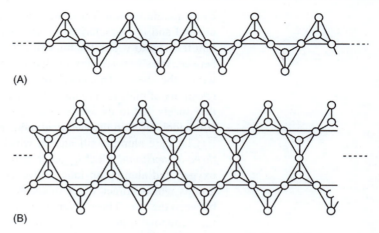

Figure 21.2 *This shows how the structure of chain-type 'giant ions' can be formed from 'SiO₄ groups'. In fact, because the oxygen atoms are shared between two units in a chain, the basic formula of the chain is $n[(SiO_3)^{-2}]$. 'Unused' oxygen valencies along the chain attach themselves to metallic ions.*

reflected in its mechanical properties. Asbestos is a naturally occurring mineral consisting of giant chains based on Si_4O_{11} units ionically bonded to Ca^{+2}, Mg^{+2} and $-OH$ ions. It is mined in a rock-like form but a series of crushing and milling operations can separate the mineral in a form of fibres, because of the basic structure described earlier. Fracture occurs of the weaker bonds which hold the strong fibres together.

21.2.2 'Sheet' type arrangements

In another very large group of naturally occurring silicates, the tetragonal silicon–oxygen units are covalently linked in a two-dimensional *sheet* form (Figure 21.3) and the basic formula becomes $n[(Si_2O_5)^{-2}]$. This type of structure is found in naturally occurring minerals such as *talc* and some *clays*.

Other minerals of a similar type contain aluminium as a constituent of the giant sheet-like units in which a silicon atom in the basic unit is replaced

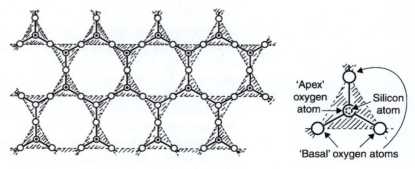

Figure 21.3 *The 'sheet' structure of silicate tetrahedra.*

by an aluminium ion. These are classed as alumino-silicates. *Mica* is such a mineral and is characterised by the extremely thin sheets (less than 0.025 mm thick) into which the mineral can be cleft because the bonding forces between the sheets are very small compared with the strong covalent bonding within the sheets. This covalent bonding also explains why the electrical resistivity of mica is very high. This fact, along with the ability to have very thin sheets, led to its use in electrical capacitors as the dielectric. Mica is still widely used in the electrical industries as an insulating material.

Clays are alumino-silicates in which the units can best be regarded as *three-dimensional* blocks, rather than sheets, and built up from silicon, oxygen and aluminium ions. However, their outer surfaces contain −OH groups which attract water molecules by van der Waal forces operating between the two. These layers of water molecules (Figure 21.4) separating the alumino-silicate blocks gives to the raw clay its plasticity, since one block can slide over another with the water molecules acting rather like a film of lubricating oil. When the clay is dried, the water molecules evaporate as they receive enough thermal energy to break the weak van der Waals bonds and so escape. The clay then ceases to be plastic. On heating to a higher temperature, further chemical changes occur in the clay and strong cross-links are formed between alumino-silicate blocks so producing a continuous hard, brittle structure. As with the reaction which occurs in thermosetting plastics materials, this chemical change is *non-reversible*. Clay, once fired, cannot be softened again with water.

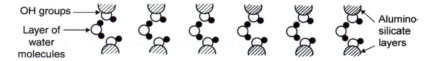

Figure 21.4 *Water molecules acting as a weak link between alumino-silicate blocks in clay due to van der Waals forces operating between them.*

21.3 Asbestos

A number of different minerals containing varying amounts of silicon, oxygen, calcium and magnesium qualify as forms of 'asbestos' but all have one feature in common, a structure based on long-chain molecules containing the repeating unit $(Si_4O_{11})^{-6}$. This is responsible for the fibrous structure (see Section 21.2.1) common to all compositions of asbestos, which is a flexible, silky material from which fibres as fine as 1.8×10^{-6} mm diameter can be separated. The fibres are strong and, with suitable compositions, tensile strengths in the region of 1500 MPa can be obtained. Strength begins to fall at temperatures above 200°C, though the mineral does not melt until a temperature of about 1500°C is reached.

Asbestos is chemically inert and is not attacked by chemical reagents. It is little wonder therefore that once its dust has reached the lungs it remains there to do its deadly work during the lifetime of the sufferer.

21.3.1 Asbestos as a health hazard

Evidence began to accumulate at the beginning of the twentieth century pointing to a connection between exposure to asbestos dust and the incidence of some diseases of the human body. It became firmly established that the inhalation of the dust can lead to lung cancer and other serious respiratory diseases such as asbestosis – a chronic inflammation of the lungs. Likewise, the ingestion (in food and water) of asbestos particles may lead to serious deterioration of other organs.

21.3.2 Obsolete products containing asbestos

Naturally, the use of asbestos in the raw fibrous form has been forbidden for many years, but such large amounts of it were used in the past that it may be prudent to mention situations where it is likely to have been used. These are products used mainly where flame-resistance, thermal and electrical insulation, or resistance to chemical attack are involved. They consist principally of asbestos to which small quantities of a suitable binder were added to effect cohesion.

The object of listing these obsolete materials here is to alert the young engineer – particularly if he is engaged on maintenance work – to the possibility *that there may still be some of these materials* around and to deal with them carefully as outlined below.

1 *Raw asbestos fibres* were used as insulation, in particular as packing for walls and floors and in underground conduits. Since it resists corrosion, it was also used for the insulation of batteries.

2 *Asbestos textiles* were produced by spinning and weaving processes basically similar to those used for other fibres. They were used in belting for conveying hot materials, friction materials and as electrical and thermal insulation. Other products included rope for caulking, gaskets, theatre safety curtains, fire-fighting suits and blankets.

3 *Asbestos paper* was used as electrical and thermal insulation and for fire protection in military helmets, armoured car roofs, automobile silencers and linings for filing cabinets. Asbestos millboard was virtually a thick form of asbestos paper and could be drilled, nailed and screwed. It was used in the linings of stoves and ovens, fireproof linings of switch boxes, safes and doors, as thermal insulation in incubators and as fire-proof wall boards.

4 *Asbestos-cement products* contained up to 75% Portland cement. Here, the strong asbestos fibres were used to strengthen the product as well as to improve thermal insulation properties. Asbestos-cement sheets were used for roofs and walls of small buildings, fire-protection booths, linings of gas and electrical cookers and switch boards. Asbestos-cement pipes were used for carrying water, sewage and gas, whilst large quantities are still in use as domestic roof guttering. Most of the pipe material has been replaced by plastics.

5   *Asbestos reinforced and filled plastics materials* included both thermoplastic and thermosetting types to which asbestos was added as loose fibre, woven textile or paper. Typical applications included circuit breakers, pulleys, castors, bearings and washing machine agitators.

21.3.3 Precautions on encountering asbestos

When confronted by asbestos products, the following precautions should be taken:

*   If dealing with the raw fibre and woven forms of it, protective clothing should be worn, including a respirator, rubber gloves and gumboots.
*   Drilling and cutting of asbestos-cement and other composites should only be undertaken if the surface can be kept wet and continual damping down is possible to prevent the formation of air-borne dust.
*   Machining processes should only be permitted if adequate exhaust systems are available to take off the dust-laden air.

An appreciation of the dangerous nature of asbestos dust must not of course blind us to the existence of other equally lethal substances. One remembers a very acrimonious meeting held in a school hall to protest about the alleged use of asbestos fibre as an insulator in its construction – a public meeting at which the air was literally blue with tobacco smoke. But enough of this cynicism!

21.4 Clay products

'Delhi Belly', 'the Turkey Trot' and 'Montezuma's Revenge' are all epithets used to describe the transient but uncomfortable (and often inconvenient) malady from which all of us who have travelled in hot climes have suffered at one time or another. The simple remedy was to take what the local pharmacist – if there was one – wrote up as 'kaolin et morph' (a suspension of kaolin in a dilute solution of morphine hydrochloride). However, it may surprise the reader to learn that the same 'kaolin' which helps to 'soothe the troubled gut' is a purified form of the alumino-silicate which forms the basis of both pottery and other industrial clays.

The mineral *kaolinite,* from which high-quality porcelain is manufactured, was originally imported from a region in the Kaoling Mountains of China after which it was named. Such products became known as *china* and kaolinite as *china clay.* Fortunately, supplies of good-quality china clay were later mined in Cornwall. Pure kaolinite is an alumino-silicate of the formula $Al_2O_3.2SiO_2.2H_2O$, but natural clays vary widely in composition. China clay, as mined in Cornwall, contains up to 95% kaolinite whilst those clays used in the manufacture of ordinary bricks may contain only 30%.

21.4.1 Fireclay

Fireclay used for furnace linings, firebricks and some crucibles is high in kaolinite. Such bricks can be used at temperatures up to 1500°C but increased amounts of other constituents like silica, lime and iron oxides all reduce the melting point by chemical reaction.

Since raw fireclay is relatively expensive, crushed used firebrick (known as *grog*) is added to new clay, partly for reasons of economy but also to reduce shrinkage of the product during firing. Once fired, clay will not shrink a second time.

21.4.2 Shaping clay products

Whilst modified forms of the potter's wheel – known as *jolleys* and *jiggers* – are still used to make a variety of domestic and ornamental products, articles such as dinner plates and the like are pressed in shaped moulds.

Lavatory basins, laboratory porcelain, pipe channels, as well as vases, porcelain figures and similar objects of re-entrant or complex shape, can be made by *slip-casting* (Figure 21.5). Here a split mould, made from some water-absorbent material such as plaster of Paris, is filled with liquid *slip* – a suspension of clay in water. Water is absorbed by the porous plaster mould, the previously suspended clay particles collecting as a uniform layer on the mould surface. When this layer has attained the required thickness, the excess slip is poured away. The mould is allowed to dry out and the halves of the mould separated when the clay form is strong enough to be handled.

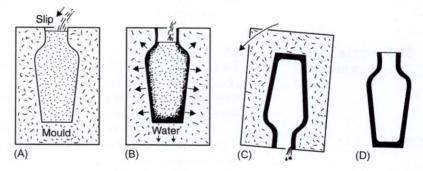

Figure 21.5 *Slip casting of pottery. (A) The plaster mould is filled with slip, (B) a layer of clay particles collects on the mould surface as water is absorbed by the mould, (C) when the shell is thick enough, excess slip is poured away, (D) the resultant 'casting' is removed from the dry mould before being further dried and 'fired'.*

21.4.3 Hydroplastic forming

Hydroplastic forming of wet clay is employed in the manufacture of simple shapes such as bricks. The clay, softened with water, is extruded using a 'screw pump' of similar design to that shown in Figure 20.10 for the extrusion of plastics materials. The extruded stock is 'parted off' as individual bricks.

21.4.4 The heat-treatment of clay products

The heat-treatment of clay products is, of necessity, a slow process since it must avoid cracking and distortion of the clay which would occur with rapid or uneven loss of water vapour. Two stages are involved:

1 *Drying* in which the work is loaded onto racks in large ovens and the temperature then raised to about 120°C in an atmosphere *saturated with water vapour*. The humidity is then very gradually reduced so that water evaporates slowly and evenly from the work.

2 *Firing* is usually carried out in long gas- or electrically heated tunnel kilns through which the work moves very slowly towards the hottest zone (about 1500°C), then into the cooling zone through which the work again moves very slowly before being discharged. The slow, uniform temperature changes avoid cracking and distortion of the product.

During the initial drying operation, *absorbed water* is removed but during the firing stage definite chemical changes occur. First, –OH ions are lost so that they are no longer there to attract (by van der Waals forces) water molecules which would cause raw clay to be plastic and 'mouldable'. Second, some vitrification occurs between the metallic oxides and silica, present in the original clay, forming a glassy phase which cements together the alumino-silicate particles, forming a strong though rather brittle structure. Though brittle under conditions of shock, porcelain has a high crushing strength as was illustrated by Messrs Josiah Wedgwood and Sons Ltd., in a photograph showing an 11-tonne double-decker bus supported by six of their fine bone-china coffee cups.

21.5 Engineering ceramics

A few ceramics materials, for example carbon and glass fibres, are important for their high tensile strengths, but most ceramics used in engineering are chosen for their great hardness. Only one of these, diamond, the hardest of them all, is of purely natural origin though small synthetic diamonds are now important as an abrasive grit.

Many readers will be familiar with 'cemented carbide' cutting tools in which particles of hard metallic carbides are held in a tough metallic matrix. Some modern cutting materials, however, are composed entirely of crystalline oxides, nitrides, carbides or borides. Thus, sintered crystalline aluminium oxide (alumina) has been used as a cutting tool material for some years. Such tool tips are bonded to an ordinary metal shank, being discarded when they become worn. Not only do alumina and other similar materials retain hardness and compressive strength at high temperatures but they generally have low friction properties coupled with a high resistance to abrasion and chemical attack. Consequently, they are often used for cutting at higher speeds than is possible with 'hard metal' tools. They are useful for cutting tough materials including plastics, rubber, aluminium and ceramics.

Yet other materials, because of their great hardness coupled with a relative density which is low compared with that of metals, are finding increasing use in military armour plating.

21.5.1 Magnesium oxide

Magnesium oxide (magnesia) MgO is obtained by heating the mineral *magnesite* $MgCO_3$ at about 1000°C:

$$MgCO_3 \rightarrow MgO + CO_2$$

The cubic structure of magnesium oxide (Figure 21.6) is basically similar to that of lithium chloride (see Figure 1.3) in that the positively charged metallic ions Mg^{+2} and negatively charged non-metallic ions O^{-2} are ionically bonded to each other. Each Mg^{+2} ion is surrounded by six O^{-2} ions. Similarly, each O^{-2} ion is surrounded by six Mg^{+2} ions and the formula of magnesium oxide is MgO. As with most ionically bonded crystalline materials, magnesium oxide is relatively brittle.

Magnesium oxide is used mainly as a refractory material, particularly in linings of furnaces used for steel-making by the now more common *basic* processes. Here, slags are chemically basic since they contain lime so that a basic furnace lining must be used to avoid any chemical reaction between slag and lining. Magnesium oxide can withstand working temperatures up to 2000°C.

21.5.2 Aluminium oxide

Aluminium oxide (alumina) Al_2O_3 is produced from bauxite, which is an impure form of alumina and the main ore from which metallic aluminium is produced.

In the crystalline structure of alumina (Figure 21.7), a mixture of ionic bonds and covalent bonds operate between the resultant ions and atoms of hydrogen and oxygen. Although the structure is basically close-packed hexagonal, one third of the aluminium sites remain vacant in accordance with the valencies of the two elements which result in a chemical formula of Al_2O_3.

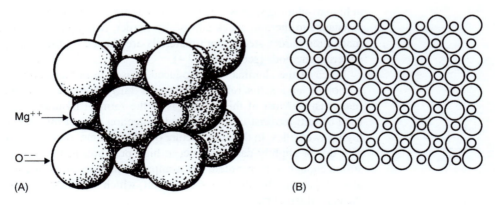

Mg^{++} →

O^{--} →

(A)   (B)

Figure 21.6 *The crystal structure of magnesium oxide ($Mg^{+2}O^{-2}$): (A) shows a single unit of the structure, (B) shows a 'plan' view of the three-dimensional structure in which each oxygen ion is shared by neighbouring units.*

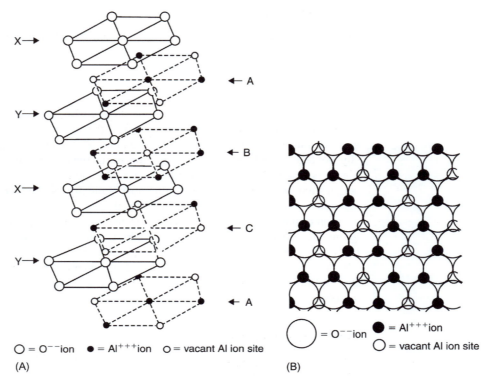

Figure 21.7 *The crystal structure of alpha alumina: (A) represents a spatial arrangement of the positions of aluminium and oxygen atom/ion sites, (B) represents a 'plan' view of the structure along plane X showing that only two-thirds of the aluminium atom/ion sites are occupied.*

Alumina is used as a cutting tool material but its poor thermal conductivity makes it susceptible to overheating and hence rapid wear when used at high machining speeds. An alumina–titanium nitride mixture has a greater resistance to thermal shock and is also harder, making it more suitable for machining harder steels. Both materials are shaped by the usual pressing-sintering methods (see Section 7.4).

Much of the alumina now produced is used in military armour plating. Although, at 3.8, its relative density is fairly high, it is comparatively inexpensive. Most of this armour is in the form of moulded breast plates used by anti-terrorist units who need protection against missiles fired from high-power rifles. In the manufacture of such armour, alumina is compacted as a dry or slightly damp powder at high pressure. In this condition, it is strong enough to be handled. The compact is then sintered for a long period at a temperature below the melting point, which densifies and consolidates the crystal structure.

The hardness, wear resistance and *chemical inertness* of alumina have led to an increase of its use in hip-joint replacements, whilst a high electrical resistivity has favoured its use for the ceramic bodies of sparking plugs.

21.5.3 Silicon nitride

Silicon nitride Si_3N_4 is a stable covalent compound. It cannot be sintered successfully since a very porous compact is the result, thus the required shapes are made in silicon. This produces porous compacts which are then heated at a temperature just below the melting point of silicon (1412°C) in an atmosphere of nitrogen to produce the nitride. As Si_3N_4 cannot be sintered, small amounts of basic metallic oxides, such as Al_2O_3 or MgO are added. On heating, a layer of SiO_2 forms on the surface of the Si_3N_4 grains due to some oxidation of the latter. This SiO_2 (being acidic) reacts with the metallic oxides (basic) to form an intergranular film of liquid which, on cooling, forms a glassy network cementing the Si_3N_4 grains into a solid mass. This glass can be devitrified to give a mainly crystalline structure either by slow cooling or further heat-treatment.

Silicon nitride tool properties depend to a large extent on the composition and amount of the intergranular phase, particularly at high temperatures.

21.5.4 Sialons

Pure Si_3N_4 can be modified in composition and properties by replacing some of the nitrogen in the structure by oxygen and at the same time substituting some aluminium for silicon. In each case, the proportions added are adjusted to keep a balance of valencies. The resulting compounds are known as *sialons* (an acronym on the element symbols of the main ingredients Si–Al–O–N).

The main properties of sialons is that they retain their hardness at higher temperatures than does alumina. They are tough but rather less so than a cemented carbide of equal hardness. Sialons are used for cutting tool materials, dies for drawing wire and tubes, rock-cutting and coal-cutting equipment, nozzles and welding shields. Because they have a good combination of high-temperature and thermal-shock resistance, sialons are used for the manufacture of thermocouple sheaths, radiant heater tubes, impellers, small crucibles and other purposes involving temperatures up to 1250°C.

21.5.5 Zirconia

Zirconia ZrO_2 has been used for many years as a high-temperature crucible and furnace refractory where temperatures up to 2500°C must be sustained. ZrO_2 is a polymorphic substance (see Section 1.4.6) which at 1170°C undergoes a change in crystal structure similar to that in steel which leads to the formation of martensite. As with steel, a sudden volume increase is involved on cooling and since ZrO_2, like martensite, is brittle then severe distortion and cracking are liable to occur, leading to spalling and disintegration. Fortunately, additions of yttria (yttrium oxide Y_2O_3) suppress the martensite type transformation to below ambient temperature – rather like the effect of nickel in 18/8 stainless steel. Lime and magnesia have a similar effect but the more expensive yttria is used because, unlike lime and magnesia, it does not lower the melting-point of zirconia.

A fully stabilised zirconia (FSZ) contains 18% yttria, but partially stabilised zirconia (PSZ) containing about 5% yttria is generally used, particularly as a temperature-resistant coating, since its *thermal conductivity* is *low,* for superalloy rotor blades in jet turbines. Here, the *high* coefficient of *thermal expansion* of PSZ matches that of the metal it is coating so that cracking and exfoliation of the coating is eliminated. PSZ is manufactured by sintering *ultrafine* grade ZrO_2 powder with 5% Y_2O_3 at 1450°C. The resultant structure is fine grained because fine-grain powder was used.

Since zirconia is 'environmentally friendly' *inside* the human body, it is finding use in implants. As it is tougher and has better 'bending strength' than alumina, it is replacing the latter in North America and France for the manufacture of artificial hip joints.

21.5.6 Some other engineering ceramics

As in the case of those engineering ceramics described earlier, these are materials which are used either because of their durability at high temperature, e.g. beryllia (beryllium oxide BeO), silicon carbide (SiC) and zirconium boride will withstand temperatures in excess of 2000°C, or because of their great hardness coupled with adequate strength and toughness.

Among cutting tool materials, the hardest and most effective of all is naturally occurring *diamond.* Its great hardness is due to the strong covalent bonds which bind all carbon atoms comprising the giant molecule (see Section 1.4.6) of this scarce and expensive precious stone. Diamonds are mined mainly in South Africa, Australia, Russia, Zaire and Botswana. *Synthetic* diamonds were first produced in the 1950s by subjecting graphite to a pressure of some 7000 MPa at a temperature of 3000°C in the presence of a catalyst. Though not the material from which diamond tiaras could be fashioned, these small black particles are very useful as a 'grit' for many cutting, grinding and polishing purposes.

Boron nitride (BN) is second only to diamond in respect of hardness. Manufactured in a similar way to synthetic diamond, it is consequently as expensive. Its main fault is that above 1000°C it will transform to a much softer crystalline form. It is therefore used to machine chill-cast irons and hardened steels at moderate speeds, rather than for high-speed processes where the tool may overheat and consequently soften.

Other very hard ceramics are used for military armour plating rather than as cutting materials. They include *boron carbide* B_4C, the low relative density of which makes it useful as armour plating for battlefield helicopters and their crew, and *titanium diboride* TiB_2 which, with a higher relative density, is confined to the protection of ground warfare vehicles against anti-tank projectiles.

21.6 Properties of ceramics

Prior to the manufacture of modern engineering ceramics, most materials of ceramic origin such as pottery, furnace refractories, bricks, tiles and electrical insulators consisted of materials containing particles of aluminosilicate origin cemented together by a brittle 'glass' network formed by chemical reactions during the high-temperature firing process. Modern processes, based on the technology of 'powder metallurgy' where high-pressure compacting is followed by sintering, produce structures which are completely crystalline. Many of these materials therefore maintain high strength at high temperatures, since the low melting-point glass network is absent. Toughness at ambient temperatures is also generally improved because of the absence of the brittle glass.

21.6.1 Strength

The high *compressive* strength of many ceramics is typified by Josiah Wedgwood's bone china cups (see Section 21.4.4) and the retention of these properties at high temperatures is a useful feature of some ceramics. For example, titanium diboride will maintain a compressive strength of 250 MPa at 2000°C, so that it is one of the strongest materials at such a temperature.

In terms of *tensile* strength, however, ceramics are generally less effective. Since many ceramics suffer from the presence of microcracks these act as stress raisers and this, coupled with a lack of ductility, means that stresses within the material cannot be relieved by the kind of plastic flow which occurs in metals. Consequently, as tensile stress increases catastrophic failure occurs precipitated by the microcracks. Table 21.1 gives the orders of magnitude of the strength of some ceramics at ambient temperature as a result of bend tests.

In most ceramics, ductility is, for all practical purposes, zero. This is largely due to the presence of small voids in the structure as well as a lack of elasticity and plasticity. These voids result from the methods of manufacture. In general, most ceramics can only be classified as 'brittle'.

Table 21.1 *Strengths of ceramics at ambient temperature*

Strengths at ambient temperature			
50–100 MPa	*100–200 MPa*	*200–400 MPa*	*Greater than 400 MPa*
Magnesia ceramics	Coarse-grained high-alumina ceramics	Sintered alumina of fine grain size	Sintered sialons
	Stabilised zirconia	Sintered silicon carbide	Sintered silicon nitrides

21.6.2 Creep

Creep only takes place in ceramics at relatively high temperatures and those which are totally crystalline in structure suffer the least creep. In those ceramics which contain glassy phases, particularly as inter-granular networks, creep will take place much more easily and at relatively lower temperatures.

21.6.3 Hardness

Most ceramics are relatively hard and it is this property which makes them useful as abrasive grits and cutting-tool tips. This hardness is largely due to the operation of strong covalent bonds between atoms in their crystal structure. Materials like silicon carbide, boron nitride and, of course, diamond are examples. Hardness tests used for ceramics use diamond indentors because of the risk of distortion with other materials. The three tests that are used are the Vickers test with a square-based pyramid indentor, the Rockwell superficial test with a diamond ball-ended cone indentor and the Knoop test with a diamond rhombohedral-based pyramid indentor. The Knoop test is used extensively for hard ceramics. Table 21.2 gives hardness values for some ceramics.

Table 21.2 *Hardness of some ceramic materials*

Material	Knoop hardness index
Diamond	7000
Boron nitride	6900
Titanium diboride	3300
Boron carbide	2900
Silicon carbide	2600
Silicon nitride	2600
Aluminium oxide	2000
Beryllium oxide	1200
(Hardened steel)	700

21.6.4 Refractoriness

Refractoriness is the ability of a material to withstand the effects of high temperatures without serious deterioration in its mechanical properties.

Fireclays, which are based on alumino-silicates, soften gradually over a range of temperatures and may collapse at a temperature below that at which fusion is expected to begin. Figure 21.8 shows only the liquidus–solidus range of the Al_2O_3–SiO_2 equilibrium diagram. The line *ABCD* represents temperatures at which 'mixtures' of pure Al_2O_3 and SiO_2 would begin to melt on being heated. In practical terms, this means that any fire-clay containing in the region of 5.5% Al_2O_3 is best avoided since it would collapse completely and suddenly to form a liquid at 1595°C. As the Al_2O_3 content increases, the amount of liquid formed at 1595°C decreases so that the brick would be less likely to collapse. However, for bricks to be usable at temperatures well *above* 1595°C, a high-alumina mixture (above 71.8% Al_2O_3) should be used so that the brick remains stable up to 1840°C. It must be pointed out, nevertheless, that this is a purely theoretical assessment – using the Al_2O_3–SiO_2 equilibrium diagram – of the expected melting range

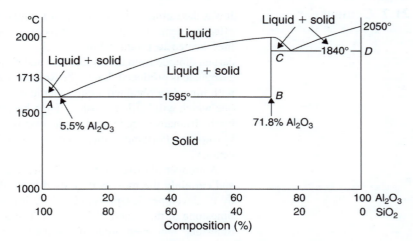

Figure 21.8 *Part of the Al_2O_3–SiO_2 equilibrium diagram showing the liquidus–solidus range.*

of *pure* clay mixture of known composition. In practice, clays are rarely mixtures of only kaolinite and either excess silica or excess alumina. Natural clays will contain other metallic oxides, all of which will *lower* its melting range to a greater or lesser degree depending upon the type and quantity of metallic oxide present. Consequently, some fireclays are unsuitable for use above 1200°C, whilst the more expensive clays containing above 70% Al_2O_3 can be used in the range 1700–1800°C.

Special ceramics for use at very high temperatures are far more expensive than fireclays. A few such materials are listed in Table 21.3.

Table 21.3 *Melting-points of some high-temperature ceramics*

Material	Melting point (°C)
Hafnium carbide	3900
Tantalum carbide	3890
Thorium oxide	3315
Magnesium oxide (magnesia)	2800
Zirconium oxide (zirconia)	2600
Beryllium oxide (beryllia)	2550
Silicon carbide	2300 (decomposes)
Aluminium oxide (alumina)	2050
Silicon nitride	1900 (sublimes)
Fused silica glass	1680 (approximately)

21.7 Cement

It was the Romans, those pioneers of engineering, who manufactured the first effective cement from small lumps of limestone found in clay beds. These they heated to produce lime, which, when mixed with water, will absorb carbon dioxide from the atmosphere and slowly harden as crystalline calcium carbonate is produced. Later, the Romans found that by mixing volcanic ash with lime they obtained a quick-hardening strong hydraulic cement, that is, one which will harden in water. Time-based cements of this type developed by the Romans were in use into the twentieth century, being slowly replaced by modern Portland cement which was introduced during the nineteenth century.

A number of different types of cement exist, but possibly the best known and certainly the most widely used is Portland cement, so named in 1824 by its developer Joseph Aspdin since it resembles Portland Stone (a white limestone of the Island of Portland). The principal ingredients are ordinary limestone – or even dredged sea shells – and clay-bearing materials comprising suitable clays or shales. The raw materials are pulverised separately and then mixed in the correct chemical proportions before being fed as a paste into a long rotary kiln where the mixture is calcined at 1500°C. The resultant 'clinker' is then ground along with a small amount of gypsum (calcium sulphate) to produce the fine greenish-grey powder – the well-known Portland cement – a typical composition of which is given in Table 21.4.

When mixed with water, Portland cement 'sets'. The chemistry of the process is quite complex but is one in which a reaction – usually termed *hydration* – between water and the silicates, aluminates and gypsum produces a hard rigid crystalline mass. Portland cement is rarely used as ceramic material by itself but generally mixed with sand to give mortar or with stone 'aggregate' to give concrete.

Table 21.4 *A typical Portland cement composition*

	Percentages by weight
Calcium silicates	73
Calcium aluminate	10
Calcium alumino-ferrite	8
Gypsum (calcium sulphate)	3
Magnesium oxide (magnesia)	3
Other metallic oxides	Balance

21.7.1 Cement as an engineering material

In building, cement and its products are widely used because they are relatively cheap to produce, being made from the most abundant constituents of the Earth's crust. Cement can be readily moulded into effective shapes without heating by simply mixing with water and pouring into a mould.

Solidified cement has very useful properties; it does not burn, dissolve or rot and possesses a moderate compressive strength. Nevertheless, in other respects, its mechanical properties are rather poor unless it is reinforced with steel (see Section 24.1). Ordinary hydraulic cement is very weak in tension or bending, particularly under impact loading when cracks propagate readily through the material. These cracks are formed from minute cavities which are present in ordinary commercially cast cement. Such cavities can be seen when a micro-section of such a cement is examined under a microscope and their action as 'stress raisers' is comparable to that of the graphite flakes in ordinary grey cast iron (see Section 15.6), another relatively brittle material.

Recent research into the production of Macro-Defect-Free (MDF) cement, in which the number and size of the cavities are greatly reduced, indicates that much improved mechanical properties are available. For example, a 3 mm thick sheet of MDF cement could not be broken by hand; and for cladding purposes 0.7 mm thick steel sheet could be replaced by 2 mm thick MDF cement *of the same weight but of vastly improved mechanical properties.* A coiled spring – similar in form to that used in automobile suspension – moulded from MDF cement was found to withstand a strain of 0.5% before failure. The replacement of plastics materials for such applications as bottle tops is also suggested. MDF cement would neither melt nor burn and is twice as stiff as reinforced plastics materials.

MDF cement is prepared by mixing large and fine particles of Portland cement with a gel of polyacrylamide in water. Thorough mixing produces a stiff dough which is then moulded to shape under pressure.

21.8 Semiconductors

Metals, which are *good conductors* of electricity, have a structure of atoms with valence electrons which are so loosely attached that they drift off and can move freely between the atoms and so, when a potential difference is applied, there are large numbers of free electrons able to respond and give rise to a current. *Insulators,* however, have a structure in which all the electrons are tightly bound to atoms and so there is no current when a potential difference is applied because there are no free electrons able to move through the material. *Semiconductors* can be regarded as insulators at a temperature of absolute zero. However, the energy needed to remove an electron from an atom in such a material is not very high and at room temperature there has been sufficient heat energy supplied for some electrons to have broken free. As a consequence, the application of a potential difference will result in a current. Increasing the temperature results in more electrons being shaken free and hence an increase in conductivity. At about room temperature, a typical semiconductor will have about 10^{16} free electrons per cubic metre and 10^{16} atoms per cubic metre with missing electrons.

Silicon is a semiconductor, being a covalently bonded solid with, at absolute zero, all the outer electrons of every atom being used in bonding with other atoms but some being readily freed when the temperature rises. When a silicon atom loses an electron, we can consider there to be a hole in its valence electrons so that, when a potential difference is applied, the movement of the freed valence electrons can be thought of as hopping from a valence site into a hole in a neighbouring atom, then to another hole, etc. Not only do electrons move through the material but so do the holes, the holes moving in the opposite direction to the electrons.

21.8.1 Doping

The conductivity of a semiconductor can be very markedly changed by impurities and so the purity of semiconductors for use in the manufacture of semiconductor devices used in electronics must be very carefully controlled, e.g. an impurity level less than one atom in a thousand million silicon atoms. However, foreign atoms can be deliberately introduced in controlled amounts into a semiconductor in order to change its electrical properties, this being termed *doping*. Phosphorus, arsenic or antimony atoms when added to silicon add easily released electrons and so make more electrons available for conduction; such dopants are called *donors* and such semiconductors with more electrons available for conduction than holes are called *n-type semiconductors*. Boron, gallium, indium or aluminium atoms when added to silicon add holes into which electrons can move; such dopants are called *acceptors* and such semiconductors with an excess of holes are called *p-type semiconductors*.

21.8.2 Production of doped silicon chips

The following is an outline of the types of processes used to produce the doped silicon chip elements in the integrated circuits which are an essential feature of modern life, whether as the microprocessor in a computer or the controller in a domestic washing machine.

1 *Purification of silica*
   Silica needs to be very pure for use in electronics. The required purity can be achieved by the use of *zone refining,* this being based on the fact that the solubility of impurities in the material is greater when the material is liquid than when solid. Thus, by moving the zone of melting along the length of a silicon rod, so the impurities can be swept up in the molten region.

2 *Crystal growth*
   For a semiconductor to be used in electronics, a single crystal is required, rather than a polycrystalline structure, since the dislocations at grain boundaries and the change in grain orientation can affect the electrical conductivity. A single crystal of silicon can be produced by the *Czochralski technique* in which a polycrystalline piece of silicon is heated to melting and an existing seed crystal with the required orientation is then lowered into the silicon melt. The seed crystal is then rotated as it is slowly drawn upwards. As the melt freezes on the seed crystal, it does so in the same orientation as the seed crystal and a single crystal rod is thus created.

3   *Slice preparation*

The single rod crystal is cut up into slices of about 500–1000 μm thickness by means of a diamond saw. The surface damage, resulting from the cutting, is removed by rubbing slices between two plane parallel rotating steel discs while using a fine abrasive, usually alumina. Residual damage is then removed by etching the surfaces with a suitable chemical to give slices of thickness about 250–500 μm.

4   *Epitaxial growth*

The surface of the slices still, however, contains defects and, as it is the surface layers in which the circuits are 'fabricated', *chemical vapour deposition (CVD)* is used to improve the surface by depositing silicon vapour onto the surface. With the temperature above about 1100°C, the deposited layer has the same orientation as the substrate. This process is termed *epitaxy* and the deposited layer the *epitaxial layer*.

5   *Silicon dioxide mask*

Silicon dioxide SiO_2 has a resistivity of about 10^{15} Ωm and so is a good insulator. It can be grown on the surface of silicon, bonding extremely well to it and providing an electrical insulating layer. Because dopants such as boron and phosphorus do not diffuse so readily through it, silicon dioxide can be used as a mask for the selective diffusion of dopants into the underlying silicon. Silicon dioxide layers can be produced by passing oxygen or water vapour over silicon slices at about 1000°C or by chemical vapour deposition of the silicon dioxide onto the silicon.

To use the silicon dioxide as a mask and so enable different areas of the underlying silicon to be doped differently, the silicon dioxide layer is coated with a negative photoresist, a monomer or short-chain polymer which is soluble in a solvent. The photoresist is then covered by the pattern to be transferred to the photoresist and then exposed to ultraviolet light. Exposure of the negative photoresist to ultraviolet light results in polymerisation or cross-linking of the short chains and a considerable reduction in solubility in the solvent. The pattern in the photoresist is then 'developed' by dissolving the non-exposed resist in a solvent. An etch is then applied to remove the silicon dioxide layer which is not covered by the photoresist; hydrofluoric acid is the standard etch used for silicon dioxide. The result is a window in the silicon dioxide through which we might introduce dopant.

6   *Doping*

The introduction of dopant atoms into a silicon slice can be by *thermal diffusion* as a result of exposing the surface to a vapour containing the dopant atoms or depositing on the surface a solid layer of silicon dioxide containing the dopant and then heating so that diffusion can occur from this layer into the silicon. An alternative to thermal diffusion is *ion implantation* in which the dopant atoms are fired at high speeds into the silicon slice.

22 Glasses

22.1 Introduction

Probably, the first glass to be 'seen' by Man was in the form of 'stones' produced by lightning strikes but little is known of its early manufacture, though bright blue glass ornaments dating from 7000 BC have been discovered. The Roman writer Pliny the Elder in the first century AD, tells us that soda glass was made accidentally by the Phoenicians from the fusion of sea-sand and soda. By Pliny's time, glass making was a profitable industry and Phoenician glass blowers were producing beautiful vases, jugs and phials in decorated glass. By AD 300, the first window-panes were being made by pouring molten glass onto a flat stone, but the glazing of windows was not widespread in England until the sixteenth century, though window glass had reached us in Norman times. Progress in this field, as in so many others involving early science and engineering, had been halted by the Fall of Rome and the Dark Ages which followed, culminating in the Black Death. In those days, the homes were either very draughty – or very dark.

Most of us still think of 'glass' as the stuff from which window-panes and beer-bottles are manufactured. We are all aware of its brittleness under mechanical shock – for some of us the first time we heard Mother use 'unacceptable' language was when she poured hot marmalade from her 'jam pan' into an inadequately heated glass jar, only to see it shatter.

Now that the fundamental nature of glass is clearly understood, materials scientists and engineers use the term *glass* to describe the structure of a whole range of substances from boiled sweets and toffee to some plastics materials, as well as the material of the well-known beer container we automatically describe as 'glass'.

22.2 Composition and structure of glass

It was mentioned earlier that Pliny describes how, centuries before his time, the Phoenicians had developed glass manufacture by melting together a mixture of sea-sand and soda. Common glass is still manufactured from roughly the same ingredients today. The chemical reaction upon which this process relies is of a simple 'acid–basic' type. Non-metallic oxides (many of which combine with water to form acids) are 'acidic' in nature whilst metallic oxides are generally strongly 'basic'. Acidic oxides usually react with basic oxides to form neutral 'salts'. Thus silicon dioxide (silica) SiO_2, generally in the form of sand, will combine with a number of metallic oxides when the two are heated strongly together. In addition to the sodium oxide Na_2O, derived from 'soda ash' (commercial sodium carbonate) when the latter is heated, lime CaO is also used in the manufacture of ordinary 'lime-soda' glass. These substances react together to give a mixture of sodium and calcium silicates – neutral salts – which will melt at temperatures much lower than those of silica or any of the metallic oxides used.

When the liquid mixture of silicates is allowed to cool, the viscous 'melt' becomes increasingly viscous and finally very hard and brittle – but, *crystallisation does NOT take place,* as it does with most materials such as metals. To explain this situation, we will deal with a very simple 'glass' made from pure silica (SiO_2) only. This would need to be heated to a temperature above 1720°C in order to produce a viscous liquid which would flow. If we then wished to produce *crystalline* silica, we would need to cool it very slowly indeed to a temperature just below its freezing-point of 1730°C in order to allow sufficient time for the SiO_2 molecules to link up forming the three-dimensional giant molecules which are present in the crystalline structures of quartz and other rocks containing silica.

'Geological rates' of cooling when Mother Earth was born were of course very slow indeed and so vast amounts of crystalline quartz (Figure 22.1A) were formed in the Earth's crust. The action of waves and weather has ground much of this to tiny particles – grains of sand. However, if we allow molten silica to cool more quickly, there is insufficient opportunity for the complex SiO_2 network to 'arrange itself' completely. As the temperature continues to fall, SiO_2 units lose more thermal energy and their movement becomes so sluggish that at ambient temperature it has ceased altogether. Thus we have failed to produce a crystalline silica structure – the material is still in its *amorphous* (or liquid) state (Figure 22.1B) and will remain so. Although still retaining its amorphous or liquid state, the silica has, at ambient temperature, become so viscous that it is extremely hard and brittle.

Many polymer chains also retain amorphous – or 'glassy' – structures. Here, the long chain-like molecules are so cumbersome that what thermal energy they possess is insufficient to overcome the van der Waals forces acting between them in order to permit them to 'wriggle' quickly enough to form a completely crystalline pattern. Hence, most polymers contain some completely amorphous regions (see Figure 20.1). If we compare this state

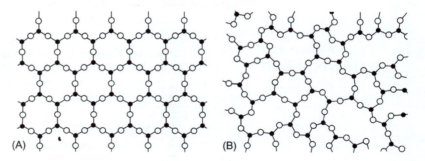

Figure 22.1 *This is a two-dimensional representation of the three-dimensional cage-like structure of silica in (A) crystalline silica (quartz) and (B) silica glass. The fourth oxygen atom is 'vertically above' the silicon atom (Figure 21.1) and will be covalently bonded in turn to another silicon atom – and so on. In (A) there is a repetitive pattern throughout the orderly crystalline structure but in (B) there is no repetitive pattern except inside the $(SiO_4)^{-4}$ units themselves – the structure is amorphous.*

of affairs in glasses and polymers with that in metals, we see that the single ions in a molten metal can move very quickly by comparison with the massive cumbersome molecules in glasses and polymers and so slip quickly and easily into their appropriate positions in the crystal lattice.

The high mobility of metallic ions in a liquid metal makes it difficult to produce amorphous glass-like structures in metals. However, it *can* be done with some metals and alloys – if we try hard enough! Generally, it is necessary to cool the molten metal at a rate of at least 10^6°C/second past its freezing point. In recent years, this has been accomplished and metallic glasses are now being manufactured as very valuable engineering materials (see Section 22.7).

22.3 Glass-transition temperature

Although silica does not crystallise at its freezing point T_m (1713°C) under normal cooling conditions, it does undergo a marked change in properties at some temperature below T_m. This is shown by a reduction in the rate at which it contracts in volume on cooling (Figure 22.2), i.e. there is a reduction in the coefficient of expansion/contraction and this coincides with a change in properties from a viscous or plastic state to one which is hard and rigid. The temperature at which this change occurs is termed the *glass-transition temperature* T_g.

There is no apparent change in the structure of the glass between T_m and T_g but clearly the glass has become a super-cooled liquid below T_m. On reaching T_g, movement of the molecules relative to one another ceases and the glass becomes hard and rigid – the molecules have become 'frozen' in their amorphous positions and a 'glassy' state has been reached. As the temperature falls below T_g, the rate of contraction becomes smaller and is due to a reduction in the thermal vibrations of the molecules.

It is not possible to give a precise value for T_g as it depends upon the rate at which the molten material is cooling when it passes through T_m. The more rapid the cooling rate the higher is the value of T_g because the molecules have less time to 'reorganise' themselves.

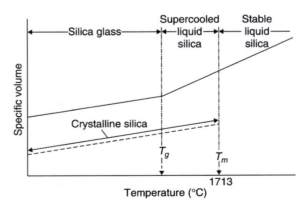

Figure 22.2 *The glass-transition temperature T_g of silica. The broken line indicates the crystallisation curve for SiO_2 which is cooled extremely slowly from above T_m.*

We have been discussing here the relationship between T_m and T_g for a simple silica glass, it having a single melting point T_m of 1713°C. Commercial glasses, which are mixtures of silicates of more than one metal, do not melt at a single definite temperature but – like metallic alloys – over a range of temperatures. Nevertheless, the relationships between T_m and T_g are basically similar.

22.3.1 Devitrification

Because glass has been cooled too rapidly to allow it to reach equilibrium by crystallisation at T_m, it is an unstable amorphous material at ambient temperatures. It does not usually crystallise because it lacks the thermal energy to allow this to happen. However, some glass will crystallise partially at ambient temperatures if given sufficient time – generally, a few hundred years. This phenomenon is called *devitrification* and leads to a loss of strength of the glass and a tendency to become extremely brittle and opaque. The surfaces of crystal grains scatters light, so reducing the light transmitted. Plans to remove some ancient stained-glass windows during the Second World War had to be abandoned because of the extreme brittleness of the glass.

22.3.2 Glass ceramics

Glass ceramics are designed so that crystallisation occurs to give a fine-grained polycrystalline material. Such a material has considerably higher strength than most glasses, retains its strength to much higher temperatures, has a very low coefficient of expansion and excellent resistance to thermal shock. Glass ceramics are produced by using a raw material containing a large number of nuclei on which crystal growth can start, e.g. small amounts of oxides such as those of titanium, phosphorus or zirconium. Most forms of glass ceramic are based on Li_2O–Al_2O_3–SiO_2 and MgO–Al_2O_3–SiO_2. The material is heated to form a glass, e.g. 1650°C, in the required product shape and then cooled and cut to size. The product might be in the

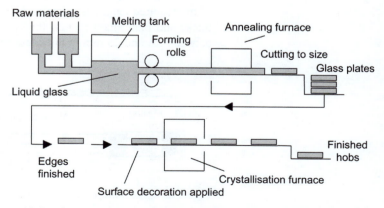

Figure 22.3 *The stages in the production of a ceramic glass cooker top.*

form of sheets, e.g. cooker tops, which are then edge finished and surface decoration applied (Figure 22.3). They are then heated to a high enough temperature, e.g. 900°C, to give controlled grain growth until the required grain size is obtained. Glass ceramics are used for cooker tops, cooking ware and telescope mirrors.

22.4 Glass manufacture

Ordinary 'soda' glass is made from a mixture of silica-sand, 'soda ash' (crude sodium carbonate) and lime (from limestone). Since glass is an easy product to recycle, large amounts of scrap glass, known as *cullet,* are used in glass manufacture. Large 'tank' furnaces, usually gas-fired and operating at 1590°C, hold up to 250 tonnes of molten glass produced from the raw materials and up to 90% cullet. The cullet comes mainly from 'bottle banks' operated by municipal authorities. Almost half of bottles find their way to these 'banks'.

As a small boy, I earned extra pocket money by collecting empty jam jars. For these, the local grocer paid me 1/2d, i.e. about 0.21p, per jar. There was a similar market for empty bottles. Apparently, it is now unprofitable, in these allegedly 'green' days, to collect, clean and reuse 'empties'. Worse still, few of these 'empties' end up even as cullet.

22.4.1 Float process

Sheet glass, such as is used for windows, is made by the float process in which a continuous ribbon of molten glass up to 4 m wide issues from the tank furnace and floats on the surface of a bath of molten tin (Figure 22.4). The forward speed of the ribbon is adjusted so that the molten glass remains in the high-temperature zone of the float tank long enough for a smooth flat surface to be obtained. The surfaces of the glass will then be parallel since both glass and tin retain horizontal surfaces. By the time that the glass leaves the float tank, it will have cooled sufficiently so that it is stiff enough to be drawn into the annealing *lehr* without the use of rollers which might disfigure the surface of the glass. Slow uniform cooling in the lehr limits internal stresses which might arise from rapid cooling. The glass is then cut into sheets of the required length as it leaves the annealing lehr. Window glass (3 mm thick) and plate glass (6 mm thick) are made by this process. Flat glass between 2.5 and 25 mm in thickness can be produced by this process.

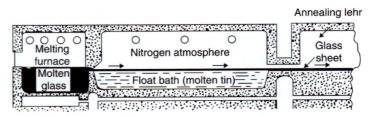

Figure 22.4 *An outline of the float process for the manufacture of sheet glass.*

22.4.2 Glass blowing

Although glass blowing was probably introduced by the Phoenicians, modern glass-blowing processes were developed in Venice in the twelfth century from whence it moved to the nearby island of Murano in 1292 to avoid 'industrial espionage'. By the middle of the nineteenth century, Flemish glass blowers had settled in the West Bromwich – Smethwick area of England to ply their craft at the Spon Lane works of Messrs Chance Brothers – sadly now long gone. In those days, the requirements of a successful glass blower were very powerful lungs and the acquired skill, but rewards were high. As a glass blower at 'Chances' my maternal great-grandfather was able to retire at 40 and buy his own pub!

Mouth-blowing is now restricted to the making of high-quality tableware and some artistic glass, but most glass hollow-ware is machine blown into metal moulds using a press-and-blow sequence of operations. A 'gob' of molten glass is first pressed into a simple mould to form a cup-shaped blank. This blank is then machine-blown to shape in a mould using a set-up very similar to that shown in Figure 20.14 which illustrates the shaping of a plastics container.

22.5 The properties of glass

In an explanation of the differences between crystalline and amorphous structures as they apply here (Section 22.2), pure silica was used as an example. Common glass, however, contains sodium (added as soda ash) and calcium (as lime) in addition. These form Na^{+1} and Ca^{+2} ions which become bonded to oxygen atoms in the silicate structures. The 'silica network' then loses its continuity to an even greater degree and in general this makes the hot glass even more malleable and easily formed than is the case with pure silica glass.

Most metals are ductile and malleable because dislocations are able to move through the *crystals* comprising the metallic structure. Glasses have no such crystal structure and any deformation can therefore only occur as a result of viscous flow of the molecules. However, a 'glass' is, by definition, a substance existing below the glass transition temperature so that very little movement of molecules is possible. This means that a glass at ambient temperature is extremely brittle. Nevertheless, some movement of molecules in a glass can occur under stress at ambient temperatures but such movement is extremely slow. If a glass rod is clamped at one end, cantilever fashion, and a weight placed at the free end, deflection of the rod of course occurs immediately. This is elastic deformation which would disappear immediately if the load were removed. If, however, the load were allowed to remain acting over a very long period and then removed, the elastic deformation would again disappear but the rod would probably be found to have acquired a 'permanent set' or deflection due to some movement of molecules into new positions having taken place during the long period of time in which the rod was in stress.

When glass is drawn to a fine fibre and cooled quickly, a high tensile strength is produced. Special glasses used in fibre-reinforced composites (see Section 24.3) can, under ideal conditions, reach strengths of up to 15 000 MPa, but in practice a lower strength of about 3500 MPa would be obtained since surface damage of the fibre is caused by contact with other materials. These microscopic surface scratches act as stress-raisers.

22.6 Glasses and their uses

The coefficient of thermal expansion of ordinary soda glass is relatively high, whilst its thermal conductivity is low, and this combination of properties makes it unsuitable for use in the majority of domestic situations involving sudden contact with boiling water. If boiling water is poured into a cold soda-glass container, it usually shatters as the surface of contact expands quickly whilst the colder glass beneath does not. This sets up higher stresses within the structure than soda glass can cope with. Table 22.1 gives the composition and uses of soda glass and a range of other glasses.

22.6.1 Pyrex

'Pyrex' is a well-known heat-resistant glass. It is of boro-silicate composition, containing boron oxide (B_2O_3), which produces a material of very *low* coefficient of thermal expansion. This makes it suitable for kitchenware, laboratory and other uses, particularly since its resistance to chemical attack is also good.

22.6.2 Glass ceramics

Some modern alumino-silicate glasses are heat-treated after fabrication to give them a partly crystalline structure (see Section 22.3.2). Such glasses have physical properties intermediate between those of true glasses and a ceramic like alumina.

Table 22.1 *Composition and uses of some glasses*

Type of glass	Typical composition (%)	Properties and uses
Soda glass	SiO_2 72, Na_2O 15, CaO 9, MgO 4	Window panes, plate glass, bottles, jars, etc.
Lead glass	PbO 47.5, SiO_2 40, Na_2O + K_2O 7.5, Al_2O_3 5	High refractive index and dispersive power. Lenses, prisms and other optics. 'Crystal' glass tableware.
Boro-silicate glass (Pyrex)	SiO_2 70, B_2O_3 20, Na_2O + K_2O 7, Al_2O_3 3	Low coefficient of expansion and good resistance to chemicals. Used for heat-resistant kitchenware and laboratory apparatus.
Alumino-silicate glass (glass ceramic)	SiO_2 35, CaO 30, Al_2O_3 25, B_2O_3 5, MgO 5	High softening temperature (T_g up to 800°C). A glass-ceramic ('Pyrosil' and 'Pyroceram') – cooking ware, heat exchangers, etc.
High silicon glass	SiO_2 96, B_2O_3 3, Na_2O 1	'Vycor' – low coefficient of expansion. Missile nose cones, windows for space vehicles.
Silicon-free glass	B_2O_3 36, Al_2O_3 27, BaO 27, MgO 10	Sodium-vapour discharge lamps.

22.7 Metallic glasses

Although it is relatively easy to retain a glassy amorphous state in many materials which exist as large ungainly molecules, it is very difficult to preserve a glassy structure in metals for reasons stated earlier (Section 22.2). Nevertheless, a number of crafty methods have been developed to achieve the very high cooling rates (in the region of 10^{6}°C/second) necessary to retain metals in an amorphous state. The most successful is based on some form of 'metal spinner' (Figure 22.5). This consists of a copper wheel rotating at a peripheral speed of up to 50 m/second and a jet of molten metal under pressure which impinges against the rim of the wheel. The 'pool' of metal which forms on the rim cools extremely rapidly, since copper is a good conductor of heat, forming a continuous ribbon which passes to a coiling unit.

In a metallic glass produced in this way, the structure is amorphous and there is no regular arrangement of the atoms as exists in a normal crystalline metal below its freezing point. Here, the atoms are still in the irregular array of the amorphous liquid, except that a little less space will be occupied because thermal movement has ceased. Imagine a plastics bag filled randomly with ball bearings – that is the type of 'structure' in which atoms are present in a metallic glass.

Some metals and alloys form glasses more readily than do others. Generally, a mixture containing 80% metal and 20% 'metalloid' (e.g. boron, silicon and so on) seems to be the most effective. Like other glasses, they are very strong but in tension elongation is usually below 1%. The most important of these glasses to date is 80Fe–20B which is mechanically stronger than the best carbon fibre. However, by far the more important property is its extreme magnetic 'softness', i.e. it has a very low remanence and a high magnetic permeability. This has led to its wide use in the USA as a core material for electric power transmission transformers, resulting in a very significant energy saving whilst at the same time being cost effective in the medium term. With traditional transformer cores, there is a 3% loss in the electricity generated. This is used in magnetising the transformer cores and can be reduced to about 0.75% when 80Fe–20B cores are used. In the UK, this translates to a saving of some 1.3 million tonnes of oil (or its equivalent in gas or coal) per annum, which would lead to a reduction of 4.5 million tonnes of CO_2 and 120 kilo tonnes of SO_2 being released as 'greenhouse gases'. Surely, the universal adoption of such glass transformer cores should be an option to be taken seriously. Or is the threatened 'Carbon Tax' now overdue?

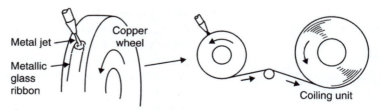

Figure 22.5 *The principles of 'metal spinning' for the manufacture of 80Fe–20B metallic glass.*

23 Composite materials

23.1 Introduction

Those engineering materials loosely referred to as *composites* include a wide range of products, ranging from those used in high-strength aircraft components to road-building tarmacadam and concrete. Generally, composites are manufactured by mixing together two separate components, one of which forms a continuous matrix whilst the other, present either as particles or fibres, provides the strength or hardness required in the composite material. Of these materials, *fibre-reinforced composites* are the most significant in the modern engineering world and the whole of the next chapter will be devoted to them. This chapter is devoted to, what can be termed, *particle composites*.

23.1.1 Particle composites

Particle composites can be divided into three groups:

- *Particle-hardened composites* containing particles of a very hard constituent embedded in a tough, shock-resistant matrix, e.g. hard metallic carbide particles in a tough metallic matrix, used for tool and die materials.
- *Dispersion-hardened composites* containing finely dispersed hard, but strong, particles which will raise the strength of the parent material, e.g. Al_2O_3 particles in specially prepared metallic aluminium.
- *'Filler' composites* containing particulate material of very low cost which has been added as a 'filler' to 'bulk-up' the matrix material. Bakelite mouldings have long been 'filled' with sawdust, wood flour or finely ground minerals such as sand and limestone.

Some cohesion between the particles and the matrix is necessary and this may be achieved by either:

- *Mechanical bonding* which will operate when the surface of the particle material is rough or irregular in texture and the matrix is added as a liquid, e.g. particles of aggregate in concrete.
- *Physical bonding* which depends upon the operation of van der Waals forces acting between surface molecules in both materials.
- *Chemical bonding* at the interface between particle and matrix; sometimes, this can have a deleterious affect if the reaction product is in the form of a brittle film.
- *Solid-solution bonding* in which the particle may dissolve in the matrix to a limited degree, forming a solid solution. Such a situation generally produces a strong positive bond.

23.2 Particle-hardened composites

These are generally the products of powder metallurgy (see Section 7.4) in which extremely hard particles of a ceramic material are held in a tough ductile matrix of some metal. Such materials are usually known as *cermets* and have been popular for many years as cutting tools and die materials. The most widely used cermets consist of particles of hard tungsten carbide held in a tough matrix of cobalt (Figure 23.1). The two components, in the form of powders, are thoroughly mixed and the mixture then compacted at high pressure in a die of the required shape. The application of high pressure causes the cobalt particles to slide over each other so that a degree of cold-welding occurs between the particles and the resultant compact is strong enough to permit handling. This stage of the process is followed by 'sintering' – that is, heating the composite at some temperature high enough above the recrystallisation temperature of the cobalt so that a continuous, tough matrix of cobalt is formed. The heating process takes place in an atmosphere of hydrogen to protect the compact from oxidation.

Figure 23.1 *The type of structure in a tungsten carbide/cobalt cermet. Particles of tungsten carbide (white) in a cobalt matrix (black) (×1000).*

The proportions of tungsten carbide and cobalt in such a cermet are varied in accordance with the properties required in the tool (Table 23.1). For machining hard materials, cutting tools containing 95% tungsten carbide and 5% cobalt are used, whilst for cold-drawing dies, where greater toughness is required, a cermet containing 75% tungsten carbide and 25% cobalt would be employed.

Some cermets are manufactured by allowing a *liquid* matrix metal to infiltrate around particles of a solid ceramic. A strong positive bond is more likely to be produced when the ceramic forms a solid solution with the matrix metal. If the result is an intermetallic compound, then the bond is likely to be weak and brittle.

Typical cermets are based on hard oxides, borides and nitrides as well as upon carbides. The more important groups are shown in Table 23.2.

23.3 Dispersion-hardened materials

A material can be strengthened by the presence of small roughly spherical particles of another material which is *stronger* than the matrix material itself. The function of these small strong particles is to impede the movement of 'dislocation fronts' which are passing through the material (see Figure 6.4). Since the dislocation fronts are unable to move through the particles – because the latter are stronger – 'loops' are formed in the fronts. These loops will provide an even more effective barrier to any further dislocation fronts following along the same plane. Thus, the strength of the material increases as the 'dislocation jam' builds up.

23.3.1 Sintered aluminium powder (SAP)

An example of such a dispersion-hardened material is 'sintered aluminium powder' (SAP) in which the dispersed strong particles – or 'dispersoid' – is aluminium oxide. SAP is manufactured by grinding fine aluminium powder in the presence of oxygen under pressure. Aluminium oxide forms on the surface of the aluminium particles and much of it disintegrates during grinding to form fine aluminium oxide powder intermingled with the

Table 23.1 *Cobalt–tungsten carbide materials*

Cobalt (%)	Tungsten carbide (%)	Tungsten carbide grain size	Uses
3	97	Medium	Machining of cast iron, non-ferrous metals and non-metallic materials.
6	94	Fine	Machining of non-ferrous and high-temperature alloys.
6	94	Medium	General purpose machining for metals other than steels, small and medium size compacting dies and nozzles.
6	94	Coarse	Machining of cast iron, non-ferrous materials, compacting dies.
10	90	Fine	Machining steels, milling, form tools.
10	90	Coarse	Percussive drilling bits.
16	84	Fine	Mining and metal forming tools.
16	84	Coarse	Mining and metal forming tools, medium and large size dies.
25	75	Medium	Heavy impact metal forming tools, e.g. heading and cold extrusion dies.

Table 23.2 *Some cermet materials*

Cermet group	Ceramic	Bonding matrix	Uses
Carbides	Tungsten carbide	Cobalt	Cutting tools.
	Titanium carbide	Cobalt, molybdenum or tungsten	Cutting tools.
	Molybdenum carbide	Cobalt (or nickel)	Cutting tools.
	Chromium carbide	Nickel	Slip gauges, wire-drawing dies.
Oxides	Aluminium oxide	Cobalt or chromium	Rocket motor and jet-engine parts – other uses where high temperatures are encountered. 'Throw-away' tool bits.
	Magnesium oxide	Magnesium, aluminium or cobalt	
	Chromium oxide	Chromium	
Borides	Titanium boride	Cobalt or nickel	Cutting tool tips.
	Chromium boride	Nickel	Cutting tool tips.
	Molybdenum boride	Nickel	Cutting tool tips.

aluminium particles. The mixture is then compacted and sintered by powder metallurgy techniques to produce a homogeneous aluminium matrix containing about 6% aluminium oxide particles. As Figure 23.2 indicates, SAP retains a much higher strength than some of the precipitation-hardened aluminium alloys at temperatures above 200°C – temperatures at which the precipitation-treated alloys lose strength rapidly as reversion (see Section 17.7) occurs.

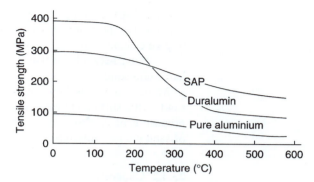

Figure 23.2 *The relationship between tensile strength and temperature for duralumin, SAP and pure aluminium.*

23.3.2 Manufacturing processes

Orthodox melting/casting processes are unsuitable for the manufacture of dispersion-hardened products, since the dispersoid is usually of lower relative density and would float to the surface of the melt. Instead, both dispersoid and matrix metal in powder form are treated in a high-speed ball mill where the high impact energy of the hard steel balls causes some 'mechanical alloying' between particles of both dispersoid and matrix metal. The final consolidation and shaping is by normal powder-metallurgy processes or by extrusion.

23.3.3 Modern superalloys

Some modern superalloys used in aerospace projects are dispersion-hardened by yttrium oxide Y_2O_3 particles. Thus, *Incoloy MA956* contains 4.5% Al, 2.0% Cr, 0.5% Ti, 0.5% Y_2O_3 with the balance being iron; it is used in combustion chambers and turbine casing sections where resistance to creep is essential. Both aluminium oxide and aluminium carbide are used as dispersoids in some aluminium alloys. *Inco MAP Al9052* is an aluminium base alloy containing 4% Mg, 1.1% C and 0.06% O and has a tensile strength of 450 MPa. It is used in military aerospace when a combination of low relative density, high strength and resistance to corrosion is required.

23.4 Mortar and concrete

These are composite materials in which a hardened paste of 'cement' and water is used to stick and hold together particles of sand, gravel, stones and other 'aggregates' to give fairly high compressive strength materials.

23.4.1 Mortar

Mortar is the adhesive material used between bricks in the building industry. Before the advent of Portland cement (see Section 21.7), mortar consisted of a mixture of slaked lime, sand and water. This is still used occasionally when a *slow* hardening process can be tolerated. The slaked lime ($Ca(OH)_2$) contained in the mortar reacts slowly with carbon dioxide present in the air to give calcium carbonate ($CaCO_3$). The calcium carbonate forms as a hard interlocking network, so that the slaked line/sand 'paste' is replaced by a hard rigid solid.

Modern mortar usually consists of a mixture of Portland cement and clean sand, with sufficient water to make the mixture workable. Ideally, the proportion of sand to cement in the dry mixture is 3:1 but higher proportions of sand are generally used.

23.4.2 Concrete

Concrete is produced from a mouldable mixture of Portland cement, stone aggregate, sand and water. When hard, it has many of the characteristics of natural stone and is extensively used in building and civil engineering. In the 'wet' state, it can be moulded easily, and is one of the cheapest constructional materials in Britain, because of the availability of suitable raw material, and also the low maintenance costs of the finished product.

The aggregate may be selected from a variety of materials. Stone and gravel are most widely used, but in some cases other substances available cheaply, such as broken brick and furnace slag, can be employed. Sand is also included in the aggregate, and, in order to obtain a dense product, a correct stone/sand ratio is essential. Aggregate materials should be clean and free from clay. The proportions of cement to aggregate used depend upon the strength required in the product, and varies from 1:3 in a 'rich' mixture to 1:10 in a 'lean' one. The cement/aggregate ratio commonly specified is 1:6, and this produces excellent concrete, provided that the materials are sound and properly mixed and consolidated (Figure 23.3). Other things being equal, the greater the size of the stones, the less cement is required to produce concrete of a given strength.

The cement and its aggregates are mixed in the dry state, water is then added and the mixing process is continued until each particle of the aggregate

Figure 23.3 *A satisfactory concrete structure: the aggregate consists of different grades (or sizes) in the correct proportions so that the smaller particles fill the spaces between the larger ones, whilst particles of sand occupy the remaining gaps, and all are held together by a film of cement.*

is coated with a film of cement paste. The duration of the hardening period is influenced by the type of cement used, as well as by other factors such as the temperature and humidity of the surroundings. Concrete made with Portland cement will generally harden in about a week, but rapid-curing methods will reduce this time. Concrete structures can be 'cast *in situ*' (as in the laying of foundations for buildings) or 'pre-cast' (sections cast in moulds and allowed to harden before being used). Such plain concrete is suitable for use in retaining walls, dams and other structures which rely for their stability on great mass. Other uses include foundations where large excavations have to be filled, whilst vast quantities of plain concrete are employed in modern motorway construction. At the other end of the scale, many small pre-cast parts are manufactured, e.g. 'reconstituted stone' for ornamental walls, as well as other sundry articles of garden 'furniture'.

23.5 Tarmacadam

Modern methods of road making were originated early in the nineteenth century by the Scots engineer, John McAdam. The method he used – coating suitable hard aggregate material with tar – is roughly similar to the process used today, except that the tar (obtained from the gasworks where coal was destructively distilled) has been replaced largely by bitumen (residues from the refining of crude petroleum). Some asphalts also occur naturally, e.g. 'Trinidad Lake'.

The bituminous material is mixed with a suitable aggregate such as crushed blast-furnace slag for the coarse foundation work, or fine gravel for the finishing layers. The resultant mixture is tough and crack-resistant because of the bituminous matrix, whilst it is hard-wearing because of the exposed surface of hard aggregate material. Its structure and properties resemble very closely those of a bearing material in which hard, low-friction particles standing 'proud' of the surface are held in a tough ductile shock-resistant matrix. However, whilst slip at a very low coefficient of friction is the objective in a bearing, the reverse is true in a road surface; the rubber tyre must be designed to provide maximum adhesion between tyre and road surface.

24 Fibre-reinforced composite materials

24.1 Introduction

This chapter follows on from the previous chapter and looks at fibre-reinforced materials. The concept of fibre-reinforced materials had its origin in nature in the structure of wood. In metallic structures, the building unit is the crystal whilst a polymer is an agglomeration of large numbers of long thread-like molecules. A glass consists of a mass of fairly large silicate units which are too sluggish in their movements to be able to crystallise. In living matter, both plant and animal, the simplest unit is the cell. As a tree grows, the wood tissue forms as long tube-like cells of varying shapes and sizes. These are known as *tracheids* but we may regard them as *fibres* which are arranged in roughly parallel directions along the length of the trunk (Figure 24.1). They vary in size from 0.025 to 0.5 mm in diameter and from 0.5 to 5.0 mm in length and are composed mainly of cellulose. The fibre-like cells are cemented together by the natural resin *lignin*; so wood can be regarded as a naturally occurring composite material in which the matrix of lignin resin is reinforced and strengthened by the relatively strong fibres of cellulose.

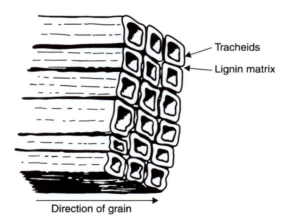

Figure 24.1 *The structure of wood – the cellulose fibres (tracheids) are cemented in a matrix of lignin resin.*

Most amateur carpenters and DIY experts soon come to realise that the strength of a piece of timber is 'along the grain', i.e. in the same direction as the cellulose fibres, whilst 'across the grain' the timber is relatively weak and brittle. Thus, wood is a very anisotropic (the term means a material in which the properties vary according to the direction in which they are measured) material, e.g. European redwood may have a tensile strength of 9 MPa along the grain, whilst across the grain, i.e. perpendicular to it, the strength may be no more than 1.5 MPa. Timber has an excellent strength-to-weight ratio and, when considered in this context, some types of wood can compete with steel. Unfortunately, a bulky cross-section is necessary if wood is to sustain a tensile load; consequently, steel with its corresponding small cross-section will generally be used instead. In compression and bending, provided there is no bulk disadvantage, wood may be preferable, as indeed it is for domestic building. Wood has a high modulus of elasticity and, when vibrated, it produces strong resonance, hence it is used for the sounding-boards of many stringed musical instruments, such as violins and guitars – the traditional or 'acoustic' types of course, not the electrically amplified gadgets 'manipulated' by the perpetuators of ear-splitting caco-phony of 'rock music'. Sorry chaps! but my youth was spent in the era of real jazz and the days of Django, Stephan and their timeless 'Quintette'.

Although Man had been using wood for a long time, it was not until the 1940s that strong fibre materials and suitable bonding resins became available so that he could hope to surpass the tensile and stiffness properties of natural wood. The development of polyester resin/glass fibre composites then began.

24.1.1 Man-made fibre-reinforced composites

In general, man-made fibre-reinforced composites include:

- Matrix materials, such as thermosetting or thermoplastics polymers and some low-melting-point metals, reinforced with fibres of carbon, glass or organic polymer.
- Polymers, usually thermosetting, reinforced with fibres or laminates of woven textile materials.
- Vehicle tyres in which vulcanised rubber is reinforced with woven textiles or steel wire.
- Materials such as concrete reinforced with steel rods.

The fibres in the composites may be all aligned in the same direction and continuous throughout the matrix, short fibres either all aligned in the same direction or randomly orientated, or the composite may have fibres orientated in two, or more, particular directions. Obviously, when the reinforcing fibres are unidirectional, as are the fibres in a tree trunk, then maximum strength is also unidirectional. In many fibre-reinforced composites where woven textiles are used instead of unidirectional fibres then maximum strength is available in more than one direction. A case in point is the woven textile or steel mat used to reinforce a vehicle tyre. However, we shall be dealing here mainly with composites containing unidirectional fibres so that we can more easily assess the advantages of such reinforcements.

In successful composites, there must be adequate bonding between fibre and matrix and this bonding may be either physical or chemical. The main function of the matrix material is to hold the fibres in the correct position so that they carry the stress applied to the composite as well as to provide adequate rigidity. At the same time, the matrix protects the fibres from surface damage and from the action of the environment. The fibres should be long enough so that the bonding force between the surface of the fibre and the surrounding matrix is greater than the force necessary to break the fibre in tension – short fibres may slip inside the matrix because the interface surface is too small so that bonding fails before the fibre breaks in tension.

The overall relative density of such a composite is important since it affects both specific strength, i.e. the strength-to-weight ratio, and the specific modulus, i.e. modulus-to-weight ratio – both particularly relevant when materials used in air or land transport vehicles are involved.

24.2 Unidirectional composites

We will now consider a composite rod consisting of long parallel fibres of some strong material held in a matrix of a rigid substance (Figure 24.2), e.g. glass or carbon fibres in an epoxy resin matrix.

24.2.1 Relative density of composite

The mass of a composite m_c is the sum of the masses of the fibres m_f and the matrix m_m, and the volume v_c is the sum of the volumes of the fibres v_f and the matrix v_m. Thus, the density of the composite ρ_c is given by:

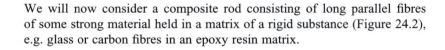

$$\rho_c = \frac{m_c}{v_c} = \frac{m_f}{v_c} + \frac{m_m}{v_c} = \frac{m_f}{v_f}\frac{v_f}{v_c} + \frac{m_m}{v_m}\frac{(v_c - v_f)}{v_c}$$

If V_f is the volume fraction of the composite that is fibre, ρ_f the density of the fibres and ρ_m the density of the matrix:

$$\rho_c = \rho_f V_f + \rho_m (1 - V_f)$$

Thus, if a glass-reinforced polyester resin contains 60% by volume of glass fibre, and the relative densities of glass and polyester are 2.1 and 1.3, respectively, then the relative density of the composite is, using the above equation:

$$\rho_c = 2.1 \times 0.6 + 1.3(1 - 0.6) = 1.78$$

Figure 24.2 *Composite rod with long parallel fibres.*

24.2.2 Tensile strength of composite

If the fibres and matrix are firmly bonded to each other then:

force acting on the composite
= force acting on the fibres + force acting on the matrix

Since stress σ = force/cross-sectional area, in each case, then:

force acting on the composite: $F_c = \sigma_f A_f + \sigma_m A_m$

where A_f and A_m are the cross-sectional areas of fibre and matrix, respectively. If we now divide throughout by the total cross-sectional area of the composite A_c, we get:

$$\frac{F_c}{A_c} = \frac{\sigma_f A_f}{A_c} + \frac{\sigma_m A_m}{A_c}$$

but F_c/A_c is the stress σ_c on the composite, whilst if we consider a unit length of the composite then V_f, the volume fraction of fibres, is A_f/A_c and V_m the volume fraction of the matrix is A_m/V_m. Therefore:

$$\sigma_c = \sigma_f V_f + \sigma_m V_m = \sigma_f V_f + \sigma_m (1 - V_f)$$

Thus, if a composite material in rod form consists of polyester fibres reinforced with 40% by volume of long parallel glass fibres and the tensile strengths of the glass and polyester are 3400 MPa and 60 MPa, respectively, the tensile strength of the composite is given by the above equation as:

$$\sigma_c = 3400 \times 0.4 + 60 \times 0.6 = 1396 \text{ MPa}$$

24.2.3 Modulus of composite

When the composite is under stress, the resultant strain ε will be the same in both fibres and matrix (since we have assumed that they are firmly bonded). Hence:

$$\frac{\sigma_c}{\varepsilon} = \frac{\sigma_f V_f}{\varepsilon} + \frac{\sigma_m (1 - V_f)}{\varepsilon}$$

But stress/strain = modulus of elasticity E in each case, therefore:

$$E_c = E_f V_f + E_m (1 - V_f)$$

Thus, if a composite material is made by adding 15% by volume of glass fibre and 5% by volume of carbon fibre to an epoxy resin matrix, the fibres being long and parallel in direction and firmly bonded to the matrix, and the values of the tensile modulus of elasticity for glass, carbon and epoxy resin are 75, 320 and 4 GPa, respectively, then the modulus of elasticity of the composite is given by the above equation as:

$$E_c = 75 \times 0.15 + 320 \times 0.05 + 4 \times 8 = 30.45 \text{ GPa}$$

24.3 Fibres

Cold-setting resins strengthened by woven textile fibres, e.g. 'Tufnol', were developed many years ago for use in the electrical trades as tough panels with high electrical insulation properties, but glass fibre was the first man-made material to be used for reinforcement purposes to produce a strong lightweight material. Now a number of other man-made fibres, in particular carbon, boron, some ceramics and polymers, all possessing the necessary high strength, stiffness and low relative density, are in use. Research continues with the object of producing materials of increasingly high strength-to-weight ratio, i.e. high *specific strength,* for the aerospace industries and, since relative density is important, many fibres are made from compounds based on the elements boron, carbon, nitrogen, oxygen, aluminium and silicon – all elements with low atomic mass which often, though *not* inevitably, means that relative density is also low. Moreover, compounds based on these elements are likely to be covalently bonded – the strongest of chemical bonds – which is generally reflected in the mechanical strength of the compound.

24.3.1 Glass fibre

Glass fibre was the first of modern fibres to be used in composites. A number of different compositions are now in production. The best known of these are S-glass (which is a high-strength magnesia-alumino-silicate composition) and E-glass (a non-alkali metal boro-silicate glass developed originally for its electrical insulation properties). Whilst glass fibre is strong and relatively cheap, it is less stiff than carbon fibre, the relative density of which is lower than that of glass.

High-quality glass fibre is spun at a high temperature and then cooled very rapidly before being given a coating of 'size'. This is a mixture, the main object of which is to protect the glass surface from contamination which would lead to a loss in strength. It also contains a derivative of silane (SiH_4) which acts as a bonding agent between glass and matrix.

24.3.2 Carbon fibre

Carbon fibre is made up of layers of carbon atoms arranged in a graphite-type structure (see Figure 1.9), but whilst graphite consists of small 'plates' of atoms which slide over each other (hence graphite is a lubricant), in carbon fibre the graphite-type structure persists to form long fibres. Whilst strong covalent bonds join the carbon atoms along a layer, only weak van der Waals forces operate *between* layers. Thus, the carbon fibre structure is anisotropic in many of its properties, e.g. in a direction parallel to the fibres the tensile modulus of elasticity reaches 1000 GPa, whilst perpendicular to it is only 35 GPa.

The bulk of carbon fibre is manufactured by the heat-treatment of polyacrylonitrile filament. The process takes place in three stages in an inert atmosphere:

1   A low-temperature treatment at 220°C. This promotes cross-linking between adjacent molecules so that filaments do not melt during subsequent high-temperature treatments.
2   The temperature is raised to 900°C to 'carbonise' the filaments. Decomposition takes place as all single atoms and 'side groups' are 'stripped' from the molecules, leaving a 'skeleton' of carbon atoms in a graphite-like structure.

3   The heat-treatment temperature is then raised to produce the desired combination of properties. Lower temperatures (1300–1500°C) produce fibres of high tensile strength and low modulus, whilst higher temperatures (2000–3000°C) provides fibres of low strength but high modulus.

The fibres are kept in tension during the heat-treatment processes to prevent curling and to favour good alignment. Carbon fibres are between 5 and 30 μm in diameter and are marketed in 'tows' containing between 1000 and 20 000 filaments. Carbon fibre is the most promising of composite reinforcements, particularly in terms of specific strength. Although initially carbon fibre was expensive, production costs have fallen as a result of improved manufacturing techniques and seem likely to fall further.

24.3.3 Boron fibre

Boron fibre has a very high modulus of elasticity but only a moderate tensile strength. It is used in the aerospace industries because of its low relative density but mainly because of its high transverse strength, which along with a large fibre diameter, enables it to withstand high compressive forces. However, since it is difficult to fabricate it is very expensive. Boron fibre is grown direct from the vapour phase onto a tungsten filament only 15 μm in diameter, the resultant boron fibre being between 100 and 200 μm in diameter (Figure 24.3). Since boron reacts chemically with tungsten, there is a continuous bond between boron and the thin tungsten core.

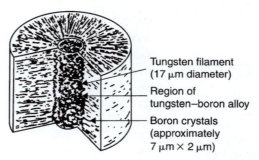

Tungsten filament
(17 μm diameter)

Region of
tungsten–boron alloy

Boron crystals
(approximately
7 μm × 2 μm)

Figure 24.3 *Section through a boron fibre grown on a tungsten filament.*

24.3.4 'Aramid' fibre

Most of the plastics materials consisting of long-chain molecules have limited strength and stiffness because the molecules are *not straight* but generally coiled or folded. Hence, under stress, extension takes place as the molecules straighten out and also slide past each other. Nevertheless, if the molecules are stretched and straightened during the fibre manufacturing process, much higher strength and stiffness can be developed. Aromatic (aliphatic organic compounds are based on a mainly chain-like structure, aromatic ones on the benzene ring) polyamide – hence 'Aramid' – fibres are the most successful of these and are manufactured under the trade name of 'Kevlar'.

After being spun, the fibres are heated *whilst in tension* in a nitrogen atmosphere at temperatures up to 550°C. The 'stretched' properties are then retained. The resultant material is strong in a direction parallel to the straightened molecular chains but, since the chains are held together only by weak van der Waals forces, the transverse and compressive strengths are relatively low. However, this does mean that Kevlar has a good capacity for absorbing energy so Kevlar-reinforced composites are useful for resisting impact damage, e.g. bullet-proof vests and the like where low relative density is important. For the latter reason, aramids are also used in aerospace industries. Their main disadvantages are a sensitivity to ultraviolet light from which they must be shielded and the fact that they absorb water.

24.3.5 Other fibres

Some ceramics, most importantly silicon carbide and alumina, are being developed as reinforcement fibres in metal matrix composites (MMCs). These are important for future developments in the aerospace industries where both cast and extruded aluminium alloys can be further strengthened by fibre reinforcement.

Table 24.1 shows the mechanical properties of a range of fibres.

Table 24.1 *Mechanical properties of some fibre materials*

Fibre type	Tensile strength (GPa)	Modulus (GPa)	Relative density	Specific strength (GPa)	Specific modulus (GPa)
Aramid (Kevlar 4)	2.75	125	1.44	1.9	87
Silica glass	5.9	74	2.2	2.2	34
E-glass	1.7	70	2.5	2.5	28
S-glass	4.6	84	2.6	2.6	32
Carbon (high strength)	3.5	200	1.7	1.7	118
Carbon (high modulus)	2.5	400	2.0	2.0	200
Boron	3.9	400	2.6	2.6	154
Silicon carbide	7.0	400	3.2	3.2	125
Alumina	1.7	380	4.0	4.0	96
Bulk materials for comparison					
Steel	1.3	260	7.8	0.2	26
Wood (spruce)	0.1	10	0.46	0.22	22

24.4 Matrix materials

Polymers are the most widely used of matrix materials in fibre-reinforced composites. The main functions of the matrix is to provide a rigid structure which holds the strong fibres in position. To provide a successful composite, the matrix material must bond adequately with the fibre surface, either frictionally or, as is more usual, chemically. When a chemical bond is formed, any reaction between the matrix material and the fibre must not be so vigorous as to damage the surface of the latter. The matrix material must flow easily during the investing process, so that the spaces between the fibres are filled, and then 'set' quickly.

The main *service* requirements of a matrix material are:

* It should be stable to a temperature at which the properties of the fibre begin to deteriorate.
* It must be capable of resisting any chemical attack by its environment.
* It should not be affected by moisture.

24.4.1 Thermosetting resins

These are polymers which cross-link during 'curing' to form a hard glassy solid. The most popular are polyesters and epoxy resins. Polyesters were the first thermosetting resins to be used, with glass-fibre reinforcement, during the Second World War in the manufacture of housings for aircraft radar antennae, the composite being 'transparent' to radio transmissions. These resins are still widely used since they are relatively cheap and are easy to work with and will 'cure' at relatively low temperatures. Unfortunately, polyesters do not bond well with fibres so that the transverse strength of the composite depends mainly on weak van der Waals forces. Moreover, since polyesters shrink considerably on curing, this further reduces the adhesion to the fibres. For both reasons, polyesters are not generally used in high-strength composites.

For high-performance composites based on either continuous carbon or aramid fibres, epoxide resins are extensively used. The principles of the manipulation of these resins will be familiar to readers who have used DIY epoxide glues. The resin is mixed with a 'hardener' and heated for up to 8 hours at about 170°C to form a hard, insoluble, cross-linked structure with a tensile strength of 80–100 MPa and a modulus of elasticity of 2–4 GPa. Such a material can be used at temperatures up to about 160°C.

Other thermosetting matrices include the well-known and inexpensive phenolic resins which are fire resistant but unfortunately rather brittle. Some high-performance polyimides can withstand temperatures to 315°C continuously but are very expensive and require long curing times following the fibre-impregnation stage.

24.4.2 Thermoplastic polymers

Thermoplastic polymers used as matrices are usually long-chain polymers of high molecular mass which develop strength at ambient temperatures, even though they remain amorphous, because the bulky units in the chains become entangled with each other, e.g. polycarbonates. Others used to develop strength by crystallising partially, e.g. nylon. Many thermoplastics have the advantage over thermosets of greater toughness and also shorter fabrication times (since 'curing' is not involved). Their main disadvantage is that high viscosity at the moulding temperature makes impregnation of the fibres more difficult.

Many thermoplastics can be used as matrices, but the most popular are nylon 66, polyethylene terephthalate (PET), polyamideimide (PAI), polysulphone (PS) (maximum working temperature 150–170°C) and polyether ether ketone (PEEK) (maximum working temperature 250°C). PEEK-based composites have good strength and excellent resistance to failure under impact.

24.4.3 Metals

Since the main object in making a composite is to produce a material of high *specific strength,* the most suitable metals are those of low relative density, e.g. aluminium, magnesium, titanium and their respective alloys. Metal matrix composites (MMCs) are of interest mainly because of the higher temperatures at which some of them can operate (around 500°C). Higher impact, transverse and shear properties are also important, whilst they can be joined by most metallurgical processes. Their main disadvantages are high relative density and difficulty of fabrication.

In the aerospace industry, silicon carbide fibre-reinforced aluminium composites have been used for wings and blades, carbon fibres with aluminium has been used for structural members, silicon carbide fibres with titanium has been used for housings and tubes, aluminium oxide fibres with magnesium has been used for structural members.

24.5 Mechanical properties

A composite containing fibres in a single direction will be extremely anisotropic. In fact, in a direction transverse to that of the fibres the strength may be only 5% or less of that measured in the direction of the fibres. Often the transverse strength is less than that of the matrix itself, because of the presence of discontinuities and incomplete bonding between fibres and the matrix material. Nevertheless, when measured in the fibre direction, many fibre-reinforced composites have greater *specific* strength and *specific* modulus than can be found in other materials.

Properties of some *unidirectional* composites are shown in Table 24.2 which indicates that, although strengths and stiffnesses of composites are not much different from those of metals, these composites have much higher *specific* properties because their *relative densities are lower.* Savings in weight of up to 30% are usually achieved over aluminium alloys.

Table 24.3 shows the properties of typical fibre-reinforced metals. These are able to give significant increases in tensile strength.

Table 24.2 *Mechanical properties of some unidirectional fibre composites*

Material (60% by volume of fibre in epoxy resin matrix)	Relative density	Tensile strength (MPa)		Modulus of elasticity (GPa) longitudinal	Specific strength (MPa) longitudinal	Specific modulus (GPa) longitudinal
		Longitudinal	Transverse			
Aramid	1.35	1600	30	80	1185	59
S-glass	2.0	1800	40	55	900	28
High-strength carbon	1.55	1770	50	174	1140	112
High-modulus carbon	1.63	1100	21	272	675	167
Bulk materials for comparison						
Duralumin	2.7	460		69	170	26
High-tensile (Ni–Cr–Mo) steel	7.83	2000		210	255	27

Table 24.3 *Tensile strengths of reinforced metals*

Composite	Tensile strength (MPa)
Aluminium + 50% silica fibres	900
Aluminium + 50% boron fibres	1100
Nickel + 8% boron fibres	2700
Nickel + 40% boron fibres	1100
Copper + 50% tungsten fibres	1200
Copper + 80% tungsten fibres	1800

24.6 Fibre-composite manufacture

Fibre materials are supplied in the following commercial forms:

- *Rovings*. A 'roving' of glass fibres, which may be several kilometres in length, consists of 'strands', or bundles of filaments wound on to a 'creel'. A 'strand' contains some 200 filaments, each about 10 μm in diameter. Bundles of continuous carbon fibres are known as *tows*.
- *Woven fabrics* in various weave types.
- *Chopped fibres*, usually between 1 and 50 mm long.

Continuous fibres give the highest tensile strength and tensile modulus but with a high directionality of properties, the strength might be as high as 800 MPa along the direction of the fibres but only 20 MPa at right angles to them. Randomly orientated fibres do not give such a high strength and modulus but are not directional in their properties. Table 24.4 illustrates this for reinforced polyester.

Table 24.4 *Properties of reinforced polyester*

Composite	Percentage by weight of glass fibre	Tensile modulus (GPa)	Tensile strength (MPa)
Polyester alone		2–4	20–70
Polyester with short random fibres	10–45	5–14	40–180
Polyester with plain weave cloth	45–65	10–20	250–350
Polyester with long fibres	50–80	20–50	400–1200

Composites are manufactured either as:

• Continuously produced sections (rod, tube or channel), or sheet, from which required lengths can be cut. Such a process can only produce composites which are anisotropic in their properties, strength being in a direction parallel to the fibre direction.

• Composites manufactured as individual components. Here, the fibre may be woven into a 'preform' which roughly follows the mould or die contour. In this case, the mechanical properties will tend to be multi-directional.

24.6.1 Poltrusion

This process is used in the manufacture of continuous lengths of composites of fixed cross-section – rods, tubes, channel profiles and sheet. Window-framing, railings, building panels and the like are also produced. Most poltrusions are of glass-fibre/polyester composites. Such materials provide good corrosion resistance at low cost as well as satisfactory mechanical properties.

In this process (Figure 24.4), the fibre material is drawn through the liquid resin, the surplus of which is removed from the fibres during the 'carding' process. It then passes through the hot-forming die which compacts the composite. Curing then follows this stage. The finished poltrusion is cut into the required lengths as it leaves the 'haul off' unit.

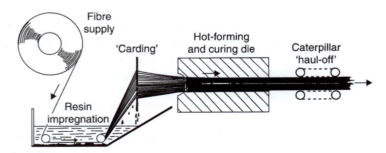

Figure 24.4 *The principles of the poltrusion process. The reinforcement fibres virtually pull the composite through the system.*

24.6.2 'Hand-and-spray' placement

This method probably accounts for a large proportion of glass-reinforced plastics (GRP) made in Britain at present. It is a labour-intensive process in which, in its simplest form, woven mats of glass-fibre rovings are laid over a mould which has previously been coated with a non-stick agent. Liquid resin is then worked into the fibre material by hand, using either a brush or roller. Polyester resins are most commonly used. This type of process is widely used in the manufacture of boats and other pleasure craft, but larger vessels such as minesweepers up to 60 m long, where a non-magnetic hull is required, have been 'hand layered' by this process.

24.6.3 Press moulding

This is the most popular process for mass producing fibre-reinforced plastics components. It is similar in principle to that shown in Figure 20.11, the charge to the mould cavity consisting of fibres and resin.

24.6.4 Resin-transfer moulding

This follows the same principle as the process shown in Figure 20.13. The fibre, usually as a woven preform, is placed in the die cavity and the resin then injected.

24.6.5 Metal matrix composites

These are relatively difficult to manufacture because high temperatures and/or pressures need to be used. Both may cause damage to the fibre reinforcement. The bulk of MMCs are based on aluminium alloy matrices which are reinforced with fibres of alumina, silicon carbide, carbon or boron. A few low melting-point metals can be gravity die cast to form the matrix, but generally it is necessary to use pressure in order to get adequate infiltration of the fibres.

In *powder metallurgy* processes, a metal powder is mixed with chopped fibres, pressed to shape and then sintered using basically the method described in Section 7.4.

24.7 Uses of fibre-reinforced composites

The most important of these materials commercially are polymer matrix composites reinforced with either glass, carbon or aramid fibres. Of these, the greater volume produced consists of glass-reinforced-polymer (GRP), mainly because of its low cost relative to the others. Such composites constitute a large proportion of the interior fittings and trim of most airliners, along with some parts of their primary structure. Sporting and leisure goods – gliders, boats, sailboards, skis and rackets – are now made largely of fibre-reinforced composites. The plastics bodies of many domestic appliances are reinforced with short clipped fibres incorporated during the moulding process.

In addition to the aerospace industries, large quantities of GRP are used in land transport vehicles in the interests of fuel economy. In motor-car manufacture, the use of fibre-reinforced plastics continues to grow, e.g.

panels, bumpers and many interior fittings. Leaf springs made in GRP have good fatigue properties and lead to a saving in weight.

E-glass was developed originally as an electrical insulation material and is now used as a fibre reinforcement for much equipment in the electrical industries – switch casings, cable and distribution cabinets, junction boxes, etc. are manufactured in GRP (glass reinforced polymers). Since these materials are also very corrosion-resistant, they are used in the chemical industries for storing aggressive chemicals and also in silos, wine vats, and pipelines for water and sewage.

Where more sophisticated applications are involved, requiring maximum specific strength and specific modulus, either carbon or aramid fibres are used for reinforcement. Aerospace and military aircraft design, helicopter blades and the like employ CFRP (carbon fibre reinforced polymers).

The following characteristics of fibre composites commend their use:

- Low relative density and hence high specific strength and modulus of elasticity.
- Good resistance to corrosion.
- Good fatigue resistance, particularly parallel to the fibre direction.
- Generally, a low coefficient of thermal expansion, especially with carbon or aramid fibres.

24.8 Reinforced wood

The development of strong synthetic resin adhesives some years ago resulted in much progress in the use of timber as a constructional material.

24.8.1 Laminated wood

Glued lamination of timber ('gluelam') has many advantages over conventional methods of mechanical jointing, such as nailing, screwing and bolting. Thin boards or planks can be bonded firmly together in parallel fashion, using these strong adhesives, and members be thus manufactured in sizes and shapes which it would be impossible to cut directly from trees. Timber members can thus be made indefinitely long by gluing boards end to end on a long sloping 'scarf' joint, and laminae can be 'stacked' to produce any necessary thickness and width. Moreover, thin laminae can be bent to provide curved structural members which are extremely useful as long-span arches, to cover churches, concert-halls, gymnasia and similar large buildings. Arches a hundred metres or more long, with members up to 2 m deep, are not uncommon. The main problem is in transporting them, since assembly 'on site' is unsatisfactory.

Adhesives used in glued laminations include casein and urea formaldehyde for interior work, and the more durable resins phenol formaldehyde and resorcinol formaldehyde for timbers exposed to weather or other damp conditions. All synthetic-resin adhesives, many of which are available from 'do-it-yourself' stores, consist of two parts – the resin and an acid hardener, supplied separately or premixed in powder form. Setting time is often several hours, though special fast-setting hardeners are available.

24.8.2 Plywood, blockboard and particleboard

An inherent weakness of natural timber is its directionality of properties, as shown by a lack of strength perpendicular to the grain. This fault is largely overcome in laminated board – or *plywood,* as it is commonly called – by gluing together thin 'veneers' of wood, so that the grain direction in each successive layer is perpendicular to that in the preceding layer (Figure 24.5). These veneers are produced by turning a steam-softened log in a lathe, so that the log rotates against a peeling knife. Most softwoods and many hardwoods can be peeled successfully, provided that the log is straight and cylindrical. Plywood varies in thickness from 3 to 25 mm; the thinnest is three-ply, whilst thicker boards are of five-ply or multi-ply construction. An odd number of plies is necessary, so that the grain on the face and the back will run in the same direction, otherwise the product will be unbalanced, and any change in the moisture-content during service may lead to warping.

Blockboard (Figure 24.5) resembles plywood, in that it has veneers on both face and back. The core, however, consists of solid wood strips up to 25 mm wide, assembled and glued side by side.

Adhesives used in the production of these materials are chosen to suit service conditions. Plywood and blockboard for interior use are glued with casein, which is resistant to attack by pests, but which will not withstand moisture for long periods. Consequently, plywood and blockboard designed for exterior use are bonded with urea-formaldehyde or melamine formaldehyde resins. Better still, a mixture of phenol formaldehyde and resorcinol-formaldehyde is used to produce material which is 'weather- and boil-proof'.

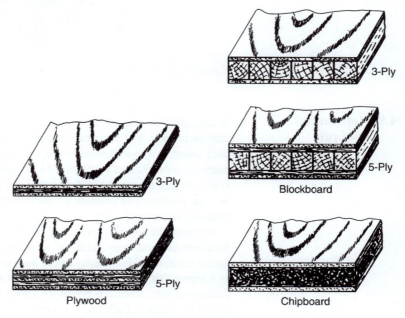

Figure 24.5 *Plywood, blockboard and chipboard.*

Another important modern introduction in timber sheet material is *particleboard* or *chipboard* (Figure 24.5). This consists of a mass of wood chips, ranging from coarse sawdust to flat shavings, bonded with a resin, and sandwiched between veneers. Here, the strength is dependent mainly on the resin bond, but there is no directionality of properties, since the particles constituting the wood filling are orientated at random.

24.8.3 Corrugated cardboard

Corrugated cardboard is derived from wood pulp and is a laminated product of corrugated cardboard sandwiched between outer layers of paper (Figure 24.6). This introduces increased stiffness along the direction parallel to the corrugation ridges. Corrugated cardboard is still widely used as a packing material for a large amount of domestic and other equipment, often in conjunction with foamed polystyrene. The latter is much lighter and stiff enough to offer good protection against mechanical shock. Foamed PS is in a sense, a composite consisting of a mass of air bubbles cemented together by thin films of PS – which brings to mind a schoolboy 'howler' of my childhood describing a fishing net as 'a lot of holes tied together with string'.

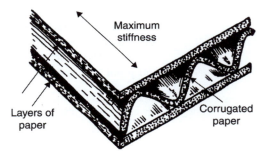

Figure 24.6 *The structure of corrugated cardboard.*

24.9 Reinforced concrete

In common with many ceramic materials, concrete is stronger in compression than in tension. For structural members such as beams, where stresses are both compressive and tensile, the use of plain concrete would be uneconomical, as indicated in Figure 24.7A and B. In reinforced concrete, the tensile forces are carried by fine steel rods. Full advantage can then be taken of high compressive strength of the concrete, and the cross-section can be reduced accordingly. The rods are so shaped (Figure 24.7C) that they are firmly gripped by the rigid concrete. Thus, reinforcement allows concrete members to be used in situations where plain concrete of adequate strength would be unsuitable because of the bulk of material necessary.

I began this chapter by indicating how the principles of fibre-reinforced composite structures are based on those which have existed in trees for millions of years. Perhaps, it is appropriate to finish with one of the early fibre-reinforced composites to be developed by Man – reinforced concrete.

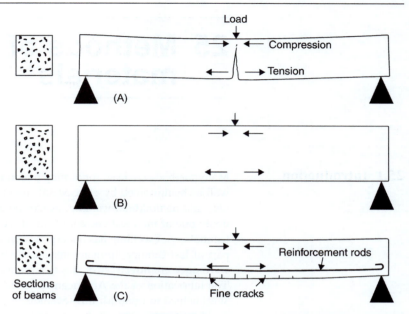

Figure 24.7 *The advantage of reinforcement in concrete. In (A), a plain concrete beam fails at the edge which is in tension. A beam strong enough in this respect (B) would be uneconomically bulky. In (C), steel reinforcement rods support the portion of the beam which is in tension.*

However, I am reminded of Pharaoh's directives to the taskmasters of the Israelite slaves some 3400 years earlier: 'Ye shall no more give the people straw to make brick, as heretofore let them go and gather straw themselves' (Exodus V, 7). In those days, bricks made from sun-dried mud were strengthened by the incorporation of straw. One may still see such bricks in use in some of the more remote mountain regions of the Balkans.

25 Methods of joining materials

25.1 Introduction

Many methods are used for joining materials. Wood is screwed or nailed, as well as being joined by glue. Metals might be welded, riveted or joined by nuts and bolts. Metal welding, as carried out by the blacksmith, is without doubt one of the most ancient of metal-working processes; yet, the modern technology of welding has undergone revolutionary change during the latter part of last century. In many instances where expensive riveting processes were once employed, steel sheets are now successfully joined by welding. The fabrication of the American 'Liberty' ships, during the Second World War, helped to establish this technique.

Joining processes can be grouped into three main groups, according to how the joint is achieved:

- Adhesives
- Soldering, brazing and welding
- Mechanical joining by means of fasteners

In this chapter, we shall consider only adhesives and soldering/brazing/welding.

25.2 Adhesives

Almost any material, including dissimilar materials, can be joined by adhesives. Such joints depend upon forces of attraction between molecules in the surface of the adhesive and molecules in the surfaces of the materials being joined. Generally speaking, the larger the molecules involved, the greater are the forces of attraction, and hence the greater the adhesion. Consequently, most adhesives are organic compounds, being composed of very large and complex molecules.

Another important feature is that a successful adhesive must mould itself perfectly to the surfaces being joined, in order that its molecules will remain in close contact with the molecules of the surfaces being joined. Adhesives bond two *substrates* (the term signifies the surfaces and those regions near to the surfaces of the work-pieces being joined) together either by chemical attraction for the surfaces involved, or by a mechanical intertwining action, by which the adhesive is carried *into* the two substrates, and sets up a tough film. Adhesives used in industry generally work on a combination of both methods, and they normally consist of solutions or suspensions. In these forms, the adhesive material can conveniently be applied to a substrate. Such adhesives are 'cured by solvent loss', that is they harden as the solvent evaporates or is absorbed by the substrates. Some of the solvents employed have become notorious by their abuse in 'glue sniffing'. Direct heat is sometimes used to accelerate solvent loss. Unfortunately, this means that, to heat the glue-line, the whole assembly must be heated, and then cooled before further handling; a decrease in production-rate thus results. One way round this problem is to use radio-frequency waves. These apply heat only to the point where it is required, namely the glue-line. The high-frequency waves act on the molecules of the adhesive and try to vibrate them; as a result, considerable heat can be generated from the friction caused by the molecules resisting the process. This energy is converted to heat and so dries off the solvent from the adhesive.

In many modern adhesives, e.g. epoxy resins such as Araldite, a two-part system is used. Two liquids, stored separately, are mixed and immediately

Table 25.1 *Adhesives*

Group	Adhesive substance (or raw material)	Materials joined
Animal glues	Animal hides or bones, fish, casein (from milk), blood albumen	Wood, paper, fabrics and leather.
Vegetable glues	Starch, dextrine	Paper and fabrics.
	Soya beans	Paper-sizing.
Natural resins	Bitumens (including asphalt)	Laying floor blocks, felt.
	Gum arabic	Paper and fabrics.
Inorganic cements	Sodium silicate	Foundry moulds.
	Portland cement, plaster of Paris	Building industries.
Elastomer materials	Natural rubber (latex/solvent)	Rubber, sealing strips.
	Synthetic rubbers (neoprene, nitrile)	Footwear industries, polythene, PVC.
Synthetic polymer materials	Polyvinyl acetate and vinyl copolymers	Wood, paper, fabrics, bookbinding.
	Cellulose derivatives (solvent release)	Glass, paper, balsa wood.
	Acrylics	Acrylics, polycarbonates.
	Anaerobic acrylics	Metals.
	Cyanoacrylates – cure in presence of moisture ('super glue')	Metals, rubbers, PVC, polycarbonates, polystyrene, polyimide.
	Epoxy/amine and epoxy/polyamide	Metals, glass, ceramics, wood, reinforced plastics, very wide range of uses.
	Phenol, urea, melamine and resorcinol fomaldehydes	Weatherproof plywoods, fabrics and paper.
	Polyurethane – hot or cold-curing liquid	Polyurethane, PVC, polycarbonates, paper and fabrics.
	Polyimide – hot-curing film	Metals, glass, ceramics, polyimide.
	Silicones	Silicone rubbers, sealing seams and joints in other materials.

begin to polymerise so that cross-linking occurs between adjacent molecules, leading to solidification of the adhesive film and, consequently, bonding of the substrates. This reaction sometimes takes place in the cold but in some cases the application of heat is necessary. With DIY materials of the 'Araldite' variety, a stronger bond is usually achieved if the bonded assembly is heated. Other such adhesives involve a blending of materials such as synthetic resins and rubbers. For example, well known adhesives like Bostik and Evo-stick are basically rubber–resin mixtures. Table 25.1 lists a range of adhesives and the materials that they can be used to join. Adhesive joints can resist tension, compression and shear better than tear or peel.

25.2.1 Service requirements

Service requirements must be considered in the choice of adhesive. The most important requirements include the joint strength required, temperature range in service and resistance to water or moisture. The working properties of the adhesive are also important, and include the method of preparation and use, the storage-life, drying-time, odour, toxicity and cost of bonding. Many adhesives require some type of solvent in order to make them fluid: the solvent should evaporate reasonably quickly. Other adhesives, such as animal glues, need to be heated to make them fluid.

As with other joining processes, surface preparation is important. Surfaces, must, above all, be thoroughly degreased, whilst the removal of dust and loose coatings is essential. Sand-blasting, wire-brushing or grinding may have to be used. Joints need to be designed so that the adhesive-bonded substrates are as large as possible, thus lap joints are preferred to butt joints.

25.3 Soldering and brazing

Soldering and brazing can be considered as essentially being 'gluing with metals'. Joints are effected by inserting a metal between the parts to be joined, the metal having a lower melting point than the parts being joined. In welding, the joint is achieved by the application of heat or pressure to the parts being joined.

25.3.1 Soldering

A solder must 'wet' – that is, alloy with – the metals to be joined, and, at the same time, have a freezing range which is much lower, so that the work itself is in no danger of being melted. The solder must also provide a mechanically strong joint.

Alloys based on tin and lead fulfil most of these requirements. Tin will alloy with iron and with copper and its alloys, as well as with lead, and the joints produced are mechanically tough. Suitable tin–lead alloys melt at temperatures between 183°C and 250°C, which is well below the temperatures at which there is likely to be any deterioration in the materials being joined.

Best-quality tinman's solder contains 62% tin and 38% lead and, being of eutectic composition, melts at the single temperature of 183°C. For this reason, the alloy will melt and solidify quickly at the lowest possible temperature, passing directly from a completely liquid to a completely solid state, so that there is less opportunity for a joint to be broken by disturbance during soldering.

Because tin is a very expensive metal as compared with lead, the tin content is often reduced to 50%, or even less. Then the solder will freeze over a range of temperatures, between 183°C and approximately 220°C. Solders are sometimes strengthened by adding small amounts of antimony.

When the highly toxic nature of lead was realised, lead piping used in plumbing was replaced by copper. Copper pipes should be joined using a *lead-free* solder, e.g. 98% tin–2% silver, which is molten between 221.3°C and 223°C (the tin–silver eutectic contains 3.5% silver and melts at 221.3°C). However, the operative who instals our copperpipe work is still known as a *plumber* (Latin: 'plumbum' – lead (Pb)). Table 25.2 lists examples of solders, their composition and uses.

For a solder to 'wet' the surfaces being joined, the latter must be completely clean, and free of oxide film. Some type of flux is therefore used to dissolve such oxide films and expose the metal beneath to the action of the solder. Possibly, the best-known flux is hydrochloric acid ('spirits of salts'), or the acid zinc chloride solution which is obtained by dissolving metallic zinc in hydrochloric acid. Unfortunately, such mixtures are corrosive to many metals, and, if it is not feasible to wash off the flux residue after soldering, a resin-type flux should be used instead. Aluminium and its alloys are particularly difficult to solder, because of the very tenacious film of oxide which always coats the surface.

Table 25.2 *Solders*

BS specification: BS EN ISO 9453	Composition (%)			Freezing range (°C)	Uses
	Sn	Pb	Others		
1 (S–Sn63Pb37)	63	37		183 (eutectic)	Tinman's solder – mass soldering of printed circuits.
3 (S–Pb50Sn50)	50	50		183–215	Coarse tinman's solder – general sheet-metal work.
14 (S–Pb58Sn40Sb2)	40	58	Sb 2	185–213	Heat exchangers, automobile radiators, refrigerators.
–	15	85		227–288	Electric lamp bases.
28 (S–Sn96Ag4)	96		Ag 2	221 (eutectic)	For producing capillary joints in all copper plumbing installations, particularly when a lead-free content is required in domestic and commercial situations.
29 (S–Sn97Ag3)	97		Ag 3	221–230	
23 (S–Sn99Cu1)	99		Cu 1	230–240	
24 (S–Sn97Cu3)	97		Cu 3	230–250	
27 (S–Sn50In50)	50		In 50	117–125	Soldering glazed surfaces.

Table 25.3 *Brazing alloys and 'silver solders'*

BS specification: BE EN ISO 1044	Composition (%)			Freezing range (°C)	Uses
	Cu	Zn	Other		
CZ6	60	Balance	Si 0.3	875–895	Copper, steels, malleable irons, nickel alloys.
Ag7	28	–	Ag 72	780	'Silver solders'– copper, copper alloys, carbon and alloy steels.
Ag1	15	Balance	Ag 50, Cd 19	620–640	'Silver solders'– copper, copper alloys, carbon and alloy steels.

25.3.2 Brazing

Brazing is fundamentally similar to soldering, in that the jointing material melts and the work-pieces remain in the solid state during the joining process. The process is carried out above 450°C and is used where a stronger, tougher joint is required, particularly in alloys of higher melting-point. Most ferrous materials can be brazed successfully and copper–zinc alloys are widely used. A borax-type flux is used, though, for the lower temperatures involved in silver soldering, a fluoride-type flux may be employed.

For aluminium and its alloys, the aluminium–silicon eutectic composition (12.6% silicon) is the basis of the brazes generally used. For copper and its alloys, the braze generally used is the eutectic composition silver–copper alloy (72% silver) with a melting point of 780°C. A silver–copper–zinc alloy, melting point 677°C, enables components to be joined at lower temperatures and the addition of cadmium gives even lower temperatures. See Table 25.3 for details of some brazing alloys.

25.4 Welding

With soldering and brazing, metal is inserted between the two surfaces being joined, the inserted metal having a lower melting point than the surfaces being joined and alloying with the surfaces. With welding, the surfaces of the parts being joined together fuse together.

The term *fusion welding* is used when the joint is produced directly between the parts being joined by the application of heat to melt the interfaces and so cause the materials to fuse together. There are four main types of fusion-welding processes:

- Arc-welding
- Electrical resistance welding
- Thermo-chemical welding
- Radiation welding

With 'solid-state welding', pressure is used to bring the interfaces of the surfaces being joined into such close contact that the two fuse together.

25.5 Arc-welding processes

Early arc-welding processes made use of the heat generated by an arc struck between carbon electrodes, or between a carbon electrode and the work. 'Filler' metal to form the actual joint was supplied from a separate rod. The carbon arc is now no longer used in ordinary welding processes and has been replaced by one or other of the metallic-arc processes.

25.5.1 Metallic-arc welding

Metallic-arc welding using hand-operated equipment is the most widely used fusion-welding process. In this process, a metal electrode serves both to carry the arc and to act as a filler-rod which deposits molten metal into the joint (Figure 25.1). In order to reduce oxidation of the metal in and around the weld, flux-coated electrodes are generally used. This flux coating consists of a mixture of cellulose materials, silica, lime, calcium fluoride and deoxidants such as ferro-silicon. The cellulose material burns, to give a protective shield of carbon dioxide around the weld, whilst the other solids combine to form a protective layer of fusible slag over the weld. Either A.C. or D.C. may be used for metallic-arc welding, the choice depending largely upon the metals being welded.

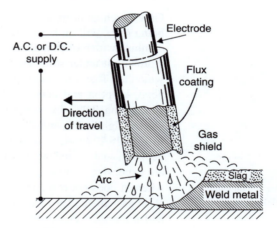

Figure 25.1 *Metallic-arc welding.*

25.5.2 Submerged-arc welding

This is essentially an automatic form of metallic-arc welding which can be used in the straight-line joining of metals. A tube, which feeds powdered flux into the prepared joint, just in advance of the electrode, is built into the electrode holder (Figure 25.2). The flux covers the melting end of the electrode, and also the arc. Most of the flux melts and forms a protective coating of slag on top of the weld metal. The slag is easily detached when the metal has cooled.

This process is used extensively for welding low-carbon, medium-carbon and low-alloy steels, particularly in the fabrication of pressure-vessels, boilers and pipes, as well as in shipbuilding and structural engineering.

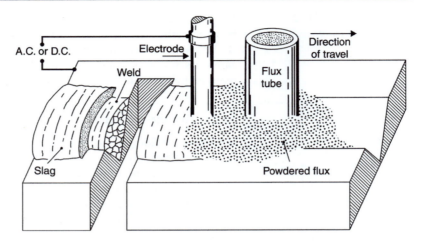

Figure 25.2 *Submerged-arc welding.*

25.5.3 Gas-shielded arc-welding

Gases such as nitrogen and carbon dioxide, which are often referred to as being *inert,* will, in fact, react with some molten metals. Thus, nitrogen will combine with molten magnesium, whilst carbon dioxide will react with steel, oxidising it under some circumstances. The only true inert gases are those which are found in small quantities in the atmosphere, viz. argon, helium, neon, krypton and xenon. Of these, argon is by far the most plentiful – comprising 0.9%, by volume, of the atmosphere – and is used for filling electric light bulbs, and also as a protective atmosphere in inert-gas welding. In the USA, substantial amounts of helium are derived from natural gas deposits, so it is used there as a gas-shield in welding. Since argon and helium are expensive to produce, carbon dioxide (CO_2) is now used in those cases where it does not react appreciably with the metals being welded.

In gas-shielded arc-welding, the arc can be struck between the work-piece and a tungsten electrode, in which case a separate filler-rod is needed, to supply weld metal. This is referred to as the *tungsten inert gas* (or TIG) process. Alternatively, the filler-rod itself can serve as the electrode, as it does in other metallic-arc processes. In this case, it is called the *metallic inert gas* (or MIG) process.

1   *The TIG process* (Figure 25.3)
    This process was developed in the USA during the Second World War, for welding magnesium alloys, other processes being unsatisfactory because of the extreme reactivity of molten magnesium. It is one of the most versatile methods of welding, and uses currents from as little as 0.5 A, for welding thin foil, to 750 A, for welding thick copper and other materials of high thermal conductivity. Since its inception, it has been developed for welding aluminium and other materials, and, because of the high quality of welds produced, TIG welding has become popular

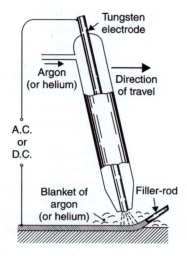

Figure 25.3 *The TIG process.*

for precision work in aircraft, atomic engineering and instrument industries.

2   *The MIG process* (Figure 25.4)
     This uses a consumable electrode which is generally in the form of a coiled, uncoated wire, fed to the argon-shielded arc by a motor drive. Most MIG equipment is for semi-automatic operation, in which the operator guides the torch or 'gun', but has little else to do once the initial

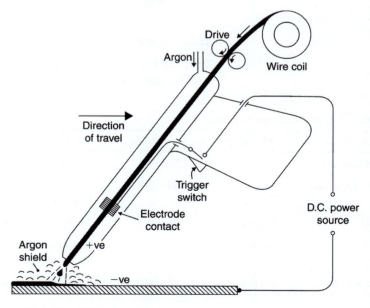

Figure 25.4 *The MIG process.*

controls have been set. In addition to the advantages arising from semi-automatic operation, the MIG process is generally a clean welding process, due to the absence of flux. Consequently, the MIG process is one of the most diversely used welding methods, in terms of the number of different jobs with which it can cope successfully. Some industries where it is used include motor-car manufacturing, shipbuilding, aircraft engineering, heavy electrical engineering and the manufacture of tanks, pressure-vessels and pipes.

3   *The CO_2 process*
This is reasonably effective, provided the filler-rod used contains adequate deoxidising agents to cope with any oxidation which may arise due to the weld metals reacting with carbon dioxide. The normal CO_2 process is a modified MIG process, in which argon has been replaced by carbon dioxide, and in which the electrode wire is rich in deoxidants.

25.5.4 Plasma-arc welding

In this relatively new process, a suitable gas, such as argon, is passed through a constricted electrical arc. Under these conditions, the gas ionises, i.e. the atoms split up into electrons and positively charged particles, the mixture being termed *plasma*. As the ions recombine, heat is released, and an extremely hot 'electric flame' is produced. The sun's surface consists essentially of high-temperature plasma, though even higher temperatures of up to 15 000°C can be produced artificially.

The use of plasma as a high-temperature heat source is finding application in cutting, drilling and spraying of very refractory materials, like tungsten, molybdenum and ceramics, as well as in welding.

25.5.5 Summary of arc-welding processes

Table 25.4 gives a summary of the main characteristics and uses of arc-welding processes.

Table 25.4 *Arc-welding processes*

Process	Shielding	Current (A)	Uses
Metallic arc	Flux	25–350	Pressure vessels, pipes, ships.
Submerged arc	Flux	350–2000	Thick plate, pressure vessels, boilers, pipes, ships, bridges, low- and medium-carbon steels.
MIG	Inert gas	60–500	Medium gauge items, car body repair.
TIG	Inert gas	10–300	Aircraft and instrument industries, Al, Cu, Ni and stainless steel sheet.
Plasma arc	Inert gas	115–240	Thick plate, stainless steel.

25.6 Electric resistance welding

In these processes, heat is produced by the passage of an electrical current across the interface of the joint or is induced within the metal near the joint.

25.6.1 Spot-welding

In this process, the parts to be joined are overlapped, and firmly gripped between heavy metal electrodes (Figure 25.5). An electric current of sufficient magnitude is then passed, so that local heating of the work-pieces to a plastic state occurs. Since the metal at the spot is under pressure, a weld is produced. This method is used principally for joining plates and sheets, but in particularly for providing a temporary joint.

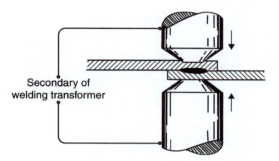

Figure 25.5 *Spot-welding.*

25.6.2 Projection-welding

This is a modified form of spot-welding, in which the current flow, and hence the resultant heating, are localised to a restricted area, by embossing one of the parts to be joined. When heavy sections have to be joined, projection-welding can be used where spot-welding would be unsuitable because of the heavy currents and pressures required. Moreover, with projection-welding, it is easier to localise the heating to that zone near the embossed projection. As Figure 25.6 shows, the projection in the upper work-piece is held in contact with the lower work-piece (A). When the current flows, heating is localised around the projection (B), which ultimately collapses under the pressure of the electrodes, to form a weld (C).

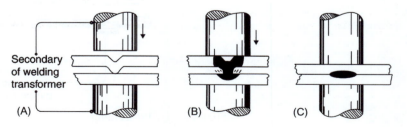

Figure 25.6 *Projection-welding.*

25.6.3 Seam-welding

This resembles spot-welding in principle, but produces a continuous weld by using wheels as electrodes (Figure 25.7). The work-pieces are passed between the rotating electrodes, and are heated to a plastic state by the flow of current. The pressure applied by the wheels is sufficient to form a continuous weld.

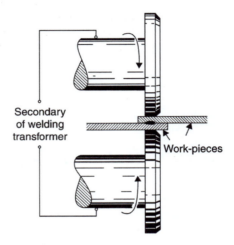

Figure 25.7 *Seam-welding.*

25.6.4 Butt-welding

This process is used for welding together lengths of rod, tubes or wire. The ends are pressed together (Figure 25.8) and an electric current is passed through the work. Since there will be a higher electrical resistance at the point of contact (it is most unlikely that the two ends will be perfectly square to each other), more heat will be generated there. As the metal reaches a plastic state, the pressure applied is sufficient to lead to welding.

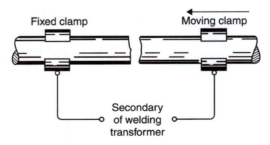

Figure 25.8 *Butt-welding.*

25.6.5 Flash-welding

The process starts with the joint faces in contact and a large current passed through the joint. This results in molten metal being produced and, as a consequence, limited arcing occurs, this arcing being known as *flashing*. This is an automatic process which is used for welding bar, tube and sections.

25.6.6 Electro-slag welding

The main feature of this process, which was developed in the USSR, in 1953, is that heavy sections can be joined in a single run, by placing the plates to be welded in a *vertical* position, so that the molten metal is delivered progressively to the vertical gap, rather as in an ingot-casting operation. The plates themselves form two sides of the 'mould', whilst travelling water-cooled shoes dam the flow of weld metal from the edges of the weld, until solidification is complete (Figure 25.9). In this process, the arc merely initiates the melting process, and thereafter heat is generated by the electrical resistance of the slag, which is sufficiently conductive to permit the current to pass through it from the electrode to the metal pool beneath. The process was originally developed for joining large castings and forgings, but its use has been extended to cover many branches of the heavy engineering industry. It can be used for welding a wide range of steels, and also titanium.

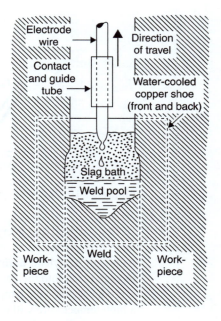

Figure 25.9 *Electro-slag welding.*

25.6.7 Induction welding

In this process, the parts to be joined are pressed together, and an induction coil placed around the joint. The high-frequency current induced in the work heats it to welding temperature.

25.7 Thermo-chemical welding

Some welding processes make use of heat obtained by a chemical reaction. Oxyacetylene welding is probably the best known of these, though the Thermit process also continues to have its uses.

25.7.1 Oxyacetylene welding

Formerly, gas welding (Figure 25.10) ranked equally important with metallic-arc welding, but the introduction of argon-shielding to the latter process placed gas welding at a disadvantage for welding metals since a flux coating is necessary. Consequently, the use of oxyacetylene welding has declined in recent years. However, it is still widely used where maintenance or general repair work is involved.

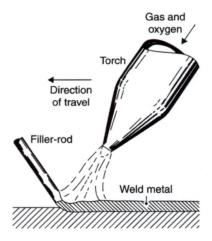

Figure 25.10 *The principle of gas-welding.*

Both oxygen and acetylene can conveniently be stored in cylinders, and their flow to the torch can easily be controlled by using simple valves. Although acetylene normally burns with a smoky, luminous flame, when oxygen is fed into the flame in the correct proportions an intensely hot flame is produced, allowing temperatures in the region of 3500°C to be attained. Such a high temperature will quickly melt all ordinary metals, and is necessary in order to overcome the tendency of sheet metals to conduct heat away from the joint so quickly that fusion never occurs. As with most other fusion-welding processes, a flux-coated filler-rod is used to supply the weld metal.

Bronze-welding is the term applied to the joining of metals of high melting-point, such as mild steel, by the use of copper-alloy filler-metals.

An oxyacetylene flame is generally used to supply the heat. Bronze-welding differs from true welding in that little or no fusion of the work-pieces takes place.

25.7.2 Thermit welding

Thermit welding is used chiefly in the repair of large iron and steel castings. A mould is constructed around the parts to be joined, and above this is a crucible containing a sufficient quantity of the Thermit powder. The parts to be welded are first preheated, and the powder then ignited. Thermit powder consists of a mixture of powdered iron oxide and aluminium dust, in calculated proportions. The chemical affinity of aluminium for oxygen is greater than the affinity of iron for oxygen; hence, the powders react:

$$Fe_2O_3 + 2Al \rightarrow Al_2O_3 + 2Fe + heat$$

The heat of the reaction is so intense that molten iron is produced and this is tapped from the crucible into the prepared joint. This process has been used for many years for on-site welding of rails.

25.8 Radiation welding

When a stream of fast-moving electrons strikes a target, the kinetic energy of the electrons is converted into heat. Since the electron-beam can be focussed sharply to impinge at a point, intense heat is produced there. To be effective, the process must be carried out in a vacuum, otherwise electrons tend to collide with the molecules in the air between the electron gun and target and become scattered from their path. Consequently, electron-beam welding is at present used mainly for 'difficult' metals which melt at high temperatures, e.g. tungsten, molybdenum and tantalum, and also for the chemically reactive metals beryllium, zirconium and uranium, which benefit from being welded in a vacuum.

25.8.1 Laser welding

Lasers emit beams of light of extremely high intensity and such beams of light are used for cutting metals and non-metals. High-quality welds can be made in many ferrous metals. At present, laser technology is still evolving.

25.9 Solid-state welding

There are a limited number of solid-state welding processes, these being processes where there is no melting of materials. The two sides of a joint are brought into intimate atomic contact with each other so that cross-joint bonding occurs.

25.9.1 Cold-pressure welding

Pressure is used to force two joint surfaces into intimate contact, the pressure being such that the surfaces deform, breaking up surface contaminants and bringing clean areas of metal into intimate contact. This method is only applicable to ductile materials, such as aluminium and copper, and can be used for lap welds with sheet and butt welds in rod and bar. This method is used for cladding sheets with a thin layer of some other metal, e.g. aluminium alloy sheets clad with a thin layer of aluminium to improve corrosion resistance. It is also a common process for sealing the ends on aluminium alloy cans.

25.9.2 Friction-welding

Friction-welding can be used to join work-pieces which are in rod form. The ends of the rods are gripped in chucks, one rotating and the other stationary (Figure 25.11). As the ends are brought together, heat is generated, due to friction between the slipping surfaces. Sufficient heat is generated to cause the ends to weld. Rotation ceases as welding commences.

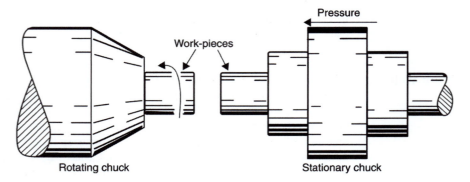

Figure 25.11 *Friction-welding.*

25.9.3 Explosive welding

With this method, the two work-pieces are impacted together by an explosive charge. It has been used for cladding slabs with metal, and tube-plate welding. It can be used with dissimilar metals.

25.10 Structure of welds

Since most welds are made at high temperatures, this inevitably leads to the formation of relatively coarse grain in the metal in and around the weld. A weld produced by a fusion process will have an 'as-cast' type of structure and will be coarse-grained, as compared with the materials of the work-pieces which will generally be in a wrought condition. Not only will the crystals be large (Figure 25.12), but other as-cast features, such as segregation of impurities, may be present, giving rise to intercrystalline weakness.

When possible, it is of advantage to hammer a weld *whilst it is still hot*. Not only does this smooth up the surface, but, since recrystallisation of the weld metal will follow this mechanical working, a tough fine-grained structure will be produced. Moreover, the effects of segregation will be significantly reduced.

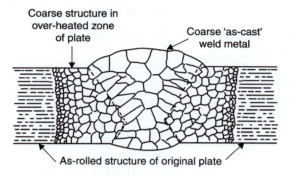

Figure 25.12 *The crystal structure of a fusion-weld.*

Some alloys are difficult to weld successfully, because of structural changes which accompany either the welding process or the cooling which follows. Thus, air-hardening steels, particularly high-chromium steels, tend to become martensitic and brittle in a region near to the weld, whilst 18–8 austenitic stainless steels may ultimately exhibit the defect known as *weld-decay* (see Section 13.4), unless they have been 'proofed' against it.

25.11 Welding of plastics

Thermoplastic materials can be welded together by methods which are fundamentally similar to those used for welding metals, all depending upon the application of heat, and sometimes pressure.

25.11.1 Hot-gas welding

In this process, a jet of hot air from a welding torch is used (Figure 25.13). In other respects, this method resembles oxyacetylene welding of metals. A thin filler-rod is used, and is of the same material as that of the work-pieces. This process is used widely in the building of chemical plant, particularly in rigid PVC.

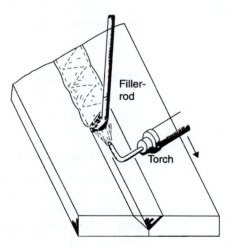

Figure 25.13 *The hot-gas welding plastics.*

25.11.2 Seam- and spot-welding

These methods are very similar to those used in metal-joining, except that, since plastics are non-conductors of electricity, the 'electrodes' must be heated by high-frequency coils.

25.11.3 Electrofusion-welding

This is a novel process used by British Gas for joining yellow polythene pipes up to 180 mm diameter (Figure 25.14). The abutting pipe ends are inserted into a connecting polythene collar which contains a moulded-in electric-heater element. Resistance heating causes a continuous joint to form in about half-an-hour.

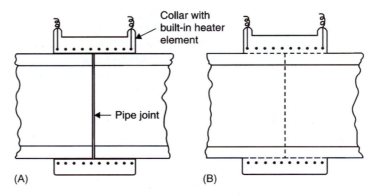

Figure 25.14 *The principles of electrofusion-welding of polythene gas pipes.*

25.11.4 Stitch-welding

This is a method of welding thermoplastics, using a device similar to a sewing-machine, but fitted with two electrodes, which weld the material progressively. Again, the electrodes are heated by a high-frequency current supply.

25.11.5 Jig-welding

In this method, the work-pieces are gripped in jigs. These jigs act as electrodes which are heated by a high-frequency field.

25.11.6 Friction-welding

This is also known as *spin-welding* and is similar in principle to friction-welding of metals (see Section 25.9.2). One part is rotated at speed, whilst the other is pressed against it. This heats both of the work-pieces. Motion is then stopped, and the parts are held together until the joint hardens.

26 Causes of failure

26.1 Introduction

We have all come across examples of materials failure, whether it be the broken rail which caused trains to be delayed or the cup that broke when dropped on the floor. A material can be considered to have failed when its ability to completely satisfy the original design function ceases. Failure can occur as a result of a number of causes, e.g. the material may fracture or perhaps degrade as a result of corrosion. In this chapter, we consider the causes of such failures and the methods that can be adopted to prevent, or at least, delay failure.

26.2 Causes of failure

Material failure can arise from a number of causes and the following are some of the more common causes:

1   *Overstressing*
    The material is overstressed, yields and then breaks. With a ductile fracture there will be significant yielding before the fracture occurs, with brittle fracture virtually none. Thus, for example, the car that hits the gatepost is likely to show a dent, as a result of yielding, in its mild steel bodywork rather than shatter into pieces. Brittle failure is a common form of failure with plastics below their glass transition temperature, i.e. amorphous plastics. With crystalline plastics, the material shows necking, as a result of yielding, before breaking. Brittle fracture is the main mode of failure for glass and ceramics – who has not dropped a cup and seen it shatter into fragments.

    If you want to tear a piece of paper, it helps if the paper contains a series of perforations so that you can tear along the 'dotted-line'. Likewise, if a metal contains a notch in its surface, or there is a sudden change in section, failure can readily occur at such points. This is because a hole, notch or sudden change in section disturbs the normal stress distribution in the material when it is loaded and gives local stress concentrations which can result in the initiation of cracks. The amount by which the stress is raised by a notch depends on its depth L and the radius r of the tip of the notch, the greater the depth and the smaller the radius the greater the amount by which the stress is increased. The factor by which the stress is increased, termed the *stress concentration factor,* is approximately given by:

$$\text{Stress concentration factor} = \sqrt{\frac{L}{r}}$$

A crack in a brittle material will have quite a pointed tip because there has been little in the way of yielding to blunt its tip. There is thus a large increase in stress at the tip. With a ductile material, there will be yielding which blunts the tip of a crack and so increases its radius and hence reduces the stress concentration. One way of arresting the progress of a crack is to drill a hole at the end of the crack to increase its radius.

The temperature of a metal can affect its behaviour when subject to stress. Many metals which are ductile at high temperatures are brittle at low temperatures. The ductile–brittle transition temperature is thus of importance in determining how a metal will behave in service. For example, alloy 817M40 (0.4% C, 0.8% Mn, 1.7% Ni, remainder Fe) when quenched and tempered at 540°C has a ductile–brittle transition temperature of –90°C. In 1965, the offshore drilling rig operating in the North Sea collapsed during operations prior to kicking down the pontoon, so that the rig could be moved and 19 people died. Investigation showed that the disaster was initiated by brittle fracture of tie bars. On the day concerned, the temperature was about 3°C, prior to this the rig having been used in the Gulf of Mexico and the Middle East where the temperatures were much higher. The material which had been ductile when tested in those conditions had become brittle in the colder conditions of the North Sea.

2   *Fatigue*

A material can be subject to an alternating stress which results in fatigue failure at a stress significantly lower than its yield stress. If you flex a metal strip, e.g. a paper-clip, back and forth repeatedly, then it is an easier way of causing the material to fail than applying a direct pull. The chance of fatigue failure occurring is increased the greater the amplitude of the alternating stresses. Stress concentrations produced by holes, surface defects and scratches, sharp corners, sudden changes in section, etc. can all help to raise the amplitude of the stresses at points in a product and reduce its fatigue resistance. Residual stresses (these are stresses arising from processing which rather than causing deformation of a material remain stored in the structure) are produced by many fabrication and finishing processes, e.g. surface hardening of steels by carburising results in surface compressive residual stresses. When surfaces have compressive residual stresses then the fatigue properties are improved, when the residual stresses are tensile it deteriorates.

In 1953 and 1954 crashes occurred of a relatively new aircraft, the Comet. This was one of the earliest aircraft to have a pressurised fuselage constructed from aluminium. Investigation showed that the failure of the fuselage was as a result of the initiation of fatigue initiated cracks at a small countersunk hole near the corner of a window. The small hole had acted as a stress raiser.

Fatigue is also a problem with plastics, though here there is a factor which is not present with metals which is that plastics when subject to alternating stress become significantly warmer. This causes the elastic modulus to decrease and at high enough frequencies of alternating stress can soften a plastic sufficiently for the plastic to no longer support the design load. Polypropylene has good fatigue resistance and so is widely used where cyclic loading might be expected, e.g. as the flip-top on a bottle.

3   *Creep*

A material can be subject to a load which causes creep, i.e. the material initially does not fail as a result of the load but the material gradually extends over a period of time until it eventually fails. For most metals, creep is negligible at room temperature but can become pronounced at high temperatures, the temperatures at which a metal becomes significantly susceptible to creep depending on the metal and alloy concerned, e.g. aluminium alloys will creep and fail at quite low stresses when the temperature rises above 200°C, while titanium alloys can be used at much higher temperatures before such a failure occurs. For plastics, creep is often quite significant at ordinary temperatures and even more pronounced at higher temperatures.

4   *Sudden loads*

A suddenly applied load can cause failure when the energy at impact is greater than the impact strength of the material. Brittle materials have lower impact strengths than ductile materials.

A drop in temperature can often cause the impact strength to drop and make failure more likely, e.g. the temperature of a steel may drop to a level at which the material turns from being ductile to brittle and then easily fails when subject to an impact load.

5   *Expansion*

An increase in temperature with a product might result in temperature gradient being produced which causes part of the product to expand more than another part. The resulting stresses can result in failure. For example, because glass has a low thermal conductivity, if you pour hot water into a cold glass then the inside of the glass tries to expand while the outside which is still cold does not, with the result that the glass can crack.

If a product is made of materials with differing coefficients of expansion, then a similar situation can arise with, when the temperature rises, different parts expanding by different amounts. The internal stresses set up can result in distortion and possible failure.

6   *Thermal cycling*

Thermal cycling is the term used for when the temperature of a product repeatedly fluctuates, so giving cycles of thermal expansion and contraction. If the material is constrained in some way, then internal stresses will be set up and as a consequence the thermal cycling will result in alternating stresses being applied to the material which can result in fatigue failure.

7   *Degradation*

Sooner or later, all engineering materials decay in their surrounding environment. Much of this degradation is due to chemical reactions typified by the 'rusting' of steel and the corrosion of metals in general. The corrosion of steel is a problem which in some degree must be faced by all of us, and it is estimated that the annual cost of the fight against corrosion on a world-wide basis runs into some thousands of millions of pounds. Other engineering metals corrode when exposed to the atmosphere, though generally to a lesser extent than do iron and steel. Because of their high resistance to all forms of corrosion, plastics are replacing metals for many applications where corrosion-resistance is more important than mechanical strength. However, many plastics materials also degenerate when in contact with the environment. Such 'weathering' is due in some degree to simple oxidation but, more seriously, to the action of ultraviolet radiation from the sun which, of course, can have disastrous effects on our own *epidermis* if we are foolish enough to court a suntan in the belief that it makes us appear more glamorous!

26.2.1 Types of fracture surfaces

The following are types of fracture surfaces found with metals, polymers, ceramics and composites:

1   *Ductile failure with metals*

When a gradually increasing tensile stress is applied to a ductile material then, when yielding starts, the cross-sectional area of the material becomes reduced and necking occurs. Eventually, after a considerable reduction in cross-sectional area the material fails. The resulting fracture surfaces show a cone and cup formation. Materials can also fail in a ductile manner in compression, such fractures resulting in a characteristic bulge and series of axial cracks around the edge of the material.

2   *Brittle failure with metals*

When a brittle material fractures, there is virtually no plastic deformation. The surfaces of the fractured material appear bright and granular due to the reflection of light from individual crystal grains within the fracture surface which have cleaved along planes of atoms.

3   *Fatigue failure with metals*

Fatigue failure often starts at some point of stress concentration and this point can be seen on the failed material as a smooth, flat, semicircular or elliptical region, often referred to as the *nucleus*. Surrounding the nucleus is a burnished zone with ribbed markings. This smooth zone is produced by the crack propagating relatively slowly through the material and the resulting fractured surfaces rubbing together during the alternating stressing of the component. When the component has become so weakened by the steadily spreading crack that it can no longer carry the load, the final abrupt fracture occurs and this region has a crystalline appearance.

4   *Failure with polymers*
In an amorphous polymer, when stress is applied it can cause localised chain slippage and an orientation of molecule chains which results in small voids being formed between the aligned molecules and fine cracks, termed *crazing,* occurring. As a result, failure gives fracture surfaces showing a mirror-like region, where a crack has grown slowly from the initial crazing, surrounded by a region which is rough and coarse where the crack has propagated at speed.

With crystalline polymers, the application of stress results in the folded molecular chains becoming unfolded and aligned and so considerable, permanent deformation and necking occurs. Prior to the material yielding and necking starting, the material is quite likely to begin to show a cloudy appearance as a result of small voids being produced within the material. Further stress causes these voids to coalesce to produce a crack which then travels through the material by the growth of voids ahead of the advancing crack tip.

5   *Failure with ceramics*
Typically, a fractured ceramic shows around the origin of the crack a mirror-like region bordered by a misty region containing numerous micro-cracks. In some cases, the mirror-like region may extend over the entire surface.

6   *Failure with composites*
The fracture surface appearances for composites depend on the fracture characteristics of the matrix and reinforcement materials and on the effectiveness of the bonding between the two. For example, for a glass-fibre reinforced polymer, the fibres may break first and then a crack propagates in shear along the fibre–matrix interface until the load, which had been mainly carried by the fibres, is transferred to the matrix which then fails. Alternatively, the matrix may fracture first and the entire load is then transferred to the fibres which carry the increasing load until they break; the result is a fracture surface with lengths of fibre sticking out from it, rather like bristles of a brush.

26.3 Non-destructive testing

Components which are produced individually, such as castings and welded joints, may vary in quality, for even in these days of rigorous on-line inspection faulty components *can* arise since the production methods for casting and weldments fall under the influences of many variable factors such as working temperature, surrounding atmosphere and the skill of the operator. If the quality of such components is very important – it may be necessary to test each component individually using some type of non-destructive examination. Such investigation seeks to detect faults and flaws either at the surface or below it and a number of suitable methods are available in each case. Since tests of this type give an overall assessment of the quality of the product, the term *non-destructive testing* (NDT) is often replaced by *non-destructive evaluation* (NDE).

26.3.1 The detection of surface cracks and flaws

Surface cracks may arise in a material in a number of ways. Some cracks show up during inspection using a simple hand magnifier, whilst others are far harder to detect. Thus, steel tools, which have been water quenched, may develop hairline cracks which are not apparent during ordinary visual inspection. Castings and weldments may crack due to contraction during the period of solidification and cooling. Detection methods used include:

1 *Penetrant methods* (Figure 26.1)
   In these methods, the surface to be examined is first cleaned adequately to remove grease, and then dried. The penetrant liquid is then sprayed or swabbed onto the surface, which should be warmed to about 90°C. Small components may be immersed in the heated penetrant. After sufficient time has elapsed for the penetrant to fill any cracks, the excess is carefully washed from the surface with warm water (the surface tension of water is too high to allow it to enter the fine cracks). Alternatively, other solvents may be used to remove excess penetrant, whilst some oily penetrants are treated with an emulsiier which enables them to be washed off with water. The test surface is then carefully dried and coated with a 'developer' such as powdered chalk. This developer can be blown on as a powder or sprayed on as a solution or suspension. In some cases, the component is dipped into a suspension or solution of the developer which is allowed to dry on the surface. The component is then set aside for some time. As the coated surface cools, it contracts and penetrant tends to be squeezed out of any cracks, so the chalk layer becomes stained, thus revealing the presence of the cracks. Most penetrants of this type contain a scarlet dye which renders the stain immediately noticeable.

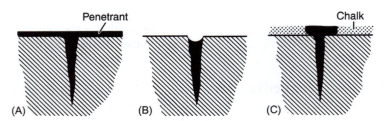

Figure 26.1 *The penetrant method of crack detection. (A) The cleaned surface is coated with the penetrant, which seeps into any crack. (B) Excess penetrant is cleaned from the surface. (C) The surface is coated with chalk. The penetrant is expelled from the crack by the contraction of the cooling metal and stains the chalk.*

In some cases, the penetrant contains a compound which becomes fluorescent under ultraviolet light. The use of chalk is then unnecessary. When the prepared surface is illuminated by ultraviolet light in a darkened cubicle, the cracks containing the penetrant are revealed as bright lines on a dark background.

Penetrant methods in general are particularly useful for the examination of non-ferrous metals and austenitic (non-magnetic) steels. Aluminium-alloy castings are frequently examined in this way.

2   *Magnetic dust methods*

These methods can be applied only to magnetic materials, but nevertheless provide a quick and efficient method of detecting cracks. An advantage over the penetrant method is that flaws immediately *below* the surface are also detected; so the magnetic dust method is particularly suitable for examining machined or polished surfaces, where the mouth of the crack may well have become 'burred' over.

The magnetic method involves making the component part of a 'magnetic circuit' (Figure 26.2A). A magnetic field is induced in the component, either by permanent magnets or by electromagnetic means. Alternatively, using probes, a heavy current at low e.m.f. is passed through the component so that a magnetic field is produced in it (Figure 5.2(C)). The magnetic lines of force that are produced pass easily through a magnetic material but on meeting a gap or other discontinuity at, or just below the surface, they spread outwards (Figure 5.2(B) and

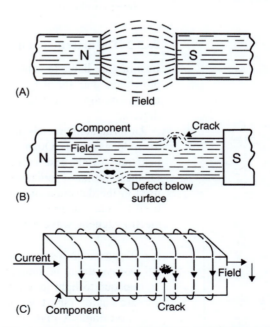

Figure 26.2 *Magnetic methods of crack detection. (A) The principle of the process – lines of force tend to 'spread' in an air gap. (B) A magnetic circuit. (C) Using a high current to produce a magnetic field.*

(C)). If some iron dust is now sprinkled on the surface of the component, it will stick to the surface where the lines of force break out, thus revealing the site of the fault. Alternatively, the magnetised component can be placed in paraffin containing a suspension of tiny iron particles. The particles will be attracted to the surface of the component at any points where, due to the presence of a fault, lines of force cut through it. The sensitivity of this method can be improved by coating the surface to be treated with a carrier liquid containing magnetic particles which have been treated with a fluorescent compound. The surface is then examined under ultraviolet light in a darkened cubicle.

3   *Acid-pickling methods*

Surface defects in steel castings can generally be revealed by pickling the casting in a 10% solution of sulphuric acid at about 50°C for up to 2 hours. Alternatively, a 20–30% solution of hydrochloric acid at ambient temperature for up to 12 hours can be used, whilst stainless steels require mixtures of concentrated nitric and hydrochloric acids. This latter mixture is *extremely corrosive* and tests involving this and other acids should be carried out only by a competent technician who will be aware of the very dangerous nature of these chemicals. After removal from the acid bath, the casting is washed thoroughly and, if necessary, any residual acid neutralised by immersing in a hot suspension of slaked lime, after which the casting is washed again and dried quickly. This simple procedure provides a cheap but effective way of dissolving oxide layers and revealing most surface defects.

26.3.2 The detection of internal defects

Castings are liable to contain unwanted internal cavities, in the form of gas blowholes and shrinkage porosity. Wrought materials may contain slag inclusions and other flaws, whilst welded joints can suffer from any of these defects.

Unfortunately, metals are opaque to visible light, so these internal faults are hidden from us. Other forms of electromagnetic radiation, however, will penetrate metals, and so enable us to 'see' into the interior of the material. Of these forms of radiation, X-rays and gamma rays are the most widely used in the detection of internal faults. Another type of wave that can be used to detect internal cavities is, however, quite different in not being an electromagnetic wave but a sound wave.

1   *X-ray methods*

X-rays travel in straight lines, as do light rays, but while metals are opaque to light they are transparent to X-rays provided that the latter are 'hard' (short wavelength X-rays have more energy than long wavelength X-rays and are termed *hard*) enough and that the metal is not so thick in cross-section that the X-rays are completely absorbed. X-rays affect a photographic film in a manner similar to that of light, so the most efficient method of detecting faults in a body of metal is to take an X-ray photograph of its interior. The reader may be familiar with a similar application of X-rays in medicine. There is, however, an important difference in the type of X-rays used since those employed for the radiography of metals need to be much 'harder' so that they will penetrate the metal successfully. These hard X-rays would cause very serious damage to our body tissues if exposed to them and for this reason such X-ray equipment must be well shielded to prevent stray radiation from reaching the operator. For this purpose, a shielding wall of 'barium cement' is often used. Barium and its compounds are very effective in absorbing X-rays. The 'barium meal' used in diagnostic medicine serves a similar purpose. The radiation from the X-ray tubes passes through the casting and forms an image on a photographic film placed behind the casting (Figure 26.3). X-rays will penetrate that region of the casting containing the cavity much more easily and so produce a *denser* image on the film. The barium meal of diagnostic medicine has the reverse effect in that it absorbs X-rays more effectively and so will cause a *less dense* image to be formed on the film. A fluorescent screen may be substituted for the photographic film, so that the result may be viewed instantaneously.

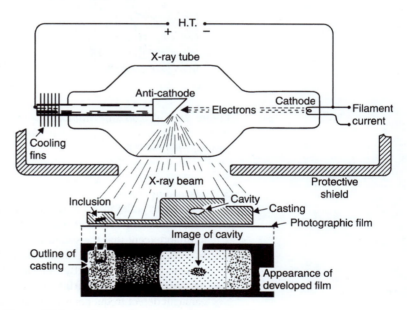

Figure 26.3 *Radiography of a casting using X-rays.*

2   *Gamma-ray methods*

Gamma rays can also be used in the radiography of metals. They are 'harder' than X-rays, and are therefore able to penetrate a greater thickness of metal or, alternatively, a more dense metal. Hence, they are particularly useful in the radiography of steel, which absorbs radiation more effectively than do the light alloys. As might be expected, exposure to gamma radiation is extremely dangerous. Originally, radium was used as a source of gamma rays but now artificially activated isotopes, e.g. cobalt-60, are used.

Manipulation of an isotope as a source of gamma rays in radiography is in many respects simpler than when using X-rays, though security arrangements need to be even more stringent. Not only are gamma rays harder but, unlike X-rays, cannot be 'switched off'. Gamma rays can be used to radiograph considerable thicknesses of steel and have the further advantage over X-rays that the equipment is more mobile and less cumbersome to transport (Figure 26.4). Thus, gamma-radiography is useful in positions difficult to access or where portable equipment needs to be used, as in examination of steelwork in motorway bridges.

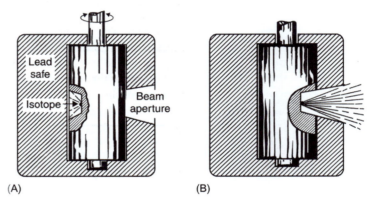

Figure 26.4 *A lead safe suitable for storing and transporting radioactive isotopes: (A) an isotope in the shrouded position, (B) the isotope in transmitting position.*

3   *Ultrasonic testing*

It is fun to yodel in the mountains, and listen to a succession of echoes from mountainsides both far and near. Ultrasonic testing is somewhat similar, except that the sound waves generated by a yodel are replaced by very high-frequency vibrations which are beyond the acoustic range which can be received by our ears. In ultrasonic testing, frequencies between 500 000 and 10 000 000 Hz are commonly used, whereas our ears can detect frequencies only between 30 and 16 000 Hz. Figure 26.5 represents the principle of ultrasonic testing. A probe containing a quartz crystal, which can both transmit and receive high-frequency vibrations, is passed over the surface of the material to be tested. The probe is connected to an amplifier, which converts and amplifies the signal, before it is recorded on a cathode-ray tube (Figure 26.5).

Under normal conditions, the vibrations will pass from the probe, unimpeded through the metal, and be reflected from the bottom inside surface at B back to the probe, which also acts as a receiver. Both the transmitted pulse and its echo are recorded on the cathode-ray tube, and

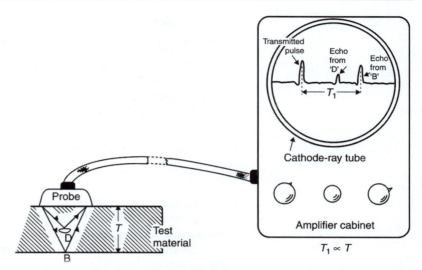

Figure 26.5 *The use of ultrasonic vibrations in detecting flaws below the surface in a metal plate.*

the distance T_1 between the 'blips' is proportional to the thickness T of the test material. If any discontinuity is encountered, such as the blowhole D, then the pulse is interrupted and reflected back as indicated. Since this echo returns to the receiver in a shorter time, an intermediate 'blip' appears on the cathode-ray tube. Its position relative to the other 'blips' indicates the distance of the fault below the surface. Different types of probe are available for materials of different thickness. This method is particularly suitable for detecting faults in sheet, plate and strip materials more than 6 mm thick, whilst modified equipment is used for detecting faults in welds.

26.4 Degradation of metals by oxidation

Metals may corrode by a process in which oxygen, ever present in the atmosphere, combines with some metals to form a film of oxide on the surface. Most metals react, at ambient temperatures, only slowly. If this film is porous, or if it rubs off easily, the process of oxidation can continue, and the metal will gradually corrode away. Aluminium oxidises very easily, but fortunately the thin oxide film so formed is very dense and sticks tightly to the surface, thus effectively protecting the metal beneath. Titanium, tantalum and chromium 'seal' themselves from further attack in this way.

At higher temperatures, this process of oxidation takes place more rapidly and as the temperature increases so does the rate of oxidation. Thus, iron oxidises ('scales') readily at temperatures above 650°C. This process of oxidation involves a transfer of electrons from the metallic atom to the oxygen atom, so that positively charged metallic ions and negatively charged oxygen ions are formed (Figure 26.6). Since these oppositely charged ions attract each other, a crystalline oxide layer is formed on the metallic surface.

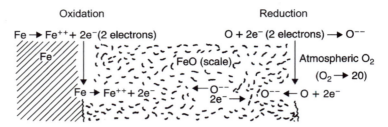

Figure 26.6 *Reactions at and near the surface of iron heated in contact with the atmosphere.*

Often, this scale is not coherent with the metal surface so that it flakes away, exposing the metal to further oxidation.

Both metal and oxygen ions are able to diffuse through the scale and so thicken the oxide layer. When the crystal lattice of this layer is not coherent – that is, does not 'match up' – with the crystal lattice of the metal beneath, it flakes away so that the metal surface is exposed to more rapid oxidation. If the oxide layer is coherent, as with aluminium, titanium, etc., the metal is protected by this closely adherent oxide layer.

26.4.1 Attack by sulphur

Some metals, notably nickel, are readily attacked at high temperatures by gases containing sulphur. Therefore, heat-resisting steels used in the presence of sulphurous vapours should not contain large amounts of nickel. Although in this case oxygen itself is not involved in the chemical attack, nickel is said to have been 'oxidised'. In fact, the term *oxidation* is applied in chemical jargon to any reaction in which metallic atoms lose electrons and so become ions.

26.5 Degradation of metals by electrolytic corrosion

Most readers will be familiar with the principle of a simple electric cell. It consists of a plate of copper and a plate of zinc, both of which are immersed in dilute sulphuric acid. If the plates are not touching each other in the solution, when they are connected to each other outside the solution (Figure 26.7), a current of electricity, sufficient to light a small bulb, flows through the completed circuit. At the same time, bubbles of hydrogen form at the copper plate, whilst the zinc plate begins to dissolve in the acid to form zinc sulphate. The zinc is 'corroding'.

When the circuit is closed, atoms on the surface of the zinc plate form ions Zn^{++} as a result of losing electrons. These electrons pass through the external circuit to the copper plate which, as a result, becomes negatively charged. The dilute sulphuric acid contains hydrogen ions (H^+) and SO_4^{--}. The hydrogen ions (H^+) in the dilute sulphuric acid will be attracted towards the negatively charged copper plate where they combine with the electrons to form hydrogen atoms which immediately combine to form hydrogen molecules (H_2), hence small bubbles of hydrogen gas form on the copper

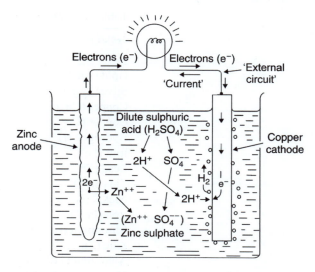

Figure 26.7 *The 'corrosion' of zinc in a simple voltaic cell.*

plate. In the meantime, the Zn^{++} ions 'pair up' with the SO_4^{--} ions to form zinc sulphate. The 'current of electricity' which causes the bulb to light up is the stream of liberated electrons from the dissolving zinc.

The dilute sulphuric acid is termed an *electrolyte*. An electrolyte is any substance which in the liquid state – in this case, as a solution in water – contains separate positive and negative ions. If the reader examines Figure 26.7, he will notice that the direction of the flow of electrons is in the opposite direction to the conventional direction used for the electric current; this apparent confusion arises from the fact that in the early days of electricity, the true nature of an electric current was not really understood. The electrode which supplies electrons to the external circuit is called the *anode,* whilst the electrode which receives electrons from the external circuit is called the *cathode.*

The zinc is anodic towards copper, so that when these metals are connected and immersed in any electrolyte, the zinc will dissolve – or corrode – far more quickly than if immersed in the electrolyte by itself. This phenomenon of electrolytic action is not confined to copper and zinc but will apply to any pair of metals, one of which will be anodic to the other. Zinc atoms lose their outer-shell electrons more easily than do copper atoms and so zinc is anodic to copper.

26.5.1 The Electrochemical (or Galvanic) Series

In the Electrochemical (or Galvanic) Series (Table 26.1), any metal in the table will go into solution and replace any metal above it (note that some metals can form more than one type of ion). This explains why, in the simple cell described earlier, zinc ions go into solution and replace the hydrogen ions from the electrolyte. Similarly, if a steel (iron) penknife blade is

Table 26.1 *The Electrochemical (Galvanic) Series*

Metal ion	Electrode potential (V)
Gold (Au^{+++})	+1.5
Silver (Ag^{+})	+0.8
Copper (Cu^{+})	+0.52
Copper (Cu^{++})	+0.34
Hydrogen (H^{+})	0.00
Iron (Fe^{+++})	−0.05
Lead (Pb^{+++})	−0.13
Tin (Sn^{++})	−0.14
Nickel (Ni^{++})	−0.25
Iron (Fe^{++})	−0.44
Chromium (Cr^{+++})	−0.74
Zinc (Zn^{++})	−0.76
Aluminium (Al^{+++})	−1.66
Magnesium (Mg^{++})	−2.37
Lithium (Li^{+})	−3.04

immersed in a copper sulphate solution, the blade becomes coated with metallic copper because iron atoms have formed ions which replace the copper ions in the solution.

The potential is measured with respect to hydrogen which is the reference, + values indicating cathodic and − values anodic with respect to hydrogen. The term *noble metals* is used for those at the top of the table and base metals for those at the bottom.

It follows that any pair of metals in the table, when immersed in an electrolyte and also in electrical contact with each other, will form a voltaic cell, and the metal which is lower in the table will go into solution as its atoms form ions and release electrons into the external circuit; it will be anodic to the other. The further apart these metals are in the table, the greater the potential difference between them and the greater the tendency of the anode to dissolve, i.e. corrode.

Often, the rate of electrolytic action is extremely slow and consequently the flow of electrons, i.e. the current, between the two metals so small as to be undetected. Nevertheless, electrolytic action will be taking place and will lead to the accelerated corrosion of that member of the pair which is anodic to the other. *Pure* water is only very slightly ionised, but in industrial regions the rainwater dissolves sulphur dioxide from the atmosphere to form the 'acid rain' we hear so much about these days. This dilute acid is very highly ionised and so forms an electrolyte which allows corrosion to proceed quite rapidly.

If the reader has been long engaged in practical engineering, he may have learned that it is considered bad practice to use two dissimilar metals in close proximity to each other and in the presence of even such a weak electrolyte as rainwater. In spite of this, many examples of such bad practice are encountered, especially in domestic plumbing. Figure 26.8A illustrates such

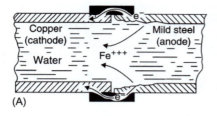 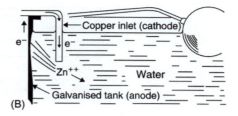

Figure 26.8 *Examples of bad plumbing practice.*

a case. Here, a section of mild steel pipe has been attached to one of copper, the system being used to carry water. Since the mild steel will be anodic to copper, it will rust *far more quickly than if the pipe were of mild steel throughout.* Rusting will be the more severe where it joins the copper pipe as Fe^{+++} ions go into solution more quickly there. These Fe^{+++} ions in solution will combine with OH^{--}, formed by the ionisation of water, and produce ferric hydroxide which is the basis of the reddish brown deposit we call *rust*.

A further example of bad practice was all too frequently encountered in domestic installations (Figure 26.8B). Here, the inlet pipe was invariably of copper whilst the storage tank was of galvanised (zinc coated) mild steel. Since most mains water contains sufficient dissolved salts to make it an electrolyte, the inevitable corrosion took place. As a result, the zinc coating in an area adjacent to the copper inlet pipe went into solution as Zn^{++} ions and the exposed mild steel, also anodic to copper, then dissolved rapidly as a result of the close proximity of the cathodic copper pipe. The disastrous leak which followed generally occurred during the annual holiday or on Christmas Day, and always just after the ground-floor rooms had been decorated. Enlightened plumbers used inlet pipes of inert plastics materials. Fortunately, modern domestic installations use plastics tanks which are free from such problems, but if your house was built before 1970, take a quick look at the 'cold-water' tank. If it's of galvanised iron – start worrying!

26.5.2 Cladding of metal sheets

Many pure metals have a good resistance to atmospheric corrosion. Unfortunately, these metals are usually expensive, and many of them are mechanically weak. However, a thin coating of one of these metals can often be used to protect mild steel. Pure tin has an excellent resistance to corrosion, not only by the atmosphere and by water, but by very many other liquids and solutions. Hence, tinplate – that is, mild steel sheet with a thin coating of tin – is widely used in the canning industry. Figure 26.9 illustrates what happens if a tin coating on mild steel becomes scratched.

The mild steel is anodic to tin and so it will go into solution as Fe^{+++} ions more quickly in the area exposed by the scratch than if the *tin coating were not there at all.* The electrons released by the ionisation of iron atoms travel to the cathode (the tin coating) where they attract hydrogen ions H^+ from the ionised water, forming hydrogen molecules which may appear as minute

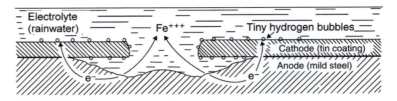

Figure 26.9 *The corrosion of scratched tinplate.*

bubbles but are more likely to remain dissolved in the water. In the meantime, Fe^{+++} ions combine with OH^- ions, also present in the ionised water, leading to the formation of rust. Consequently, to be of any use in protecting mild steel, a tin coating must be absolutely sound and unbroken.

Figure 26.10 illustrates the effect of a zinc coating in similar circumstances. Here, the zinc is anodic to mild steel, so in this case it is the zinc which goes into solution (as Zn^{++} ions) in preference to the iron (mild steel). These Zn^{++} ions then combine with OH^- ions present in the water to form a white corrosion deposit based on zinc hydroxide. In fact, *the mild steel will not rust as long as any zinc remains in the vicinity of the original scratch.* Thus, by being dissolved itself, the zinc is in fact protecting the mild steel from corrosion. Such protection offered to the mild steel by the zinc is known as *sacrificial protection*. Naturally, the zinc will corrode fairly quickly under these circumstances, and the time for which this sacrificial protection lasts will be limited. For this reason, every attempt must be made to produce a sound coating, whatever metal is used, but, if the coating is anodic to the mild steel beneath, it will still offer some protection, even after the coating is broken. Thus, galvanising is more useful than tin-coating in protecting steel. Tin-coating is used only in canning and in food-preparing machines because it is not attacked by animal or fruit juices. Such juices would attack zinc coatings, and possibly produce toxic compounds.

There is always a danger of electrolytic corrosion taking place in the steel hull of a ship, due to the proximity of the bronze propeller (steel is anodic to bronze and sea water is a strong electrolyte). However, if a zinc slab is fixed to the hull (Figure 26.11), near to the propeller, this will sacrificially protect the steel hull, since the zinc is more anodic than steel. Naturally, the zinc slab must be replaced from time to time, as it is used up.

The reader may verify the foregoing principles by means of a very simple experiment. Obtain three identical clean iron nails, a small strip of tin foil and a strip of zinc sheet. The latter may be obtained from an old dry-battery,

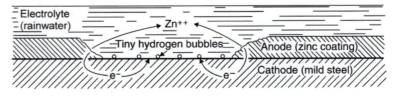

Figure 26.10 *The 'sacrificial' protection offered by zinc to mild steel.*

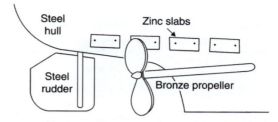

Figure 26.11 *Sacrificial protection of a ship's hull by zinc slabs.*

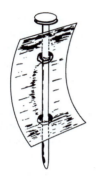

Figure 26.12
Simple experiment.

provided it is cleaned, and washed free of any electrolyte. To obtain tin foil is not so easy these days, since most metal foil used in food and cigarette packaging is of aluminium. Pierce two holes in each metal strip, and thread it onto a nail, as shown in Figure 26.12. Take three separate glass containers, and fill each with a very weak salt solution (a pinch of table salt to a litre of water will be enough). This solution serves as an electrolyte. Then drop a nail/metal strip 'couple' into each of two vessels, and a nail by itself into the third vessel. Examine them each day and note the extent to which corrosion has taken place by the end of about a week. It will be found that the nail accompanied by the tin strip has rusted most, whilst the one accompanied by the zinc strip has not rusted at all. Some white corrosion product from the zinc will, however, be found to be present. The unaccompanied nail serves as a 'control' and will be found to be less rusty than that accompanied by the tin foil.

26.5.3 Rusting of iron and steel

Galvanic cells can be established in a number of ways. In section 26.5.2, we looked at how such cells can be set up with two dissimilar metals. Another way is what is termed *concentration cells*. If a metal is in contact with a concentrated electrolyte solution, it will tend not to ionise to as great an extent as when it is in contact with a dilute electrolyte. Thus, for a piece of metal in contact with an electrolyte of varying concentration, that part of the metal in contact with the more dilute electrolyte will be anodic with reference to that part of the same metal in contact with the more concentrated electrolyte.

An example of a concentration cell is where a piece of metal dips into an electrolyte which contains dissolved oxygen. There will be more oxygen dissolved near the electrolyte surface where it is in contact with air than deeper below the surface. Such a cell is set up with a drop of water on the surface of a piece of steel, there being a greater amount of oxygen near the edge of the water drop compared with the centre of the drop. Thus, the centre becomes anodic with respect to the drop edge. Hence, at the drop centre, Fe^{++} ions are formed and move towards the cathodic edges of the drop and form ferrous hydroxide when they meet up with OH^- ions. This becomes deposited as ferrous hydroxide which subsequently oxidises to give what we term *rust*.

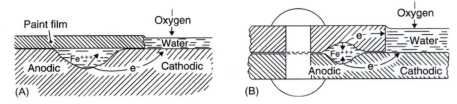

Figure 26.13 *Corrosion due to a non-uniform electrolyte.*

Thus, the rusting of iron and steel is not a case of simple oxidation (as in Section 26.4) but is associated with the presence of both *air and moisture*. Iron will not rust in dry air, nor will it rust in pure water; but, when air and moisture are present together, iron and particularly steel, will begin to rust very quickly. Rusting continues unabated, because the layer of corrosion product formed is loose and porous, so that a fresh film of rust will form beneath and lift upwards the upper layer.

As an illustration, consider the case of flaking paint work on a motor car (Figure 26.13A). Rainwater can seep beneath the loose paint film and since it becomes thus isolated from the atmosphere, it will inevitably contain less dissolved atmospheric oxygen, as time goes on, than that water which remains exposed to the atmosphere. The composition of the electrolyte is therefore non-uniform and that metal covered by the oxygen-depleted water becomes anodic to the metal covered by oxygen-rich water. Thus, Fe^{+++} ions go into solution more quickly beneath the paintwork, leading to the build up of corrosion products there. This is why blisters form and ultimately push off the paint film.

This type of corrosion can also be caused by faults in design. Consider two steel plates of similar composition riveted as shown in Figure 26.13B, with rivets of composition similar to that of the plates. Water seeping into the fissure between the plates will become depleted in oxygen as compared with water which is in continuous contact with the atmosphere. Thus, non-uniformity in the composition of the electrolyte will cause metal within the fissure to become anodic to the remainder and corrosion will take place there. This is generally termed *crevice corrosion*. Piles of leaves, sand or other forms of inert sediment collecting at the bottom of a metal water storage tank will cause corrosion of the metal beneath it in a similar manner.

26.5.4 Stress corrosion

Another type of concentration cell (see Section 26.5.3) is formed within a material which has been cold-worked and contains regions that are more highly stressed than other regions. A highly stressed material will tend to ionise to a greater extent than atoms in a less stressed state and so the stressed part becomes anodic with respect to the less stressed part. Therefore, when covered by an electrolyte, the more heavily cold-worked regions corrode more rapidly than those areas which have received less cold-work (Figure 26.14).

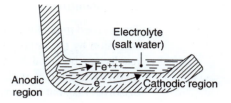

Figure 26.14 *Corrosion due to non-uniform cold-work.*

This explains the tendency of the cold-formed rim of a motor-car 'wing' to rust through quickly. Here, where the metal has been cold-worked more severely and where, incidentally, the paint is more likely to be defective, corrosion will be accelerated as the region will be anodic to the remainder of the wing. In winter, road moisture is likely to be also heavily laden with salt and this will provide a very strong electrolyte which will favour the rapid solution of iron as Fe^{+++} ions.

Throughout a cold-worked metal, locked-up strain energy is greater at the grain boundaries than elsewhere in the distorted crystals. The 'season cracking' of 70–30 brass cartridge cases was caused by intercrystalline corrosion resulting from the locked-up elastic strains of this type. Failure resulting from this form of corrosion was prevalent in cold-worked cartridges shipped to India in the days of the Raj. Strong ammonia fumes from the stables combined with the damp atmosphere of the monsoon produced a strong electrolyte which allowed intercrystalline corrosion in the anodic regions of the grain boundaries. This problem was overcome by low-temperature annealing of the cartridge cases in order to disperse the effects of cold-work and so remove the potentially anodic areas. Such a solution cannot generally be applied to the problem of the motor-car wing since some cold-work is necessary to retain the stiffness of the structure.

Locked-up stresses may be set up near welded joints as a result of non-uniform cooling of the weld, though here galvanic corrosion is more likely to be due to differences in composition between weld metal and work-piece.

26.6 The protection of metal surfaces

The corrosion of carbon steel is due partly at least to electrolytic action between different phases in the structure of the steel. Pearlite consists of microscopically thin layers of ferrite and cementite, arranged alternately in the structure. Ferrite is anodic to cementite, and so it is this ferrite which corrodes away, leaving the cementite layers standing proud. Being very brittle, these cementite layers soon break away. In order to protect the surface of steel, therefore, it must be coated with some impervious layer which will form a mechanical barrier against any electrolyte which is likely to come into contact with its surface. In all cases, the surface to be coated must be absolutely clean and rust free.

26.6.1 Painting

This is used to coat vast amounts of mild steel, not only to protect it against corrosion by the atmosphere, but to provide an attractive finish. Optimum results are obtained by first 'phosphating' the surface of the steel. This involves treating it with a phosphoric acid preparation, which not only dissolves rust, but also coats the surface of the steel with a dense and slightly rough surface of iron phosphate. This affords some protection against corrosion, but also acts as an excellent 'key' for the priming paint and the undercoat of subsequent paint.

In the automobile industry, undercoat paint is applied using a process called *electrophoresis* or *electro-painting*. The articles to be painted are made the anode in a bath which contains the paint 'resin' suspended as particles in a soap solution in water. The negatively charged ions from the dissolved soap become attached to the 'resin' particles, so that the negatively charged combination is then attracted to the anode (the article being painted). A very uniform paint film is produced because the charged particles 'seek out' any areas not already insulated by paint.

26.6.2 Stove-enamelling

This finish is used to provide a hard-wearing corrosion-proof coating for many domestic appliances, such as washing-machines, refrigerators and cooking-stoves.

26.6.3 Coating the surface with another metal

A thin coating of a corrosion-resistant metal can be applied to one which is less corrosion-resistant, in order to protect it. The aim is *always* to provide a *mechanical* barrier against possible electrolytes or corrosive atmospheres, but it must be remembered that, whilst zinc and aluminium will offer sacrificial protection should their coatings become damaged, the presence of damaged coatings of most other metals will *accelerate* corrosion.

Metallic coating can be applied in a number of different ways:

1   *Hot dipping* can be used to coat the surface of iron and steel components with both tin and zinc. Tinplate is still manufactured in South Wales, where the industry was established some 300 years ago. Clean mild-steel sheets are passed through a bath of molten tin and then through squeeze rolls, which remove the surplus tin. Galvanising is a similar process, whereby articles are coated with zinc. Buckets, dustbins, wheelbarrows, cold-water tanks and barbed-wire are all coated by immersion in molten zinc. Many of these galvanised items are now largely superseded by plastics materials products, though barbed-wire festoons the British countryside in increasing amounts as it replaces hedgerows grubbed-out or systematically killed off by the mechanical flail.

2   *Spraying* can be employed to coat surfaces with a wide range of molten metals, though zinc is most often used. In the Schoop process, an arc is

struck between two zinc wires within the spray gun, and the molten metal so produced is carried forward in an air blast. This type of process is useful for coating structures *in situ,* and was employed in the protection of the Forth road-bridge.

3   *Sherardising* is a 'cementation' process, similar in principle to carburising. Steel components are heated in a rotating drum containing some zinc powder at about 370°C. A very thin, but uniform, layer of zinc is deposited on the surface of the components. It is an ideal method for treating nuts and bolts, the threads of which would become clogged during ordinary hot-dip galvanising.

4   *Electroplating* is used to deposit a large number of metals onto both metallic and non-metallic surfaces. Gold, silver, nickel, chromium, copper, cadmium, tin, zinc and some alloys can be deposited in this way. Electroplating is a relatively expensive process, but provides a very uniform surface layer of very high quality, since accurate control of the process is possible at all stages. Moreover, heating of the component being coated is not involved, so that there is no risk of destruction of mechanical properties which may have been developed by previous heat-treatment.

5   *Cladding* is applicable mainly to the manufacture of 'clad' sheet. The basis metal is sandwiched between sheets of the coating metal and the sandwich is then rolled to the required thickness. During the process, the coating film welds onto the base metal. 'Alclad', which is duralumin coated with pure aluminium, is the best known of these products.

26.6.4 Protection by oxide coatings

In some cases, the film of oxide which forms on the surface of a metal is very dense and closely adherent. It then protects the metal beneath from further oxidation. The 'blueing' of ordinary carbon steel during tempering produces an oxide film which offers some protection against corrosion. Anodising is applied to suitable alloys of aluminium, in order to give them added protection against corrosion. The natural oxide film on the surface of these materials is an excellent barrier to further oxidation, and, in the anodising process, this film is thickened by making the article the anode in an electrolytic bath. As current is passed through the bath, atoms of oxygen are liberated at the surface of the article and these immediately combine with the aluminium and so thicken the natural oxide film. Since aluminium oxide (alumina) is extremely hard, this film is also wear-resistant; it is also thick enough to enable it to be dyed an attractive colour.

26.6.5 Metals and alloys which are inherently corrosion-resistant

Stainless steels are resistant to corrosion partly because the tenacious chromium oxide film which coats the surface behaves in much the same way as does the oxide film on the surface of aluminium. They are also corrosion-resistant, because of the uniform structure which is generally present in such steels. If a structure consists of crystals which are *all of the same composition,* then there can be no electrolytic action between them, as there is, for example, between the ferrite and cementite in the structure of an ordinary carbon steel.

Pure metals are generally corrosion-resistant for the same reason; though, of course, particles of impurity present at crystal boundaries can give rise to intercrystalline corrosion. This applies particularly to iron and aluminium, which can be obtained as much as 99.9999% pure. Unfortunately, such metals are generally so expensive in this state of purity that their everyday use is not possible.

26.6.6 Galvanic protection

The sacrificial protection of a ship's hull from electrolytic corrosion, which would be accelerated by the presence of a 'manganese bronze' propeller near by, has been described earlier in this chapter. In a similar way, underground steel pipelines can be protected from corrosion by burying slabs of zinc (or magnesium) near to the pipe at suitable intervals. Alternatively, a small e.m.f. can be used (Figure 26.15) to make the pipe *cathodic* to its surroundings. Either a battery or a low-voltage D.C. generator (operated from a small low-maintenance windmill) can be used to supply the e.m.f.

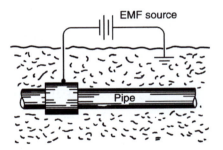

Figure 26.15 *Using an impressed e.m.f. to protect a steel pipeline from corrosion.*

26.7 Stability of plastics

Many plastics materials are relatively inert and will resist chemical attack by those reagents which would lead to the severe corrosion of most engineering metals. Thus, polythene is unaffected by prolonged contact with concentrated acids (including hydrofluoric acid which readily dissolves glass) yet it is not completely stable when exposed to outside atmospheres, tending to become opaque and brittle with the passage of time. Similarly, rubber will 'perish', that is becomes brittle and useless after some time unless steps have been taken to stabilise it.

26.7.1 Weathering of plastics materials

Almost all polymers, but in particular thermoplastic materials, deteriorate in appearance at a significant rate when exposed to the atmosphere unless they have been 'stabilised'. This deterioration in appearance, coupled with an increase in brittleness, continues to the extent that some plastics eventually become useless. At ambient temperatures, this weathering process appears to be due to the combined effects of oxygen and ultraviolet (UV) light, the effects of the oxygen by itself being negligible. Most polymers absorb UV light and this excitation causes polymer molecules to vibrate so that chemical bonds break and chains thus shorten. At the same time, the presence of oxygen leads to the formation of other chemical groups and possibly cross-links. Both events will reduce flexibility and strength of the polymer whilst increasing brittleness.

The simplest method by which plastics can be protected from weathering is to include some substance in the original moulding mixture which will 'screen' the material from UV radiation by making it opaque to UV. 'Carbon black' and other pigments will absorb UV radiation and so act as an effective screen. A large number of organic compounds are used as compatible absorbers of UV radiation when transparency to visible wavelengths of light is required in the polymer. Their function is to absorb UV radiation without decomposing. A further type of UV absorber converts UV radiation into light of longer wavelength which has no deleterious effect. Such a substance is hydroxyphenylbenzotriazole. Other stabilisers and anti-oxidants may be added. Thus, 0.1–0.2% phenol or an amine will effectively absorb oxygen and so prevent it from reacting with the polymer chains.

Other forms of short-wavelenth radiation, such as γ-rays or X-rays will cause degradation – or *scission* – of polymer molecules. High-speed particles, such as electrons and neutrons, may have a similar effect. Scission refers to a breaking of the linear 'backbone' chains. The strength is reduced because the total of van der Waals forces holding molecules together will be less if these molecular chains are now shorter.

Weathering is of course more serious with polymer materials exposed to intense UV radiation outdoors, but often the cost of stabilising the material must be balanced against the cost of frequent replacement. Weathering is inevitable with materials like the thin LDPE sheet used for covering horticultural greenhouses. Such sheet will not last for more than two seasons, but the overall cost of replacement is lower than it would be using a glasshouse which would, in addition, require a stronger frame and still involve some maintenance.

The phenomenon of weathering was previously known as *ageing* but this term is now generally used to describe an 'annealing' process whereby the structure of a polymer attains equilibrium and, consequently, stability in mechanical properties. For example, a PVC moulding which has been cooled rapidly may have a tensile strength of 40 MPa, but if immediately 'aged' (annealed) at 65°C the tensile strength will be increased to 60MPa. 'Natural ageing' at ambient temperature (20°C) would achieve a similar result but only after a number of years.

26.7.2 Perishing of rubbers

Both natural rubber and some synthetic elastomers, such as SBR, are prone to attack by ozone in the atmosphere. Ozone, an allotrope of oxygen in which

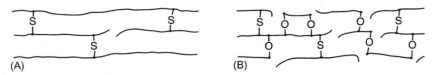

Figure 26.16 *The perishing of rubber. (A) Original sulphur (S) cross-links only; (B) scission of linear molecules and also extra oxygen (O) cross-links.*

the molecule contains three atoms of oxygen, is much more reactive than 'ordinary' oxygen in which the molecule contains two atoms, and can provide oxygen atoms which form cross-links between rubber molecules at points not used up during the vulcanisation process (Figure 26.16). At the same time, some degradation (scission) of the linear chain results. The extra cross-linking reduces elasticity whilst scission reduces strength, so that the rubber becomes weak and brittle.

In rubber, for car-tyre production, the moulding mixture contains about 3.5% amines added as anti-oxidants and anti-ozonants. The effect of both oxygen and ozone in promoting perishing of rubber is accelerated by UV radiation but the presence of some 30% carbon black in the mix acts as a screen to UV penetration.

26.7.3 Stress cracking and crazing of polymers

Some active chemical reagents have an adverse effect on the surface finish of some plastics materials when the latter are in a state of mechanical stress. Thus, water solutions of detergents can produce brittle cracking in polythene utensils which contain residual stresses from the moulding process. *Environmental stress cracking,* as it is termed, is therefore a result of the combined effects of stress and chemical attack. The surface cracking of stressed PVC gas pipes takes place in the presence of some hydrocarbon impurities, whilst some organic liquids and gases may also promote the formation of a network of minute cracks (crazes) in amorphous polymers such as clear polythene. Although such crazed material has a poor visual appearance, it generally retains its strength quite well. Nevertheless, there is a danger that these minute surface cracks may act as stress-raisers and so promote ultimate failure.

26.7.4 Stability to solvents

Some plastics materials dissolve in or absorb water. Nylon, for example, will absorb considerable amounts of water either when immersed or in a humid atmosphere. Unexpectedly, this 'wet nylon' has a much higher impact value than does 'dry nylon'. Most of the common plastics, however, are insoluble in water.

Solution occurs when the van der Waals forces operating between the solvent molecules and the polymer molecules are greater than the van der Waals forces operating between the adjacent polymer molecules within the solid polymer. Individual polymer molecules are thus pulled away from the solid and so go into solution. Many thermoplastic polymers are dissolved in various organic solvents. Thermosetting and other cross-linked polymers cannot be dissolved in this way because the van der Waals forces exerted by the solvent molecules are much too small to break the strong covalent cross-links holding the polymer chains together. For this reason, a cross-linked polymer does not dissolve or lose its original shape. Instead, it may absorb molecules of the solvent into itself because of the action of van der Waals forces. Some rubbers (cross-linked elastomers) swell when in contact with certain organic solvents though they are impervious to water.

Conversely, for a thermoplastics polymer, small quantities of solvents may lead to a softening of the polymer. This can often be utilised in *plasticisation,* as an important aid in polymer technology, but it is an unwelcome property in that it limits the range of organic liquids with which many polymers can be permitted to make contact.

26.8 Preservation of timber

In 1628, the Swedish warship *Wasa* capsized and sank at the start of her maiden voyage. Until 1961, she remained at the bottom of Stockholm harbour, when it was discovered that she was in remarkably good condition, and was subsequently raised. This great oak ship, with all its beautifully carved decorations and wooden sculptures, has since had a museum built around it. The excellent state of preservation was explained by the fact that the water in which the wreck lay was too saline to support freshwater organisms, but not sufficiently salty to allow 'sea-worms' to survive. Thus, due to the absence of any predatory organisms, the wood did not decay.

Metals are subject to corrosion by fairly simple chemical attack in the presence of moisture, but timber, which is obtained from a living organism, is attacked by other living organisms – insects and fungi. Normally, when a tree dies, it is, in the source of time, reduced to its original chemical ingredients by the scavengers of the forest – boring insects, fungi and bacteria. The chemical ingredients return to the soil, and so the cycle of nature repeats itself. When timber is used for constructional purposes, these scavengers are regarded as enemies and steps must be taken to repel them.

26.8.1 Insect pests

In Britain, various types of beetle constitute the most important insects which attack wood. In each case, the female lays her eggs on or just beneath the surface of the wood, and the larvae (or grubs), which ultimately hatch, eat their way through the timber until, considerably they become dormant and form pupae, or chrysalides. Eventually, the adult beetles emerge, find their way to the surface via bore-holes, and the life cycle repeats itself.

The *common furniture-beetle* attacks both softwoods and hardwoods, but seems to prefer old furniture. The *death-watch beetle,* on the other hand, generally attacks hardwoods, and is most frequently found in the roof timbers of old churches. It thrives in damp situations arising from a lack of ventilation and derives its name from the tapping sound it makes during the mating season – a not unusual behaviour pattern among certain higher forms of life. These two are the most common insect pests which attack wood, though there are several others encountered less frequently. They often reveal their presence by the little mounds of wood powder scattered around the scenes of their crimes, though the death-watch beetle differs from the others in that its larvae produce little bun-shaped pellets. If timber is badly affected, the only sure way to eradicate the pest is to cut out such timber and burn it. A number of proprietary preservatives are available to ward off such attack. To be effective, such preservatives should be permanent in their effect, poisonous to both insects and fungi, but non-toxic to human beings and animals.

26.8.2 Fungus attack

A very large number of fungi and mould (or micro-fungi) attack timber in their search for food and lodgings. However, the best known are those which cause the effects generally known as *dry-rot* and *wet-rot*.

The fungus which causes dry-rot thrives in damp, poorly ventilated situations where a temperature between 16°C and 20°C prevails. It consists of a somewhat disgusting, dark, feathery mass, with branching tendrils, and is also able to penetrate brickwork. An unpleasant mouldy smell generally prevails when dry-rot is present. The timber becomes discoloured and develops a dry, shrunken appearance. This is the origin of the term *dry-rot*, but it must be emphasised that the fungus which causes it operates only in damp situations.

The fungi responsible for wet-rot cause internal rotting of the wood and thrive only in very wet conditions. A pale green scum first appears, but this soon changes to dark brown and ultimately to black.

Whilst these forms of fungal attack occur most frequently in the basements and ground floors of buildings, they may occur elsewhere, if the combination of dampness and poor ventilation produce conditions under which they can thrive. All infected timber must be cut out and burnt, and the surviving sound wood treated with a preservative.

26.9 Service life

The 'service life' of a product can be considered to be the length of time for which it performs its required function adequately under the specified operating conditions. One of the important factors affecting service life is the behaviour of the materials used for the product. Thus, factors that have to be taken into account are:

- External loading levels, rate of loading (impact loading), frequency of loading (fatigue), duration of loading (creep).
- Material property degradation (corrosion).
- Defects in the form of cracks, porosity (in castings), cavities (in welds) introduced during manufacturing.
- Conditions under which used, e.g. temperature, temperature cycling, humidity, chemicals, contact with other materials.

- Bad design features such as the presence of notches, sharp corners, small holes, surface roughness.
- Lack of, or inappropriate, maintenance.

Service life of a product might be improved by:

- Changing the material to a more suitable one.
- Changes in the product design to avoid notches, etc.
- Protective measures against corrosion.
- Changes in the service conditions, e.g. the level of stress or the elimination of alternating stresses.
- Preventative maintenance by which, for example, a component subject to wear is replaced before it fails, or repair maintenance in which failed components are replaced.

27 Choice of materials and processes

27.1 Introduction

At about the same time I had acquired the status of a 'teenager' it was possible to buy a brand new rather heavy 'sit-up-and-beg' pedal bicycle from the local branch of H*lf*r*s for the sum of £3 19s 11d (near enough £3.99). The 'bike' is still part of modern life and we now have 'mountain-bikes' which are encountered in all sorts of unlikely places, such as canal towpaths, bridle ways and steep mountain tracks.

The design and materials used for bicycles has changed over the years. The earliest bicycles, about 1865, had a solid wooden frame joined by adhesives; it was light but not very durable. The first all-metal machine, with solid rubber tyres, appeared in 1870. It was, however, not until much later, 1885, that improvements in metallurgy enabled fine chain and sprockets to be developed and gears to be added to bicycles to give what became termed as the *safety bicycle,* so called because the earlier bicycles had a much larger front wheel than rear wheel and were prone to tip the rider forward over the front wheel if that wheel was stopped by a stone or perhaps a dog running across the road. Think of the hard work involved in pedalling uphill without any gears! The pneumatic tyre was developed by the British physician John Dunlop in 1885, so making for a less 'bone-shaking' ride.

The bicycles introduced after the First World War were 'made-to-last' and had every feature possible – the result being a very heavy bicycle. Modern bicycles have changed to become as light as possible – less effort to pedal up hills. They can even have, what would have been until very recently, an unheard of number of gears! While steel is used for the tubular framework of a cheap (?) bicycle, costing under £100, lighter and more 'with-it' bicycles have aluminium alloy frames and cost a few hundred pounds. Bicycles for the fanatics – sorry, 'enthusiasts' – who ride these machines are willing to pay in excess of £600 for the frame alone, provided that it offers an acceptable 'strength-to-weight ratio'. Currently, it seems that about half-a-dozen different materials are competing for this trade. The main contenders are:

1   A metal matrix composite in which aluminium is dispersion strengthened by the presence of aluminium oxide particles (see Section 23.3.1, this composite being a development of SAP). Frames made with this are TIG welded.
2   A metal matrix composite, again with an aluminium matrix, but this time stiffened with boron carbide in amounts up to 15%. This composite is also TIG welded.

3   Another metal matrix composite in which aluminium is stiffened by the incorporation of 40% beryllium, giving a material of low relative density about 2.3. The metal beryllium, though scarce and very expensive, has now come onto the market for civil uses as a result of the end of the Cold War. Recently, a cycle with a frame constructed entirely from this 'modern miracle metal' (in fact beryllium was discovered in 1797 and industrial production on a small scale began in 1916!) was marketed in the USA for around $25 000. The use of beryllium under all but the most carefully controlled conditions is made more difficult because its dust (from machining operations) and vapour (from welding) are extremely toxic.

4   A type of maraging steel (see Section 13.2.7) based on iron, nickel, cobalt, chromium, molybdenum and carbon from which frame tubes with a wall thickness of only 0.5 mm can be drawn. This makes them extremely light and strong but also difficult to weld due to ensuing heat distortion. Silver soldering (see Table 25.3) is used instead. The 'sit-up-and-beg' bicycle of my youth had a frame made of tubular steel but the wall thickness was much thicker and so the bicycle was much heavier and harder work to cycle uphill. The strength of a tube to the bending forces it experiences in a bicycle frame depends on a combination of the wall thickness and the yield stress of the material. If you have a higher yield stress then you can have a thinner wall thickness.

5   A novel carbon-fibre reinforced nylon matrix in which a woven mat of mixed carbon and nylon fibres is hot moulded to form the frame. During this process, the nylon fibres fuse to form the thermoplastic matrix holding the carbon fibres. Early carbon-fibre reinforced polymers used the material to form tubes, imitating the steel tube design of the conventional frame, but joining was difficult and the properties of the material are better utilised in a monocoque construction.

These then are processes and materials competing to manufacture very strong lightweight cycle frames. Here, we are dealing with the 'pursuit of excellence' with few, if any, restraints on the cost of the product. But how often can this situation prevail in the real world of engineering? Everyday engineering production involves a highly competitive enterprise which, in the long run, nearly always leads to some sacrifice of quality, in favour of lower cost. The manufacturer is not necessarily to blame for this state of affairs, since he is usually satisfying an existing demand, even though that demand is often stimulated by a vigorous advertising campaign. Often the cost of perfection is too high, and in most cases the effects of competition, coupled with a lack of power of discrimination by the consumer, leads to the manufacture of an article which is 'cheap and nasty' – or in many cases nasty without being cheap.

In an age when the development of new materials was extremely slow, industrial philosophy favoured the manufacture of high-quality durable products. Thus, a pocket watch in its silver case was handed down from father to son for several generations. Kettles, cooking pots and domestic equipment generally was made to last indefinitely. However, methods of mass production which followed in the wake of the Industrial Revolution meant that goods never quite reached the standard of craftsmanship attained by individual craftsmen – the petrol-driven 'tin Lizzie' of the 1920s never quite reached the standard of craftsmanship of the horse-drawn carriage of earlier days. Of course, mass production had to provide a commodity which was cheap enough to be purchased by the masses who were employed in making it. Invention, research and development now proceed so quickly that many products become obsolescent within a very short time of purchase (the market value of electronic gadgets like computers falls so rapidly that one purchased now becomes almost unsaleable in a couple of years).

Fortunately when technical development of a class of appliances stabilises, quality of production tends to improve, as there is less incentive for the consumer to buy a new model because of some minor innovation in design, and this in turn encourages him to hope that it will last longer without breaking down. Obviously, there will always be the manufacturer who seeks to make something a little cheaper – and a little nastier. It is up to Joe Public to discriminate between products offered for sale.

27.2 Selection of materials

In choosing a material for a specific application, the engineer must consider:

- The ability of the material to withstand service conditions.
- The method(s) by which it will be shaped.
- The overall cost, i.e. the cost of the material(s), with in some cases the availability of the material, and the cost of the shaping process(es).

In some cases, a number of materials may be satisfactory in respect of fulfilling the service requirement. Normally, the engineer will choose the one which, when the cost of forming and shaping is taken into account, results in the lowest overall production cost. Thus, the cheapest material may not necessarily be the one which is used; a more expensive material may be capable of being formed very cheaply and so give a lower overall production cost.

27.3 Service requirements

The service requirements of a material may involve properties which fall under one or all of the three headings:

1   *Mechanical properties*
    (a) Tensile strength
    (b) Elasticity
    (c) Toughness (impact value)
    (d) Stiffness (modulus of elasticity)
    (e) Hardness
    (f) Fatigue resistance
    (g) Resistance to creep
    (h) Frictional and wear resistance properties

2   *Physical properties*
   (a) Relative density
   (b) Melting point or softening temperature
   (c) Thermal conductivity
   (d) Coefficient of expansion
   (e) Effect of temperature changes on properties
   (f) Electrical conductivity
   (g) Magnetic properties

3   *Chemical properties*
   (a) Resistance to oxidation
   (b) Resistance to electrolytic corrosion
   (c) Resistance to degradation by electromagnetic radiation
   (d) Resistance to biological attack by plants or animals

We will now deal with the influence of some of these properties on material selection.

27.3.1 Tensile strength and specific strength

Table 27.1 is a list of important engineering materials in order of tensile strength.

Such a list (Table 27.1) is headed by maraging steels and the well-known 'nickel–chrome–moly' constructional steels, followed by other steels, fibres, non-ferrous alloys, ceramics and plastics materials in appropriate order. The choice of a material, however, will not usually be made with reference to strength alone. Such criteria as the temperature and chemical nature of its working environment must be considered, as well as the type of stress it must bear, i.e. static, impact or alternating.

The choice of dimensions of the load-bearing member may also be relevant. Thus, if the dimensions allow a material of *lower* tensile strength but *greater* cross-sectional area to be used, then a cheaper material may be available. But this may mean using a member of *greater* weight and so the factor cost/kg is involved. For example, although a medium-carbon steel is only about half as strong as a similar nickel–chromium steel, it is only about one-third of the cost of the latter in terms of cost/kg. Consequently, although a member in medium-carbon steel capable of bearing a similar load will be greater in cross-sectional area than one in nickel–chrome steel, it will cost less.

The energy required to blast a space vehicle free of the gravitational 'pull' of Mother Earth is considerable so that it is important to ensure that its mass is kept to a minimum. Therefore, constructional materials used in such a craft must have a high 'strength-to-weight ratio'. This is now referred to as *specific strength* and is the ratio (tensile strength)/(density) or, for simplicity, as it is sometimes expressed (tensile strength)/(relative density) – the difference is purely one of the units used. Thus, maraging steel with a tensile strength of 2460 MPa and a relative density of 8.2 has a specific strength of $2460/8.2 = 300$ MPa; using the density of $8.2 \times 10^3 \text{kg/m}^3$ or 8.2 Mg/m^2,

Table 27.1 *Approximate values of tensile strengths and specific strengths of some important engineering materials*

Material	Tensile strength (MPa)	Specific strength (MPa)
Maraging steel	2460	299
S-glass composite*	1800	900
High-strength carbon fibre composite*	1770	1140
Ni–Cr–Mo constructional steel	1700	215
Beryllium bronze, heat treated	1300	146
Titanium alloy TA10	1120	280
Nimonic 115	1100	134
0.5% C steel, quenched and tempered	925	119
18/8 Stainless steel, hard rolled	800	98
Phosphor bronze, hard drawn	700	80
65/35 Brass, hard rolled	695	84
0.2% C steel, quenched and tempered	620	79
Duralumin, heat treated	420	155
400 grade grey cast iron	400	52
Zinc-base die-casting alloy	300	43
Magnesium-base alloy, cast and heat treated	200	118
Aluminium alloy LM6, cast	190	73
Thermoplastic polyester	170	120
PEEK	92	70
Nylon 6:6	85	74
Epoxy resin	80	69
Bakelite	55	40
PVC, rigid form	49	36
Polythene (HD)	31	32

60% by volume in epoxy resin matrix.

rather than relative density, gives a specific strength of 2460/8.2 = 300 MNm/Mg. Of course, it is not only when dealing with space vehicles that we must take into account energy used in working against gravity. Heavy road-haulage vehicles constructed from unnecessarily heavy component parts consume extra energy – and release unacceptable amounts of 'greenhouse gases' as a result. Undoubtedly, fibre-reinforced composites, with their much higher strength-to-weight ratios, will be used increasingly in such cases.

27.3.2 Stiffness, modulus of elasticity and specific modulus

The modulus of elasticity E is derived from the ratio stress/strain for stresses applied to a material below its yield point. Thus, E is equivalent to the slope of that part of the force/extension diagram below the yield point and is a measure of the 'stiffness' of the material. For example, the slope of that part of the diagram for a steel is much 'steeper' than that for aluminium – the

Table 27.2 *Modulus of elasticity and specific modulus for some engineering materials*

Material	Modulus of elasticity (GPa)	Specific modulus (GPa)
High-tensile Ni–Cr–Mo steel	210	27
18/8 Stainless steel	205	26
High-carbon steel	203	26
Low-carbon steel	203	26
Titanium alloy (6Al-4V)	115	28
Duralumin	69	26
S-glass*	55	28
Aramid*	80	59
Boron*	210	95
High-modulus carbon*	272	167

* 60% by volume fibre in epoxy resin matrix and modulus values are for longitudinal specimens.

steel is much the stiffer material under the action of either tensile or compressive loads. The modulus of elasticity can also be related to the density, or relative density, to give the value 'specific modulus', i.e. specific modulus = (modulus of elasticity)/(density) or (modulus of elasticity)/(relative density) – the difference is purely one of the units used.

Table 27.2 gives some typical values of the modulus of elasticity and specific modulus. When considering the specific modulus of materials, note that there is little to choose between metallic alloys and it is here where composites reinforced with carbon, boron or aramid fibres score heavily.

27.3.3 Toughness and impact value

The easy way to define 'toughness' is to say that it is the opposite of 'brittleness', but to the engineer it implies the ability of a material to resist the propagation of a crack and the term *fracture toughness* is used. A simpler measure which is used as an indicator of toughness is the ability of a material to absorb energy under conditions of mechanical shock, i.e. impact, without fracture. The Izod and Charpy impact tests are a demonstration of this as they measure the mechanical energy absorbed from a swinging hammer as it fractures a standard test-piece.

Strong materials are not necessarily tough ones. Thus, soft, annealed copper with a tensile strength of no more than 200 MPa may well almost 'stop' the swinging hammer in the Izod test – so needing all the energy of the swinging hammer to break it and hence indicating a very tough material. Any treatment which increases the yield stress, or the hardness of a material, generally reduces its toughness. Thus, a quenched and tempered low-alloy steel may be stronger and harder than the same steel in the normalised condition but the Izod value will usually be lower.

Whilst most ceramics are brittle when compared with metals, there are exceptions. Thus, crystalline alumina, although used in armour plating because of its great hardness, also possesses adequate toughness. Thermo-plastic polymers are usually weak but tough whilst thermosetting polymers are generally stronger and harder but more brittle.

Consider the requirements for a container in which milk can be purchased. Until relatively recently there was only one material used and that was the glass bottle. However, this was heavy and easily broken because the glass was not tough (bottles used in the 1940s had a weight of about 500 g but this weight was steadily reduced over the years to now about 220 g, however, now more easily broken). The glass bottle did, however, have one great merit and that was that the empty bottle could be returned and, after cleaning, refilled with milk. Nowadays, milk is hardly ever delivered, as it used to be, to the doorstep but bought at the supermarket. The need is now for a light, unbreakable (tough), container and so milk now comes in plastics containers, generally polythene.

27.3.4 Fatigue resistance

In Section 27.3.1, we were comparing the tensile strengths of some materials under the action of *static* loads, in Section 27.3.3 under *impact loads.* In practice, however, we must often consider the behaviour of a material under the action of 'live' (fluctuating) or alternating stresses and we use as an indicator of such behaviour the *fatigue limit,* this being the maximum stress a material can sustain under the action of an 'infinite number' of reversals of stress.

For most metallic alloys, the fatigue limit is usually between one-third and one-half of the static tensile strength. For example, a 0.1% carbon steel in the normalised condition has a static tensile strength of 360 MPa and a fatigue limit of 190 MPa.

With many plastics materials, it is difficult to assess fatigue properties because the heat generated by the energy used in overcoming van der Waals forces acting between the large molecules is not quickly dispersed as it is in metals which have a much higher thermal conductivity. The plastics material therefore tends to overheat on test and fail at a low stress. The higher the frequency of stress alternations the quicker the build up of heat and the more likely the polymer to fail at low stress.

27.3.5 Creep resistance

Creep is the gradual extension which can take place in materials at a stress *below* the yield stress. Although measurable creep may occur in some materials such as lead and a number of polymers at ambient temperatures, it is generally associated with engineering materials at high temperatures. Hence, metallic alloys used under these conditions are usually dispersion hardened by including in the microstructure tiny, strong particles which impede the progress of dislocations along the crystallographic planes. It is also necessary for high-temperature alloys to have a good resistance to oxidation.

Table 27.3 *Rupture stresses at different temperature and for different times*

Alloy (%) and condition	Rupture stress (MPa)						Oxidation
limit (°C)	At 500°C		At 700°C		At 800°C		
	1000 hours	10 000 hours	1000 hours	10 000 hours	1000 hours	10 000 hours	
Carbon steel							
0.2 C, 0.75 Mn* Normal 920°C, temper 600°C	118	59					450
Nickel–chromium alloys							
80 Ni, 20 Cr** ½ hour at 1050°C, air cool			100	59	31	18	900
80 Ni, 20 Cr + Co, Al, Ti*** 8 hours at 1080°C, air cool + 16 hours at 700°C, air cool	925	815	370	245	140	70	900

Room temperature values for tensile strength: *430MPa, **565MPa, ***560MPa.

The need, engendered in the early 1940s, for a material suitable for the manufacture of parts for jet engines and gas turbines led to one of the epic metallurgical research projects of the twentieth century and resulted in the production of the 'Nimonic' series of alloys by Messrs Henry Wiggin. The basis of a Nimonic is an alloy containing 75% nickel and 20% chromium, giving a material of high melting-point which will remain tough at high temperatures because of the grain-growth restricting influence of nickel and at the same time be oxidation-resistant because it becomes coated with a tenacious film of protective chromium oxide. Small amounts of carbon, along with titanium, aluminium and molybdenum, form the particles of 'dispersoid' which resist creep and so strengthen the alloy at high temperatures. Other alloys based on nickel and chromium, along with dispersion-hardening additions, are now available. Some are wrought alloys whilst others are cast to shape. Table 27.3 shows the effects of temperature and time on a carbon steel and two high-temperature nickel–chromium alloys.

27.3.6 Refractoriness

Whilst most materials working at high temperatures require high creep resistance, others which are exposed to aggressive environments combining high temperatures and corrosive atmospheres need carry only light static loads. Such materials are used in furnace linings and other static parts, crucibles and the like. Materials which are used principally because of their ability to withstand high temperatures are known as *refractories*. Some of the 'heavy metals' like tungsten and molybdenum fall into this category. Table 27.4 lists some refractories along with other engineering materials of which maximum working temperatures are indicated.

Table 27.4 *Maximum working temperatures of some engineering materials*

Material	Maximum working temperature (°C)	Material	Maximum working temperature (°C)
Refractories		*Metals*	
Zirconia	2500	Nimonic alloys	900
Magnesia	2000	High Ni–Cr steels	800
Boron nitride	1800	12% Cr stainless steels	600
Silica	1700	Low Ni–Cr–Mo–V steel	550
Fireclay	1400	Titanium alloys	500
Sialons	1250	C steels (normalised)	400
Molybdenum	1000	Cast iron	300
		Copper alloys	190
Polymers		Aluminium alloys	180
Thermosetting	150		
Thermoplastic	100		

27.3.7 Friction and wear resistance

When two surfaces rub together, energy is consumed in proportion to the frictional forces which operate between them. At the same time, there is a tendency for particles of the softer material to become detached – its surface wears away. In machines, these problems are reduced to some extent by lubrication, the function of a lubricant being to *separate* the running surfaces and so reduce friction. Lubricants include both liquids (from water to heavy mineral oils) and solids (molybdenum disulphide (MoS_2) and graphite).

As far as load-bearing surfaces are concerned, materials with high hardness values usually have a low coefficient of friction and a good resistance to wear. Unfortunately, such materials are often intermetallic compounds (see Section 8.4) which are both weak and brittle. Nevertheless, the need for a material which is hard, has a low coefficient of friction, but is also reasonably tough and ductile is met by using an alloy with a complex microstructure as is found in bearing metals (see Section 18.6). Thus, the 'white' bearing metals consist of hard particles of an antimony–tin intermetallic compound (SbSn) embedded in a tough matrix of tin-base or lead-base alloy; whilst bronze bearings consist of particles of the hard intermetallic compound $Cu_{31}Sn_8$ embedded in a tough copper–tin solid solution. In both cases, the softer matrix tends to wear away leaving the hard compound standing proud. This in itself reduces friction between the running surfaces but also provides channels which assist the flow of lubricants.

Similar properties are required in the journal which is running on the bearing surface provided, i.e. a strong, tough 'body' with a hard, low coefficient of friction surface. This is provided by using one of two classes of steel:

1   A low-carbon or low-alloy, low-carbon steel, the surface of which is then either carburised and case-hardened (see Section 14.2) or nitrided (see Section 14.5).

2   A medium-carbon or low-alloy steel, the surface of which receives separate heat-treatment by either flame- (see Section 14.7) or induction-hardening (see Section 14.8).

In each case a tough, strong core is provided with a very hard wear-resistance skin.

Cast irons tend to have a good wear-resistance – 'grey' irons because the presence of graphite flakes has a lubricating effect and 'white' irons because of the presence of very hard cementite.

Wear-resistant, self-lubricating plastics are now used for many applications which previously used metals. The plastics need less maintenance and are easy to manufacture. The plastics are formulated with an internal lubricating agent or agents. Silicon and PTFE are often used as the additives for high speed and pressure applications. Table 27.5 gives some examples of such plastics and the dynamic coefficients of friction obtained when they slide on steel.

Table 27.5 *Coefficients of friction for lubricated plastics on steel*

Material	Lubricating additive	Coefficient of friction
Nylon 6:6	18% PTFE, 2% silicone	0.08
Nylon 6:6 composite with 30% carbon	13% PTFE, 2% silicone	0.11
Acetal	2% silicone	0.12
Acetal	20% PTFE	0.13
For comparison: mild steel on mild steel		0.62

27.3.8 Stability in the environment

Iron is one of the most plentiful and inexpensive of engineering metals. Furthermore, its alloy, steel, can either by cold-work (mild steel) or by heat-treatment (medium- and high-carbon steels) be made to provide us with a wider range of mechanical properties than any other material. It has one major fault – its corrosion-resistance is poor and, except when alloyed with expensive chromium, it will rust if exposed to damp atmospheres. The cost of protecting steel structures – bridges, pylons, radio masts and the like – from corrosion in the UK alone runs into more than £10 billion per year, whilst of the 'unprotected' steel some 1000 tonnes 'escapes back into nature' as rust on every single day.

Steel will rust more quickly in the annealed or normalised conditions, since in the presence of an electrolyte (moisture containing dissolved atmospheric pollutants such as SO_2 from burning coal) the layers of ferrite and cementite present in pearlite constitute an electrolytic cell (see Section 26.6) leading to the 'rusting' of the ferrite component which is anodic to cementite. A quenched (hardened) carbon steel is slightly more corrosion-resistant because most of the carbon is in solution and electrolytic action is reduced.

Thus, for carbon- and low-alloy steels which are to be used in an aggressive environment, some form of coating whether by paint, paint assisted by electrophoresis (see Section 26.6.1), enamelling, or some form of metal coating must be used. Where mild-steel presswork is exposed to corrosive conditions, it may be worthwhile considering alternative materials to painted mild steel. 'Family cars' now have bodywork where the 'skirtings' – notorious corrosion traps – are largely plastics.

Most of the engineering non-ferrous metals have much higher corrosion resistance than does carbon steel, but, weight-for-weight, they are more expensive. Titanium and its alloys are exceptionally resistant to attack by air, water and sea water but are very expensive. Nevertheless, when it can be considered in terms of cost/cubic metre instead of cost/kg, the low relative density of titanium may make it an attractive proposition (see Table 27.8). Nickel alloys too are very resistant but expensive, whilst the amounts of nickel and chromium necessary to make steel 'stainless' also render it up to 8 times more expensive, weight-for-weight, than mild steel. Copper and its alloys are moderately resistant to atmospheric corrosion but some brasses suffer dezincification in contact with some natural waters – small amounts of arsenic inhibit this fault if added to brasses used for condenser tubes. Aluminium, which in terms of cost per unit volume is less expensive than other non-ferrous alloys, has an excellent resistance to corrosion (see Section 17.3) but some of the high-strength alloys, particularly those containing zinc (2L88) require protection. Zinc-base die-casting alloys have a good corrosion resistance provided that the metal is of high purity – 'four-nines zinc', i.e. 99.99% pure is required for this purpose. Table 27.6 gives an overview of the relative corrosion resistance of a number of engineering alloys.

Whilst plastics materials have a high resistance to chemical attack and can be used underground for conveying water and gas, some are prone to damage by ultraviolet (UV) light and must be made opaque by suitable

Table 27.6 *Relative corrosion resistance of unprotected metals*

Metal	Fresh water	Sea water	Industrial atmosphere
Low-carbon steel	Poor	Poor	Poor
4–6% chromium steel	Good	Good	Good
18–8 stainless steel	Very good	Excellent	Excellent
Monel (70% Ni–30% Cr)	Very good	Excellent	Excellent
Aluminium	Fair	Poor	Very good
Copper	Very good	Very good	Very good
Red brass (85% Cu–15% Zn)	Good	Very good	Very good
Nickel	Excellent	Excellent	Very good

additives. PTFE resists attack by the most aggressive reagents but is very expensive. Natural rubber may 'perish' due to exposure to intense UV light and the presence of atmospheric ozone.

Glass has an excellent resistance to all chemical reagents except hydrofluoric acid (used for etching stained glass) as does vitreous enamel. 'Glass' coated steel tanks are widely used in the chemical industries for containing corrosive liquids. If a heavy spanner is accidentally dropped on to such a vitreous-coated surface, the chipped area can be 'drilled through' and the surface sealed by carefully hammering in a tantalum plug. Tantalum is a very soft, ductile heavy metal with an extremely high resistance to corrosion for, like aluminium and chromium, its surface seals itself with an adherent, impervious oxide film.

27.3.9 Electrical conductivity

A material has a high electrical conductivity if electrons are able to pass through it easily, the higher the conductivity the less a current dissipates energy (as heat) in passing through it. Electrical conductors such as power lines must lose as little energy as possible (the 'I^2R loss') in this way and so a low resistance is required. With electrical heater elements, the opposite is required and these elements require a high resistance so that heat is generated and so are made of alloys with a low conductivity (or high electrical resistivity). They need to generate the heat without the element melting. Hence for this purpose, high-temperature nickel–chromium alloys are used.

Copper is the most widely used metal industrially where high electrical conductivity is required and the bulk of copper produced is for the electrical trades. In respect of electrical conductivity it is second only to silver, which is slightly more conductive, but when the relative costs of the two metals are considered then copper is of course the metal used. As a standard against which the conductivities can be compared, there is an international standard specified as 'The International Annealed Copper Standard' (IACS) which relates to copper of which the resistance of a wire 1 m long and weighing 1 g is 0.153 28 ohm at 20°C. Such copper is said to have a conductivity of 100%. Since this standard was adopted in 1913, the chemical purity of commercial copper has improved and this explains why seemingly impossible conductivities of 101% or more are quoted.

The electrical conductivity of copper is adversely affected by the presence of impurities (Figure 27.1). Some impurities like zinc, cadmium and silver have little effect if present in small amounts. Thus, about 1% cadmium is added to copper destined for use in overhead telephone cables in order to strengthen them sufficiently to support their own weight. The conductivity is still of the order of 95%. The presence of only 0.1% phosphorus, however, reduced electrical conductivity by 50% so that phosphorus-deoxidised copper should not be used where high conductivity is required. The electrical conductivity of aluminium is only 60% that of copper, but in terms of conductivity per unit mass, aluminium is a better conductor, for which reason it is used in the power grid network. The electrical conductivities of

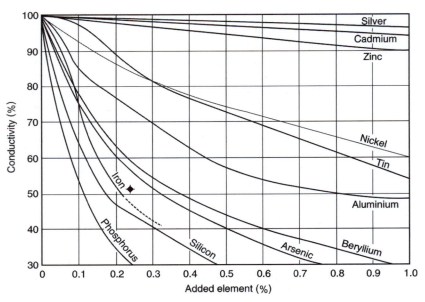

♦ To the limit of solid solubility.

Figure 27.1 *The effect of impurities on the electrical conductivity of copper.*

Table 27.7 *Electrical conductivities (IACS) of grades of copper and its alloys*

Material	Conductivity (%)
Silver	105
IACS Cu standard	100
Oxygen-free high-conductivity Cu	101.4
Hard-drawn Cu (1% Cd)	95
Phosphorus-deoxidised Cu (0.8% P)	54
Phosphorus-deoxidised arsenical Cu	5
70–30 brass	27
Tin bronzes	10–27

various grades of copper and other alloys likely to be used in electrical trades are shown in Table 27.7.

27.3.10 Relative costs of important engineering materials

Table 27.8 gives a very approximate idea of the relative costs of engineering materials. In terms of relating the cost of polymer materials to those of metals, 'weight-for-weight' polythene is very approximately 3 times as expensive as mild steel but when considered 'volume-for-volume' mild steel is roughly 3 times more expensive than polythene.

Table 27.8 *Approximate relative costs of some engineering materials*

Material	Cost per unit mass	Relative density	Cost per unit volume	Material	Cost per unit mass	Relative density	Cost per unit volume
Bar form				*Castings*			
Mild steel	1.0	7.87	1.0	Grey cast iron	1.0	7.4	1.0
Medium carbon steel	1.5	7.83	1.5	Brass	3.5	8.38	4.0
				Zinc-base die-casting alloy	4.0	6.8	3.7
Low-alloy steel	4.5	7.85	4.5				
Brass	6.5	8.45	7.0	Aluminium bronze LM6	4.8	2.65	1.7
Aluminium	8.8	2.70	3.0				
Stainless steel	9.7	7.92	9.8	Phosphor bronze	6.8	8.93	8.2
Copper	12.0	8.94	13.6				
Phosphor bronze	14.6	8.93	16.6	*Polymer materials*	*Cost/unit mass or volume*		
Titanium	24.0	4.51	13.7	Polythene	1.0		
				PVC	1.4		
Sheet form				Natural rubber	1.7		
Mild steel	1.0	7.87	1.0	Phenolics	3.8		
Lead	2.9	11.30	4.0	ABS	4.5		
Brass	4.0	8.45	4.3	Polycarbonate	11.0		
Stainless steel	6.3	7.92	6.3	Nylons	12.0		
Aluminium	6.7	2.70	2.3	PTFE	25.0		
Copper	7.0	8.94	8.0				
Titanium	20.0	4.51	11.5				

In the real world, both the *quality* and the relative *costs* of materials change and the design engineer must keep under review such factors as:

- Variations in the market value of a material relative to those of competing materials.
- Improvements in relevant properties of materials resulting from developments in production methods.

For example, until recently there was no serious competitor for copper as a prestigious roofing material, but in Japan, where corrosion from marine atmospheres is prevalent, it was decided to roof the Municipal Aqualife Museum in Kobe with titanium sheet. On the face of it, this would seem to be an expensive project since, despite a gradual fall in the price of titanium as production methods improve, it is still several times more expensive than is copper when considered on a weight-for-weight basis. However, the relative density of titanium is only 50.7% that of copper, so, assuming equal thickness of cladding sheet are involved the price difference is already halved; but the comparison does not end there. Since the use of titanium cladding reduces by half the load to be carried by roof members, modifications in the design of the latter on the side of economy are also possible. Since the price of titanium is likely to fall further relative to that of other metals, and particularly copper, we can expect to see increased use of titanium as a roofing material, particularly in buildings of high prestige.

Improvements in quality of some of the old well-established materials continue to take place. We may not regard prosaic mild steel as a technically 'glamorous' material but in the years since the Second World War considerable improvements have been achieved in its chemical 'cleanliness' and a reduction in the quantity of dissolved atmospheric nitrogen it contained hitherto. These achievements have led to a big improvement in the overall ductility of those high-quality mild steels produced in this way. So much so, that they have largely replaced far more expensive materials like 70–30 brass, cupro-nickel and aluminium for the manufacture of deep-drawn components. This 'revolution' has of course been assisted by improvements in tool design and methods of lubrication. Much of the practice ammunition used by the military under conditions where corrosion is no problem uses mild steel cases to replace those formerly deep drawn from 'cartridge' (70–30) brass.

27.4 Choice of shaping process

The properties of a metallic material which affect its suitability for forming and shaping operations include:

- Malleability
- Ductility
- Strength
- The effects of temperature on the above properties
- Castability
- Machinability
- Capacity, if any, for heat-treatment
- Methods by which the material can be joined.

Forging processes can be carried out on metals and alloys which are malleable in either the hot or cold state but whilst a metal may be malleable at all temperatures those which are ductile lose their ductility at high temperatures where strength falls so that the metal tears apart in tension. Thus ductile metals can be drawn or deep drawn, spun or stretch formed. Table 27.9 gives some values of the percentage elongation, a measure of ductility, of metals used in such processes. Some materials are neither malleable nor ductile and can only be cast to shape or shaped by powder metallurgy techniques.

Frequently, the shaping process is also used to develop strength in a component. Thus, an aluminium cooking-pot or the bodywork of an automobile becomes sufficiently rigid as a result of work-hardening which accompanies the shaping process.

Some metals and alloys have a limited scope for shaping. These alloys, such as Nimonics and the Incoloy series, destined for service at high temperatures in gas turbines, need of course to be both strong and stiff at those temperatures if they are not to suffer distortion. This hardness and stiffness in turn makes such alloys very difficult to forge, even at high temperatures where they are intended to be stiff. Thus a slow, expensive

Table 27.9 *The ductility of some metals and alloys (as measured by percentage elongation in the tensile test) for use in cold-pressing or deep-drawing processes*

Material	Percentage elongation
70–30 brass	70
5% aluminium bronze	70
3% tin bronze	65
65–35 brass	65
Aluminium (99.9%)	65
Stainless steel (17% Cr–10% Ni)	65
Copper (OFHS)	60
Cupro-nickel (5% Ni)	50
Mild steel	40
Stainless iron (13% Cr)	40
Aluminium alloy (1.2% Mn)	40

processing schedule is involved with frequent reheating stages. Sometimes investment casting is the alternative in such cases but, whichever process is chosen, shaping is an expensive process, the cost of which must be added to the initial high cost of such materials.

Thermoplastic polymers and glasses can be shaped by hot pressing and injection moulding but, since both of these materials retain some ductility at high temperatures, they can also be blown to shape.

27.4.1 Processes

The selection of process, or processes, to produce a particular product depend on such factors as:

- Number of components required
- Equipment, tooling and labour costs, i.e. the capital costs to set up a process and then the running costs
- Processing times
- Material costs and availability
- Component form, detail such as holes required, and dimensions
- Dimensional accuracy and surface finish required.

Very often the choice of shaping process will depend upon the number of components required. Thus, it may be desirable to use a die-casting process to produce a component, and thus eliminate machining operations which would otherwise be necessary if sand-casting were employed instead. Unfortunately, the use of a die-casting process will only be economical if a large number of castings, say 5000 or more, is required, thus spreading the high initial cost of the steel dies over the large number of castings. Similarly, drop-forging is only economical if very large numbers of components are required, since expensive die-blanks must be used. Sometimes a compromise may be struck, at the expense of some reduction in mechanical properties, by using a malleabilised-iron sand-casting in cases where only a few components are required.

If the component form required is, say, a tube then such a product could be produced by centrifugal casting, drawing or extrusion. If the component form had a hollow section then forging cannot be used. The type of material to be used for a product will also affect the choice of process. For example, if the material to be cast has a high melting point then the process can be either sand-casting or investment casting.

The dimensional accuracy required from a process will also affect the choice of process, some processes more easily giving high dimensional accuracy than others. Generally, processes which give good dimensional accuracy also give better surface finishes.

27.4.2 Changing conditions

Conditions are continually changing, as new materials become available. During the last few decades, plastics have replaced metals and glass for innumerable applications. As an example, most curtain-rail material was extruded from a 60–40 type brass. Now this material is produced in a thermoplastic polymer which is extruded on to a steel core which provides the necessary strength. Many components which were once made in the form of metal die-castings are now manufactured as plastics mouldings, die-casting being retained as the shaping process where higher strength, rigidity and temperature resistance are required. In disposable tubes for toiletries, such as toothpaste, aluminium has been replaced by plastics materials but unfortunately it is now less easy to squeeze out the last 'blob' – though perhaps it is my inborn suspicion of manufacturers' motives that makes me ponder this fact!

The main barrier to the greater use of polymers is their low range of softening temperatures. Nevertheless, the corrosion-resistance of a plastics moulding is often very much higher than was that of the former die-casting, particularly if the latter was made from an alloy in a poor state of purity.

Without doubt, the field in which most engineering materials development will take place in the near future is that of fibre-reinforced composites. Having spent a long life working mainly as a metallurgist, I am of the opinion that the properties of metals and alloys have been exhaustively investigated and that there is now limited scope for outstanding developments in metallic alloys to take place – but I wonder if perhaps I made a similar remark, mistakenly, as a young whipper-snapper in the 1930s?

27.5 Developments in materials

Over recent years an increasing amount of development is occurring in types of materials referred to as smart materials and nanomaterials. The following is a brief outline of such developments.

27.5.1 Smart materials

Materials which are becoming increasingly used are the so-termed smart materials. The term *smart material* is used for materials that can sense and respond to the environment around them and change their properties to respond to the environmental conditions. For most materials, properties cannot be significantly altered by changes in the environment, e.g. an oil becomes a little thinner if the temperature increases, but smart materials have properties which change dramatically, e.g. a liquid might change to a solid when a magnet is brought near to it. Some types of smart materials are:

- *Piezoelectric materials* are materials that produce a voltage when a mechanical stress is applied, e.g. a microphone which produced an electrical signal as a result of the changes in pressure produced by sound waves. Conversely, a stress can be produced when a voltage is applied. Thus, items can be made which bend, expand or contract when a voltage is applied. An example of such a use is as the airbag sensor in a car, sensing the force of an impact on the car and producing a voltage which deploys the airbag. The piezoelectric effect occurs naturally in quartz, tourmaline, Rochelle salt and ammonium dihydrogen phosphate, ceramics such as lead–zironate–titanate (PZT), and polymers such as polyvinyldene fluoride (PVDF).

- *Shape memory alloys* (see Sections 16.5.1 and 18.2.6) and polymers are materials which after deformation remember their original shape and return to it when heated without any mechanical aid. Applications of shape-memory polymers include medical stents, i.e. tubes that are threaded into arteries and expand on heating to body temperature to allow increased blood flow.

- *Thermochromic materials* change colour in response to changes in temperature. One application is as a bath plug which changes colour when the water is too hot. Another application is as a forehead thermometer which when put in contact with the forehead gives a colour which is an indication of the temperature.

- *Electrochromic materials* change colour as a result of an electric current or voltage being applied. Thus, an electrochromic glass might change transparency and absorb certain wavelengths of light as a result of a voltage being applied, typically the glass assumes a blue tint, and the effect is reversed when the voltage direction is reversed.

- *Photochromic materials* change colour in response to changes in light conditions. An example of the use of such a material is light-sensitive sunglasses that darken when exposed to bright sunlight.

- *Magneto-rheostatic fluids* go from a very fluid state to a highly viscous or solid state when placed in a magnetic field and the effect is reversed when the field is removed. The changes can take place very fast, in a time of the order of a millisecond. They are used to construct hydraulic-based damping systems to suppress vibration to buildings and bridges to suppress the damaging effects of high winds, e.g. the dampers attached to the cables of suspension bridges. The precise amount of damping is controlled by altering the viscosity of the hydraulic fluid as a result of changing a sensor-controlled current and hence a magnetic field.

- *Electro-rheostatic fluids* go from a very fluid state to a highly viscous or solid state when placed in an electric field and the effect is reversed when the field is removed. The changes can take place very fast, in a time of the order of a millisecond. Such materials are used in engine mounts to reduce noise and vibration in vehicles.

27.5.2 Nanomaterials

Over recent years, nanomaterials have become increasingly researched and developed as their potential for widespread application becomes apparent. *Nanomaterials* can be metals, polymers, ceramics or composite materials but they all have the defining characteristic of being very small, namely in the range 1–100 nanometres (1 nm = 10^{-9} m). To get some idea of what size this is, a pinhead is about one million nanometres.

In the 1970s a new type of materials, hollow carbon spheres, was discovered by Harold Kroto and Richard Smalley. These involved 60 carbon atoms chemically bonded together in a ball-shaped molecule (C_{60}) which became known as *buckyballs* or fullerenes in honour of the architect Buckminster Fuller (see Section 1.4.6 and Figure 1.11). This led to the development of carbon nanofibres with diameters less than 100 nm. In 1991, Sumio Ijima of the NEC laboratory in Japan discovered *carbon nanotubes* during high-resolution transmission electron microscopy observation of the soot generated from the electrical discharge between two carbon electrodes. Carbon nanotubes are graphite sheets wrapped into a cylindrical form and have a diameter of a few nanometres and a length up to a few micrometres. Such tubes are able to withstand large stresses with very little deformation (a Young's modulus of the order of 1000 GPa), withstand very large tensile stresses (30 GPa), are thermally stable to very high temperatures, have a high thermal conductivity (twice that of diamond) and have the capacity to carry an electric current a thousand times better than copper wires. Such nanotubes are now produced commercially. One application of such nanotubes is in the producing of composites by embedding them in a metal, polymer or ceramic matrix. They also offer the possibility of developing electronic devices smaller and more powerful than any previously used.

Nanopaints have been developed involving inorganic or metallic nanoparticles added to a paint. The result is an improvement in hardness and scratch-resistance and such paints are finding a use in the car industry. Nanomaterials are also being used with paints to improve their properties to resist staining and mould growth.

Another form of nanomaterial that has been developed is a composite involving the blending at the nano-level of clays with polymers. The result is a great increase in the elastic modulus, an increase in strength, improved thermal stability (typically an extra 100°C above the normal service temperature of the polymer), and improved flammability properties. In 2002, General Motors used a thermoplastic olefin nanoclay composite in running-board step-assists.

27.5.3 Graphene

Graphene is a crystal form of carbon in which the atoms are arranged in a regular hexagonal pattern in a layer which is just one-atom thick. The term graphene was first used in 1987. Graphene is one of the strongest materials with a very high breaking strength and tensile modulus. Applications being considered for its use include thin but flexible display screens, solar cells where its high conductivity and optical transparency are of merit and a component of integrated circuits.

27.5.4 Metal foams

A metal foam is a cellular structure consisting of a solid metal containing gas-filled pores. The pores can be closed-cell or open-cell. Typically for metal foams 75–95% of the volume consists of voids, so making these very light and stiff materials. Metallic foams tend to retain some physical properties of their base material but notably with a reduced thermal conductivity.

The idea of a closed-cell metal foam was first reported in 1926 by Meller in a French patent and such foams have been developed since about 1956 by John C. Elliott at Bjorksten Research Laboratories. Commercial production was started only in the 1990s. Closed-cell metal foams are light and stiff and are mainly used as an impact-absorbing material, but unlike many polymer foams they remain deformed after impact and so can only be used once. Closed-cell metal foams are able to float in water. Open celled metal foams are used in heat exchangers. Metallic foams are currently being looked at as a new material for automobiles. Metallic foams currently being developed use aluminium and its alloys. These foams have low density, high stiffness, fire resistance, low thermal conductivity, are efficient at sound dampening, can be recycled and are cheap to cast by using powder metal-lurgy. In comparison to polymer foams, metallic foams tend to be stiffer, stronger, more fire resistant, and have better weathering properties but they are heavier, more expensive, and are electrical conductors and not insulators.

27.5.5 Amorphous metals

Most metals are crystalline in their solid state, having a highly ordered arrangement of atoms. The term amorphous metal is used for a solid metallic material, usually an alloy, which has been developed to have a disordered atomic-scale structure and so is non-crystalline with a glass-like structure. However, unlike glass such as window-glass amorphous metals have good electrical conductivity and are not electrical insulators.

In general, polycrystalline metals in the molten state are very fluid and their atoms can easily move around to form the polycrystalline structure which is typical of metals. The alloys used for amorphous metals contain atoms of significantly different sizes and, as a consequence, in the molten state have a high viscosity which prevents atoms having enough time during cooling to move to form an ordered lattice. For example, amorphous metal wires have been produced by sputtering molten metal onto a spinning metal disk and the resulting rapid cooling is too fast for crystals to form and the resulting material has a disorderly atomic structure.

Compared with polycrystalline metal alloys, amorphous metals are hard and have higher tensile yield strength but lower ductility and fatigue strength. The higher strength arises because of their lack of the dislocations that limit the strength of crystalline alloys. Vitreloy, an amorphous metal, has a tensile strength that is almost twice that of high-grade titanium. Because they are glasses they soften and flow upon heating and this allows for processing in much the same way as polymers, e.g. by injection moulding. Hence amorphous alloys are being used in sports equipment and medical devices.

27.5.6 Prototyping

This a method which is used to make solid objects from 3D computer models. This involves printing thin layers of material on top of each other and so building up a 3D object. One method that is used for this process involves laser sintering. The starting material is a powdered metal or polymer. The laser is used melt a layer of the powder into a solid. When a layer has been so 'printed' the plate supporting the powder is moved down a small amount and a new layer of powder is 'printed'. This is repeated until the entire solid object is formed. In this way a digital design can very rapidly be transformed into a 3D object.

28 Selection of materials

28.1 Introduction

Chapter 27 outlined the various characteristics of materials. But there is still the problem of determining which material would be suitable for a particular application, as invariably a number of different factors are generally involved in making the decision. It is not a simple case of just looking for a material with, say, the maximum stiffness, i.e. modulus of elasticity (see Section 27.3.2), but a balance might need to be achieved between the maximum stiffness and the lowest density to give a light but stiff material. There might also be the need to consider the effect on the environment of making a particular choice. For example how the product might be disposed of when its working life is finished. Can it be recycled?

Who is not aware of the problems created by the plastic carrier bags used by supermarkets to pack your purchases in at the check-out desk. I can remember when carrier bags were made of paper. Paper, however, had the disadvantage of losing its strength when it became wet and the wet carrier bag disintegrating under the weight of the goods in it and depositing them all the ground. Plastic bags do not do this, maintaining their strength when wet. However plastic bags do not degrade when they are disposed of in land-fill sites. The solution seems to be to minimise the use of plastic bags by supplying reusable plastic bags rather than ones which are so flimsy they are thrown away after one use. So far degradable bags with the cheapness and convenience of plastic bags have not been developed. But this is not the only issue involved in choosing materials. There is also the issue to be considered of global warming and the carbon footprint of making a choice.

There has been a rise in the average temperature of the Earth's atmosphere and oceans since the late nineteenth century and it seems likely to continue. Since the early twentieth century the Earth's mean surface temperature has increased by about 0.8°C. This warming seems likely to have been primarily caused by increasing concentrations of greenhouse gases produced by human activities such as the burning of fossil fuels. Burning fossil fuels such as natural gas, coal, oil and gasoline raises the level of carbon dioxide in the atmosphere, and carbon dioxide is a major contributor to the greenhouse effect. The *greenhouse effect* is a process by which thermal radiation from the Earth's surface is absorbed by atmospheric greenhouse gases, and then re-radiated in all directions. Since some of this re-radiation is back towards the Earth's surface, it results in an increase in the average surface temperature above what it would be in the absence of such gases.

A term used to assess the effect of making a particular choice of a material or process is the *carbon footprint*, this being defined as the greenhouse gas emissions caused. The most common way to reduce the carbon footprint is to Reduce, Reuse, Recycle, i.e. consider processes that emit less greenhouse gases, use reusable materials for products and recycle materials.

28.2 Material property limits

In selecting materials for a particular application there is likely to be some property which could be critical and for which there is a limit below which a material would not be a feasible choice. For example there may be a temperature up to which the material must maintain a certain strength. For example, in selecting a polymer to make a teaspoon there would be the criterion that the spoon must be able to withstand immersion in boiling water and maintain its stiffness. There is thus a requirement for a minimum value of the modulus of elasticity, the measure of stiffness, at 100°C. There is thus a need to consider what the component that is being designed does and what non-negotiable conditions must be met. This enables property limits to be determined and so eliminate materials which cannot do the job required.

As an illustration consider the problem of determining the material that could be used to make the container for a fizzy drink liquid. One limiting property is that it must be reasonably impermeable to carbon dioxide so that the drink will not lose its 'fizz'. Table 28.1 shows the permeabilities of three plastics. As we want there to be virtually no loss of fizz then the nearest we can get with the materials in Table 28.1 is PET. Indeed this is the material generally used for plastic bottles containing fizzy liquids.

Table 28.1 *Permeability to carbon dioxide*

Polymer	Permeability to carbon dioxide 10^{-8} mol mN^{-1}s^{-1}
Low density polythene	5700
Polyvinyl chloride (PVC)	98
Polyethylene terephthalate (PET)	30

There are other limiting properties which could be taken into consideration in determining the material for the fizzy drinks container. Thus we would want immunity to corrosion by the drink of the container, easy and cheap to shape to the required format and possibly capable of being recycled. Glass it thus a contender if we did not need to take into account container weight and ease of being broken. If we did not need transparency then we could consider metals for the drinks container. Thus aluminium and steel can become contenders. But if we want immunity to corrosion by the liquid then aluminium has the advantage unless the steel is to be protected by a lacquer. There are thus generally a number of limiting properties that have to be considered in arriving at possible materials. Probably one of the greatest limiting factors is cost.

28.3 Material property indices

In considering the properties of a material that might make it suitable for a particular application we often have to deal with requirements involving a combination of properties. As an illustration, consider the requirement for a strut of a specified length which is to have the properties of low mass but withstand high axial forces. Mass m is the product of the length L of the strut, its cross-sectional area A and its density ρ. The strength that is relevant is the yield stress σ_y and since stress is force F per unit cross-sectional area, then:

$$\frac{F}{m} = \frac{\sigma_y A}{LA\rho}$$

Thus for a fixed length, the maximum force per unit mass is determined by the ratio of the yield stress to density. This ratio is termed the *performance index* and is what we have to try to maximise for this application.

$$\text{Performance Index} = \frac{\sigma_y}{\rho}$$

If we had wanted to maximise the stiffness of the strut for the minimum mass then, as the tensile modulus of elasticity E is a measure of stiffness, the performance index can be derived as:

$$\text{Performance Index} = \frac{E}{\rho}$$

As a further example of the derivation of a material property index, consider the requirement for a cantilever to be as stiff as possible for the minimum mass. For a cantilever of length L the force F producing a deflection y at the free end is given by:

$$y = \frac{FL^3}{3EI}$$

For a stiff cantilever we want the maximum force per unit deflection F/y.

$$\frac{F}{y} = \frac{3EI}{L^3}$$

For a square cross-section of side b then $I = b^4/12$ and so we can write:

$$\frac{F}{y} = \frac{3Eb^4}{12L^3}$$

The mass m of the cantilever is $b^2 L\rho$, where ρ is the density. Thus if we take the square root of the above equation we can eliminate b and so obtain:

$$\frac{(F/y)^{1/2}}{m} = \frac{1}{2L^{5/2}} \times \frac{E^{1/2}}{\rho}$$

For a constant length, we have the maximum stiffness per unit mass when $E^{1/2}/\rho$ is maximised. Thus the performance index is:

$$\text{Performance Index} = \frac{E^{1/2}}{\rho}$$

Table 28.2 gives examples of performance indices.

Table 28.2 *Performance indices*

Component	Required property	Performance index
Strut/tie rod	Strength, minimum mass	$\dfrac{\sigma_y}{\rho}$
	Stiffness, minimum mass	$\dfrac{E}{\rho}$
Beam	Stiffness, minimum mass	$\dfrac{E^{1/2}}{\rho}$
	Strength, minimum mass	$\dfrac{\sigma_y^{2/3}}{\rho}$
Column/compression strut	Strength, minimum mass	$\dfrac{\sigma_y}{\rho}$
Panel/flat plate loaded in bending	Stiffness, minimum thickness	$\dfrac{E^{1/3}}{\rho}$
Flywheel	Stored energy, minimum mass	$\dfrac{\sigma_y}{\rho}$
Spring	Stored energy, minimum mass	$\dfrac{\sigma_y^2}{E\rho}$

28.4 Materials-cost sensitivity

The cost of the materials used in a product will have an impact on the cost of the product. However, the total cost of a product involves considering not just the cost of the material but also the costs of processing it to give the required product. In some cases the cost of the processing can be a major fraction of the total cost and so an increased cost of the material will have less impact on the product price than a product where the materials cost is a major element of the total cost. It is thus necessary in considering materials for a particular product to consider the sensitivity of the product cost to the cost of the materials. For example, only about 5% of the cost of a golf club arises from the cost of the materials. Thus, using a more expensive material will have little effect on the cost of the club.

28.5 The materials life cycle

The life cycle of a material can involve the following steps (Figure 28.1):

1. Production from the Earth's natural resources and processing to give a material. This involves energy and extraction of the raw material is likely to have an impact on the environment and result in some resource depletion.
2. Product manufacture. This involves energy for the manufacturing process and generally waste material.
3. Product use. This can involve energy consumption and, in some cases, can result in contamination of the environment.
4. Product disposal. This can be to a landfill site or incineration.
5. Recycling back to give useful material which can repeat the cycle. Possibly only a fraction of the materials of a product is recycled to give material which can repeat the life cycle.

Energy and materials are consumed at each stage in the cycle. A particular item is the consumption of energy involved in the transportation of the materials/products. This can result in the release of gases such as carbon dioxide, sulphur dioxide and nitrogen. In particular, carbon dioxide is an important greenhouse gas, absorbing heat radiation from the Earth's surface which otherwise would have left the atmosphere and so resulting in an increase in the Earth's temperature.

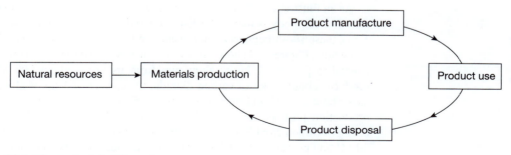

Figure 28.1 *The materials life cycle*

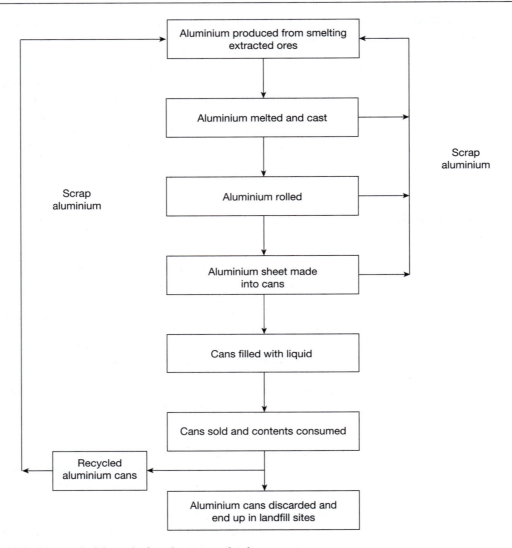

Figure 28.2 *Materials life cycle for aluminium drinks cans*

As an illustration of a materials life cycle, consider the aluminium drinks can. The can body is made of 3004 alloy and the lid 5182 alloy (see Table 17.1), these alloys being compatible enough to allow for recycling of the can as a whole. Figure 28.2 shows the cycle. It has been estimated that the energy needed to produce 1 kg of aluminum from its ore is about 200 MJ and produces about 12 kg of carbon dioxide, the casting of 1 kg of aluminium uses about 2.5 MJ and produces about 0.15 kg of carbon dioxide, the can production process uses per kilogram about 2.5 MJ and produces about 0.2 kg of carbon dioxide and recycling of aluminium takes per kilogram about 20 MJ and produces about 1 kg of carbon dioxide. At present about 50% of aluminium is recycled. It is this high rate of recycling which keeps the cost of making an aluminium drinks can down below the cost of using other materials such as glass, plastic or steel. Aluminium is one of the most abundant metals in the Earth's crust and it has been estimated that at the current rate of usage, of the order of 30 million tonnes per year, there is enough for another thousand years.

As a comparison, the polyethylene terephthalate (PET) drinks bottle (see Table 28.1 and the associated discussion) takes for the production of 1 kg of raw material about 80 MJ and produces about 2.3 kg carbon dioxide, for processing 1 kg about 10 MJ and produces about 1 kg of carbon dioxide, and recycling takes per kilogram about 35 MJ and produces about 1 kg of carbon dioxide. Currently about 20% of the drink bottles are recycled.

28.6 End of life of products

At the end of the life of a product, the material may be used for landfill, combusted for heat recovery, recycled or reconditioned. Throwing away products to landfill presents the problem that the land available to be filled by such waste is rapidly diminishing. Combustion of materials can be used to generate heat energy. However, before this can be done there is a need to remove non-combustible materials and remove those materials that can generate toxic fumes. The heat generated can be used to generate electricity. Recycling is the process of recovering material at the end of the life of a product and returning it to a condition in which it can be reused. However, not all materials can be recycled. The recycling of metals is a highly developed industry. However, the recycling of polymers can present problems. Polymeric materials generally contain fillers in addition to a polymer and so the process of recycling a polymer can be an expensive process. Reconditioning involves replacing parts that have worn out or become defective and so enabling the product to continue being useful.

Appendix A: Properties of engineering metals

The following are the properties of the important engineering metals.

Metal		Relative density	Melting point (°C)	Tensile strength (MPa)	Percentage elongation	Characteristics and uses
Aluminium	Al	2.7	660	59	60	The most widely used of the 'light metals'. Common in the Earth's crust.
Antimony	Sb	6.6	630	10	0	A bright, crystalline metal, used in limited amounts in bearing and type-metals.
Beryllium	Be	1.8	1285	310	2.3	A light metal, the use of which is limited by its scarcity. Used in beryllium bronze and nuclear-power industries.
Cadmium	Cd	8.6	321	80	50	Used for plating steel, and to strengthen copper for telephone wires.
Chromium	Cr	7.1	1800	220	0	A metal which resists corrosion, and is therefore used for plating and in stainless steels.
Cobalt	Co	8.9	1495	250	6	Used mainly in permanent magnets, in super-highspeed steels and corrosion-resistant 'super-alloys'.
Copper	Cu	8.9	1083	220	60	Now used mainly where very high electrical conductivity is required; also in brass and bronzes.
Gold	Au	19.3	1063	120	30	Of little use as an engineering metal, because of softness and scarcity. Used mainly in jewellery, and as a system of exchange.
Iron	Fe	7.9	1535	500	10	Quite soft when pure, but rarely used in engineering in the unalloyed form. As steel, our most important metal.
Lead	Pb	11.3	327	18	64	Not the densest of metals, as phrase 'as heavy as lead' suggests. Very resistant to corrosion – used in chemical engineering, but main uses in batteries and pigments.
Magnesium	Mg	1.7	651	180	5	Used in conjunction with aluminium in the lightest of engineering alloys.
Manganese	Mn	7.2	1260	500	20	Very similar to iron in many ways – used mainly as a deoxidant and desulphuriser in steels.

continued

Metal		Relative density	Melting point (°C)	Tensile strength (MPa)	Percentage elongation	Characteristics and uses
Mercury	Hg	13.6	−39	Liquid at ordinary temperatures.		The only liquid metal at normal temperatures. Used in 'mercury-in-glass' thermometers and scientific equipment.
Molybdenum	Mo	10.2	2620	420	50	A heavy metal, used mainly in alloy steels. One of the main constituents of modern high-speed steels. Also used in stainless steels.
Nickel	Ni	8.9	1458	310	28	A very adaptable metal, used in both ferrous and non-ferrous alloys. The metallurgist's main 'grain refiner'. Principal uses – stainless steels and electroplating.
Niobium	Nb	8.6	1950	270	49	Also known as 'columbium' in the USA. Used mainly in alloy steels and in high-temperature alloys.
Platinum	Pt	21.4	1773	130	35	A precious white metal. Used in scientific apparatus, because of its high corrosion-resistance, also used in some jewellery.
Silver	Ag	10.5	960	140	50	Has the highest electrical conductivity, but is used mainly in jewellery and, in a few countries, for coinage.
Tin	Sn	7.3	232	11	60	Widely used but increasingly expensive. 'Tin cans' carry only a very thin coating of mild steel. Very resistant to corrosion. A constituent of bronze.
Titanium	Ti	4.5	1667	230	55	A fairly light metal, which is becoming increasingly important as its price falls due to the development of its technology.
Tungsten	W	19.3	3410	420	16	Used in electric light filaments, because of its high melting-point. It is also the main additive of most high-speed steels and heat-resisting steels, but its main use now is in cemented carbides.
Uranium	U	18.7	1150	390	4	Now used mainly in the production of atomic energy.
Vanadium	V	5.7	1710	200	38	Used in some alloy steels as a hardener.
Zinc	Zn	7.1	420	110	25	Used widely for galvanising mild- and low-carbon steels. Also as a basic for some die-casting alloys. Brasses are copper-zinc alloys.
Zirconium	Zr	6.4	1800	220	25	Used as a grain-refiner in steels. It is also used for atomic energy applications.

Appendix B: Glossary of key terms

Additives Plastics and rubbers almost invariably contain, in addition to the polymer or polymers, other materials. These are termed additives and added to modify the properties and cost of the material.

Ageing This term is used to describe a change in properties that occurs with certain metals due to precipitation occurring, there being no change in chemical composition.

Alloy A metal which is a mixture of two or more elements.

Amorphous An amorphous material is a non-crystalline material and has a structure which is not orderly.

Annealing This involves heating to and holding at a temperature which is high enough for recrystallisation to occur and which results in a softened state for a material after a suitable rate of cooling, generally slowly. The purpose of annealing can be to facilitate cold-working, improve machinability, improve mechanical properties, etc.

Anodising This term is used to describe the process, generally with aluminium, whereby a protective coating is produced on the surface of the metal by converting it to an oxide.

Atomic number This is the number of protons in the atomic nucleus.

Austempering A heat-treatment for ferrous alloys in which quenching is used from the austenising temperature at a rate fast enough to avoid the formation of ferrite or pearlite and then held at a temperature above M_s until the transformation to bainite is complete.

Austenite This term describes the structure of a face-centred cubic iron crystalline structure which has carbon atoms in the gaps in the face-centred iron.

Bayer process A process for extracting alumina from bauxite ores.

Bend, angle of The results of a bend test on a material are specified in terms of the angle through which the material can be bent without breaking, The greater the angle the more ductile the material.

Bessemer process A process for making steel by blowing air through molten pig iron.

Brass An alloy containing mainly copper and zinc.

Brazing Welding processes that join solid materials together by heating them to a suitable temperature and using a suitable filler metal.

Brinell number The Brinell number is the number given to a material as a result of a Brinell test and is a measure of the hardness of a material. the larger the number the harder the material.

Brittle failure With brittle failure a crack is initiated and propagates prior to any significant plastic deformation. The fracture surface of a metal with a brittle fracture is bright and granular due to the reflection of light from individual crystal surfaces. With polymeric materials the fracture surface may be smooth and glassy or somewhat splintered and irregular.

Brittle material A brittle material shows little plastic deformation before fracture. The material used for a china teacup is brittle. Thus because there is little plastic deformation before breaking, a broken teacup can be stuck back together again to give the cup the same size and shape as the original.

Bronze A copper-rich copper-tin alloy.

Carbide A compound of carbon with one or more metallic elements.

Carbonitriding A case-hardening process in which carbon and nitrogen is absorbed by a surface.

Carburising This is a treatment which results in a hard surface layer being produced with ferrous alloys. The treatment involves heating the alloy in a carbon-rich atmosphere so that carbon diffuses into the surface layers, then quenching to convert the surface layers to martensite.

Case-hardening Processes in which by changing the composition of surface layers of ferrous alloys a hardened surface layer can be produced.

Casting This is a manufacturing process which involves pouring liquid metal into a mould or, in the case of plastics, the mixing of the constituents in a mould.

Cast iron Cast ferrous alloys containing 3 to 4% carbon plus 1 to 3% silicon and traces of other elements such as sulphur, manganese and phosphorus.

Cementite This is a compound formed between iron and carbon, often referred to as iron carbide. It is a hard and brittle material.

Cermit A powder metallurgy product consisting of ceramic particles bonded with a metal.

Charpy test value The Charpy test is used to determine the response of a material to a high rate of loading and involves a test-piece being struck a sudden blow. The results are expressed in terms of the amount of energy absorbed by the test-piece when it breaks. The higher the test value the more ductile the material.

Cold-working This is when a metal is subject to working at a temperature below its recrystallisation temperature.

Composite material This is a material composed of two different materials bonded together in such a way that one serves as the matrix surrounding fibres or particles of the other.

Compressive strength The compressive strength is the maximum compressive stress a material can withstand before fracture.

Copolymer This is a polymeric material produced by combining two or more monomers in a single polymer chain.

Coring A condition resulting from non-equilibrium solidification and results in differing compositions between the centre and surface.

Corrosion This is when a material suffers deterioration by reacting with its immediate environment.

Creep Creep is the continuing deformation of a material with the passage of time when it is subject to a constant stress. For a particular material the creep behaviour depends on both the temperature and the initial stress, the behaviour also depending on the material concerned.

Critical cooling rate The rate of continuous cooling needed to prevent undesirable transformations. With a steel it is the minimum cooling rate needed to give a martensitic structure.

Critical point In an equilibrium diagram it is the value of composition, temperature and pressure at which the phases are in equilibrium.

Crystalline This is when a material has a regular, orderly arrangement of atoms or molecules.

Cyaniding A case-hardening process in which a ferrous material is heated in a molten salt containing cyanide to cause the absorption of carbon and nitrogen in the surface.

Dendrite A crystal having a tree-like branching pattern.

Density The density of a material is its mass per unit volume.

Dislocation A linear imperfection in a crystalline array of atoms.

Ductile failure With ductile failure there is a considerable amount of plastic deformation prior to failure. With metals the fracture shows a typical cone and cup formation and the fracture surfaces are rough and fibrous in appearance.

Ductile materials Ductile materials show a considerable amount of plastic deformation before breaking. Such materials have a large value of percentage elongation.

Elastic limit The elastic limit is the maximum force or stress at which on its removal the material returns to its original dimensions.

Electrical conductance This is the reciprocal of the electrical resistance and has the unit of the siemen (S). It is thus the current through a material divided by the voltage across it.

Electrical conductivity The electrical conductivity is defined by L/RA, where R is the resistance of a strip of the material of length L and cross-sectional area A. Conductivity has the unit of S/m. The IACS specification of conductivity is based on 100% corresponding to the conductivity of annealed copper at 20°C, all other materials are then expressed as a percentage of this value.

Electrical resistance Resistance is the voltage across a material divided by the current through it, the unit being the ohm (Ω).

Electrical resistivity The electrical resistivity is defined in terms of the resistance R of a conductor of cross-sectional area A and length L by RA/L and has the unit of Ωm.

Electroplating This is the electro deposition of a metal on an object serving as a cathode.

Endurance This is the number of stress cycles to cause failure.

Endurance limit This is the value of the stress for which a test specimen has a fatigue life of N cycles.

Equilibrium diagram A graphical representation of the temperature, pressure and composition limits of the phase fields in an alloy system that exist under conditions of equilibrium.

Expansion, coefficient of linear The coefficient of linear expansion is a measure of the amount by which a unit length of a material will expand when the temperature rises by one degree. It is defined by

$$\text{coefficient} = \frac{\text{change in length}}{\text{length} \times \text{temperature change}}$$

It has the unit /°C or °C^{-1} or /K or K^{-1}.

Extrusion Conversion of an ingot into lengths of uniform cross-section by forcing the metal to flow plastically through a die orifice.

Fatigue life The fatigue life is the number of stress cycles to cause failure.

Ferrite This term is a structure consisting of carbon atoms lodged in body-centred cubic iron. Ferrite is comparatively soft and ductile.

Flow lines Texture showing the direction of metal flow during hot- or cold-working.

Forging Plastically deforming metal into the required shapes by compressive force.

Galvanic series Metals and alloys arranged according to their relative electrode potentials in a particular environment.

Gamma iron The face-centred cubic form of pure iron.

Glass transition temperature This is the temperature at which a polymer changes from a rigid to a flexible material. The tensile modulus shows an abrupt change from the high value typical of a glass-like material to the low value of a rubber-like material.

Grain This term describes a crystalline region within a metal, i.e. a region of orderly packed atoms.

Hardenability The relative ability of a ferrous alloy to form martensite when quenched from a temperature above the upper critical temperature.

Hardness The hardness of a material may be specified in terms of some standard test involving indentation, e.g. the Brinell, Vickers and Rockwell tests, or scratching of the surface of the material, the Moh test.

Heat-treatment The controlled heating and cooling of metals in the solid state for the purpose of altering their properties.

Hooke's law When a material obeys Hooke's law its extension is directly proportional to the applied stretching forces.

Hot-working This is when a metal is subject to working at a temperature in excess of its recrystallisation temperature.

Hypereutectic alloy An alloy with a composition which has an excess of the alloying element compared to that of the eutectic composition.

Hyper-eutectoid alloy An alloy with a composition which has an excess of the alloying element compared to that of the eutectoid composition.

IACS The International Annealed Copper Standard. This is used to report electrical conductivity.

Impact properties See Charpy test value and Izod test value.

Induction hardening A surface hardening process in which the surface layer of a ferrous material is heated by electromagnetic induction to above the upper critical temperature and then quenched.

Ionic crystal A crystal in which atomic bonds are ionic bonds, i.e. the result of electrostatic forces of attraction between positively and negatively charged ions.

Izod test value The Izod test is used to determine the response of a material to a high rate of loading and involves a test-piece being struck a sudden blow. The results are expressed in terms of the amount of energy absorbed by the test-piece when it breaks. The higher the test value the more ductile the material.

Laminate A composite composed of two or more bonded layers.

Limit of proportionality Up to the limit of proportionality the extension is directly proportional to the applied stretching forces, i.e. the strain is proportional to the applied stress.

Liquidus In an equilibrium diagram it is the line representing the temperatures at which the various compositions in the system begins to freeze on cooling or melt on heating.

Machinability The relative ease of machining a metal.

Malleable cast iron A cast iron made by heat-treatment of a white, i.e. cementite structure, iron to replace the cementite by carbon or remove it completely to give a more malleable and ductile iron.

Malleability This describes the ability of metals to permit plastic deformation in compression without rupturing.

Martensite This term is used to describe a form of atomic structure. In the case of ferrous alloys it is a structure produced when the rate of cooling from the austenitic state is too rapid to allow carbon atoms to diffuse out of the face-centred cubic form of austenite and produce the body-centred form of ferrite. The result is a highly strained hard structure.

Melting point This is the temperature at which a material changes from solid to liquid.

Modulus of elasticity The ratio of the stress to strain below the proportional limit. It is also known as Young's modulus.

Moh scale This is a scale of hardness arrived at by considering the ease of scratching a material. It is a scale of 10, with the higher the number the harder the material.

Monomer This is the unit, or mer, consisting of a relatively few atoms which are joined together in large numbers to form a polymer.

M_s **temperature** The temperature at which martensite starts to form on cooling.

Necking Reduction in the cross-sectional area of a metal as a result of stretching.

Nitriding This is a treatment in which nitrogen is diffused into surface layers of a ferrous alloy to produce hard nitrides and hence a hard surface layer.

Non-destructive testing Inspection that does not destroy the part being tested.

Normalising This heat-treatment process involves heating a ferrous alloy to a temperature which produces a fully austenitic structure followed by air cooling.

Orientation A polymeric material is said to have an orientation, uniaxial or biaxial, if during the processing of the material the molecules become aligned in particular directions. The properties of the material in such directions are markedly different from those in other directions.

Pearlite This is a lamella structure of ferrite and cementite.

Percentage elongation The percentage elongation is a measure of the ductility of a material, the higher the percentage the greater the ductility. It is the change in length which has occurred during a tensile test to breaking expressed as a percentage of the original length.

Percentage reduction in area This is a measure of the ductility of a material and is the change in cross-sectional area which has occurred during a tensile test to breaking expressed as a percentage of the original cross-sectional area.

Peritectic reaction An isothermal reversible reaction in which a liquid phase reacts with a solid phase to produce a single and different solid phase on cooling.

Pig iron High-carbon iron produced by the reduction of iron ore in a blast-furnace.

Plastic deformation Deformation that remains after the removal of the load that caused it.

Poisson's ratio The ratio of the transverse strain to the axial strain when a body is subject to a uniaxial stress.

Polycrystalline A solid composed of many crystals.

Precipitation hardening This is a heat-treatment process which results in a precipitate being produced in such a way that a harder material is produced.

Proof stress The 0.2% proof stress is defined as that stress which results in a 0.2% offset, i.e. the stress given by a line drawn on the stress–strain graph parallel to the linear part of the graph and passing through the 0.2% strain value. The 0.1% proof stress is similarly defined. Proof stresses are quoted when a material has no well-defined yield point.

Quenching This is the method used to produce rapid cooling. In the case of ferrous alloys it involves cooling from the austenitic state by immersion in cold water or an oil bath.

Radiography Non-destructive testing using X-rays or gamma-rays.

Recovery This term is used in the treatment involving the heating of a metal so as to reduce internal stresses.

Recrystallisation The process whereby a new, strain-free grain structure is produced from that existing in a cold-worked metal by heating.

Resilience This term is used with elastomers to give a measure of the 'elasticity' of a material. A high resilience material will suffer elastic collisions when a high percentage of the kinetic energy before the collision is returned to the object after the collision. A less resilient material would lose more kinetic energy in the collision.

Rockwell test value The Rockwell test is used to give a value for the hardness of a material. There are a number of Rockwell scales and thus the scale being used must be quoted with all test results.

Ruling section The limiting ruling section is the maximum diameter of round bar at the centre of which the specified properties may be obtained.

Rust A corrosion product of ferrous alloys consisting of hydrated oxides of iron.

Sacrificial protection Reducing the corrosion of a metal by coupling it with another metal which is electrochemically more active in the environment concerned and so sacrificed.

Secant modulus For many polymeric materials there is no linear part of the stress–strain graph and thus a tensile modulus cannot be quoted. In such cases the secant modulus is used. It is the stress at a value of 0.2% strain divided by that strain.

Shear When a material is loaded in such a way that one layer of it slides relative to its neighbour in a direction parallel to their plane of contact.

Sintering This is the process by which powders are bonded by molecular or atomic attraction as a result of heating to a temperature below the melting points of the constituent powders.

S/N **diagram** A diagram showing the relationship of stress and the number of cycles before fracture in fatigue testing.

Solidus In an equilibrium diagram it is the line representing the temperatures at which various compositions finish freezing on cooling or begin to melt on heating.

Solute The component that is dissolved in a solvent.

Solution treatment This heat-treatment involves heating an alloy to a suitable temperature, holding at that temperature long enough for one or more constituent elements to enter into the crystalline structure, and then cooling rapidly enough for these to remain in solid solution.

Specific gravity The specific gravity of a material is the ratio of its density compared with that of water.

Specific heat capacity The amount by which the temperature rises for a material when there is a heat input depends on its specific heat capacity. The higher the specific heat capacity the smaller the rise in temperature per unit mass for a given heat input.

$$\text{specific heat capacity} = \frac{\text{heat input}}{\text{mass} \times \text{temperature change}}$$

Specific heat capacity has the unit $J \, kg^{-1} \, K^{-1}$.

Specific stiffness The modulus of elasticity divided by the density.

Specific strength The strength divided by the density.

Stiffness The property is described by the modulus of elasticity.

Strain The engineering strain is defined as the ratio (change in length)/ (original length) when a material is subject to tensile or compressive forces. Shear strain is the ratio (amount by which one layer slides over another)/(separation of the layers). Because it is a ratio strain it has no units, though it is often expressed as a percentage. Shear strain is usually quoted as an angle in radians.

Strength See Compressive Strength and Tensile strength.

Stress In engineering, tensile and compressive stress is defined as (force)/(initial cross-sectional area); true stress is (force)/(cross-sectional area at that force). Shear stress is the (shear force)/(area resisting shear). Stress has the unit Pa (pascal) with 1 Pa $= 1$ N/m^2 or 1 N m^{-2}.

Stress relieving A treatment to reduce residual stresses by heating the material to a suitable temperature, followed by slow cooling.

Stress–strain graph The stress–strain graph is usually drawn using the engineering stress (see Stress) and engineering strain (see Strain).

Surface hardening This is a general term used to describe a range of processes by which the surface of a ferrous alloy is made harder than its core.

Temper This term is used with non-ferrous alloys as an indication of the degree of hardness/strength, with expressions such as hard, half-hard, three-quarters hard being used.

Tempering This is the heating of a previously quenched material to produce an increase in ductility.

Tensile modulus The tensile modulus, or Young's modulus, is the slope of the stress–strain graph over its initial straight line region.

Tensile strength This is defined as the maximum tensile stress a material can withstand before breaking.

Thermal conductivity The rate at which energy is transmitted as heat through a material depends on a property called the thermal conductivity. The higher the thermal conductivity the greater the rate at which heat is conducted. Thermal conductivity is defined by

$$\text{thermal conductivity} = \frac{\text{rate of transfer of heat}}{\text{area} \times \text{temperature gradient}}$$

Thermal conductivity has the unit W m^{-2} K^{-1}.

Thermal expansivity See Expansion, coefficient of linear.

Toughness This property describes the ability of a material to absorb energy and deform plastically without fracturing. It is usually measured with the Izod test or the Charpy test.

Transition temperature The transition temperature is the temperature at which a material changes from giving a ductile failure to giving a brittle failure.

Ultimate strength The maximum stress that a material can withstand.

Vickers test results The Vickers test is used to give a measure of the hardness of a material; the higher the Vickers hardness number the greater the hardness.

Work hardening This is the hardening of a material produced as a consequence of working subjecting it to plastic deformation at temperatures below those of recrystallisation.

Yield point For many metals, when the stretching forces applied to a testpiece are steadily increased a point is reached when the extension is no longer proportional to the applied forces and the extension increases more rapidly than the force until a maximum force is reached. This is called the upper yield point. The force then drops to a value called the lower yield point before increasing again as the extension is continued.

Young's modulus See Tensile modulus.

Index